Project Business
Management

Best Practices and Advances in Program Management Series

Series Editor
Ginger Levin

RECENTLY PUBLISHED TITLES

PgMP® Exam Test Preparation: Test Questions, Practice Tests, and Simulated Exams
Ginger Levin

Managing Complex Construction Projects: A Systems Approach
John K. Briesemeister

Managing Project Competence: The Lemon and the Loop
Rolf Medina

The Human Change Management Body of Knowledge (HCMBOK®), Third Edition
Vicente Goncalves and Carla Campos

Creating a Greater Whole: A Project Manager's Guide to Becoming a Leader
Susan G. Schwartz

Project Management beyond Waterfall and Agile
Mounir Ajam

Realizing Strategy through Projects: The Executive's Guide
Carl Marnewick

PMI-PBA® Exam Practice Test and Study Guide
Brian Williamson

Earned Benefit Program Management: Aligning, Realizing, and Sustaining Strategy
Crispin Piney

The Entrepreneurial Project Manager
Chris Cook

Leading and Motivating Global Teams: Integrating Offshore Centers and the Head Office
Vimal Kumar Khanna

Project and Program Turnaround
Thomas Pavelko

Project Business Management

Oliver F. Lehmann, MSc, PMP®

CRC Press
Taylor & Francis Group
Boca Raton London New York

CRC Press is an imprint of the
Taylor & Francis Group, an **informa** business
AN AUERBACH BOOK

Disclaimer

This book is intended to provide guidance on how to behave as a project manager in a contractual business context in national and International environments. It is intended to help practitioners, educators, and associated professionals to make the decisions that are needed to navigate around the impediments and menaces that commonly arise in projects under contract. It has also been written to help manage complex project supply networks and generate outcomes of cross-corporate projects that are considered successes by the organization itself and also by the business partners involved.

This book is a management book and does not provide, nor is a replacement for, the necessity to take legal advice. The author strongly recommends readers to consult legal experts such as external lawyers or corporate counsels before any decisions are made or actions are done that could have negative legal consequences for the organization or other organizations involved.

The author of the book has applied maximum diligence to give guidance that is helpful in many project situations, but he does not assume any responsibility for any claims, disputes, or legal troubles of any sort, which may be encountered by projects that use the recommendations made in this book.

Persons and organizations described in the case stories are purely fictitious, and any similarity with real persons or organizations would be coincidental.

Copyright Notes

"IPMA" and "Level B" are registered trademarks of the International Project Management Association.

"ISO" is a registered trademark of the International Organization for Standardization.

"Level B" is a registered trademark of the International Project Management Association.

"PMI," "PMP," "Project Management Professional," "*PM Network*," and "*PMBOK* ® *Guide*" are registered trademarks of the Project Management Institute.

CRC Press
Taylor & Francis Group
6000 Broken Sound Parkway NW, Suite 300
Boca Raton, FL 33487-2742

© 2019 by Taylor & Francis Group, LLC
CRC Press is an imprint of Taylor & Francis Group, an Informa business

No claim to original U.S. Government works

International Standard Book Number-13: 978-1-138-19750-3 (Hardback)

Visit the Taylor & Francis Web site at
http://www.taylorandfrancis.com

and the CRC Press Web site at
http://www.crcpress.com

Printed and bound in Great Britain by
TJ International Ltd, Padstow, Cornwall

Dedication

I dedicate this book to my wife, Silvia, who allowed me the time needed to not rush it but diligently give it value and depth.

I also dedicate it to all project managers in the world who practice project business management in complex project supply networks (PSNs), mastering the balance between the financial well-being of their own organization and its partners in business, upholding a "Mission Success First" culture for the benefit of the project.

Contents

Dedication *vii*

Contents *ix*

List of Illustrations *xv*

List of Tables *xix*

Foreword *xxi*

Preface *xxvii*

Acknowledgments *xxxiii*

About the Author *xxxv*

Chapter 1 Laying Out the Scenery: The Business Side of
Project Management 1

1.1 The Importance of Project Business Management 1

1.2 Introductory Questions

1.3 The Story So Far 4

  1.3.1 "Best Practices" or Uniqueness? 5

  1.3.2 A Research Project 7

  1.3.3 A Typology of Projects—Project Types 8

  1.3.4 Mark 1 vs. Mark *n* Projects 8

  1.3.5 Greenfield vs. Brownfield Projects 9

  1.3.6 Siloed vs. Solid Projects 10

  1.3.7 Blurred vs. Focused Projects 10

  1.3.8 High-Impact vs. Low-Impact Projects 11

  1.3.9 Customer Projects vs. Internal Projects 11

  1.3.10 Standalone Projects vs. Satellite Projects 12

  1.3.11 Predictable Projects vs. Exploratory Projects 12

  1.3.12 Composed Projects vs. Decomposed Projects 13

1.4 Further Types of Projects 13

  1.4.1 Engineers' Projects and Gardeners' Projects 13

  1.4.2 Discretionary Projects and Mandatory Projects 14

  1.4.3 Single Handover Projects and Multiple Handover Projects 15

	1.4.4	No-Deadline Projects, Single-Deadline Projects, and Multiple-Deadline Projects	16
	1.4.5	One-Shot Projects vs. Multi-Shot Projects	16
	1.4.6	Open Typologies and Closed Typologies	17
1.5	An Introduction to Project Business Management		18
	1.5.1	The Significance of Customer Projects	21
	1.5.2	Procurement as Part of the Architecture of a Project	22
	1.5.3	The Basic Nature of Customer Projects	25
1.6	Projects as Profit Centers		29
	1.6.1	Small Consultancies and Development Offices	30
	1.6.2	Large Consortia	30
	1.6.3	Major Project Providers	31
	1.6.4	Providers of Supportive Projects and Freebie Projects	33
	1.6.5	Mixed Customer Projects	35
1.7	The Players in *Project Supply Networks*		35
	1.7.1	The Buyer	36
	1.7.2	Sellers #1: The Contractor	37
	1.7.3	Sellers #2: The Prime Contractor	37
	1.7.4	Sellers #3: Subcontractors over Various Tiers	38
	1.7.5	Freelancers	38
	1.7.6	Other Players	39
Chapter 2	**The Difficult Way to the Contract**		**41**
2.1	Bringing Order Into Chaos May Not Be Enough		41
2.2	Introductory Questions		44
2.3	The Lure of the *Buy* Option		45
	2.3.1	The Dilemma of Management Attention	47
	2.3.2	Agility and Speed of Change	50
	2.3.3	Diversity of Skills	51
	2.3.4	Unlocking Growth Potentials	52
	2.3.5	Buying to Save Money or Time	54
	2.3.6	Buying as a Political Decision	54
	2.3.7	Tapping External Assets	56
2.4	Customer Project Management: Where Does the Market Stand?		57
	2.4.1	The Rationale of the Research	57
	2.4.2	The Design of the Research	59
	2.4.3	The Survey Respondents	60
	2.4.4	The Overall Result	61
	2.4.5	The Influence of the Workload Assignment	62
	2.4.6	The Influence of the Project Size	63
	2.4.7	The Influence of the Project Location	63
	2.4.8	Conclusions from the Research	64

2.5   Risks of the *Buy* Option   66
    2.5.1   Time Losses   66
    2.5.2   Unclear Scope   67
    2.5.3   Fragmentation   68
    2.5.4   Unattainable Cost Benefits   69
    2.5.5   Narrowing Strategic Options   70
    2.5.6   Remoteness of Error Fixing   71
    2.5.7   Speed Blindness   72
    2.5.8   Bringing Strange People into the Project   75
    2.5.9   Opening the Project for Corruption   77
    2.5.10   Coercing Behaviors by Contractors   78
    2.5.11   Opening New Doors for Malware   80
    2.5.12   The Reputation at Stake   80
    2.5.13   The Contractor as a Data Leak   83
    2.5.14   Other Risks   84
2.6   Finding and Approaching Sellers   85
    2.6.1   Personal Recommendation   86
    2.6.2   B2B Marketplaces   86
    2.6.3   Public Tendering   88
    2.6.4   Bidders' Conferences   89
2.7   Requesting Seller Responses   90
    2.7.1   Traps in Terminology   90
    2.7.2   Statement of Work (SOW)   93
    2.7.3   Thresholds for Procurement   94
2.8   The Offer/No-Offer Decision   95
    2.8.1   Risks of an Offer Development Process Without Input
           from Project Managers   96
    2.8.2   Templates   97
    2.8.3   TRAC: Influencing Factors for the Bid/No-Bid Decision   98
    2.8.4   A TRAC-Based Weighting System   100
    2.8.5   A TRAC-Based Force-Field Analysis   101
2.9   Winning the Contract   102
    2.9.1   AIDA—Singing for the Business   102
    2.9.2   Hit Rates and Capture Ratio   104
    2.9.3   Being an Incumbent   108
    2.9.4   Conclusion   111
2.10   Offers: Bids, Proposals, Quotations, etc.   112
    2.10.1   Developing and Submitting the Offer   112
    2.10.2   Types of Seller Responses   113
2.11   Binding and Non-Binding Offers   115
    2.11.1   Binding Offers   115

	2.11.2	The Invitation to Treat	117
	2.11.3	Bid Bonds, Performance Bonds	118
	2.11.4	Questions Forbidden	119
2.12	Submission Deadlines		120
2.13	Teaming Agreements		122
	2.13.1	Informal Relationship	124
	2.13.2	Prime/Subcontractor Relationship	125
	2.13.3	The Consortium	126
	2.13.4	Mixed and Expanded Teaming Agreements	129
2.14	Pricing		130
	2.14.1	The "Perfect Price" in a Non-Competitive Setting	130
	2.14.2	The "Perfect Price" in a Competitive Setting	134
	2.14.3	When the Offer Is Too Low	138
	2.14.4	The Customer Dictates the Price	139
	2.14.5	The Absolutely Last Price	139
	2.14.6	"Mission Success First" in Negotiations	140
2.15	Writing Complex Proposals		140
	2.15.1	Where Do You Place the Most Important Contents of the Complex Proposal?	140
	2.15.2	Familiarity of the Audience	141
	2.15.3	Behavioral Expectations by the Buyer	143
	2.15.4	The Friendly Dog Effect	143
2.16	Non-Disclosure Agreements and Non-Compete Clauses		145
2.17	Submitting and Presenting the Offer		147
	2.17.1	The Most Fundamental Consideration for the Presentation	147
	2.17.2	Preparing the Offer Presentation	149
	2.17.3	Preparing for Q&As	152
	2.17.4	Some More Don'ts of Offer Development	152
2.18	The Contract		153
	2.18.1	Binding and Nonbinding Agreements	153
	2.18.2	Signing the Contract	156
	2.18.3	Startup Meetings: On-Boarding and Kick-Off	158
Chapter 3 Contracting			**159**
3.1	Contracting as a Process		159
3.2	Introductory Questions		159
3.3	Good Faith and Mutual Obligations		161
	3.3.1	The Doctrine of "Good Faith" in International Project Contracting	162
	3.3.2	Good Faith in Common Law	164
	3.3.3	Good Faith in Civil Law	164
	3.3.4	Good Faith and Basic Trust	165

	3.3.5	Concurrent Sourcing	174
3.4		International Contracts	176
3.5		Incomplete Contracts	178
3.6		Project-Related Contract Types	181
	3.6.1	Contract Types in Civil Law Systems	182
	3.6.2	Contract Types in Common Law Systems	182
	3.6.3	Assigning Cost Risks to Contract Parties	183
	3.6.4	Motivational Price Adjustments	184
	3.6.5	The Capped Target Cost Contract	186
	3.6.6	The Rolling Award Fee Contract	189
3.7		Protective and Relational Contracting	192
3.8		Refinements and Changes	194

Chapter 4 Managing Complex and Dynamic PSNs — **197**

4.1		Change Requests in Complex Project Supply Networks	197
4.2		Introductory Questions	198
4.3		Teaming Agreements	200
	4.3.1	Direct Procurement	201
	4.3.2	Indirect Procurement	201
	4.3.3	Indirect Procurement over a Consortium	202
	4.3.4	Mixed Structures	203
	4.3.5	Customer's Involvement in Subcontracting	203
4.4		Managing PSNs Is Managing Interfaces	204
	4.4.1	The Sentiments of Industries	204
	4.4.2	The Five Dysfunctions	206
	4.4.3	How Competitiveness and Cooperation Interrelate	207
4.5		Interfaces Among Contractors	208
	4.5.1	Customer Projects with One Customer	208
	4.5.2	Customer Projects with Several Customers	209
4.6		Risks from Outsourcing Projects Under Contract	211
	4.6.1	Risks from Differences in Software Tools Used	211
	4.6.2	Risks from Interpersonal and Social Conflicts	214
4.7		Avoiding Crises in Project Business Management	221
4.8		Supportive Action, Provisions, and Enabling Services	225
	4.8.1	Understanding "Mission Success First"	225
	4.8.2	Objectives and Constraints	226
	4.8.3	Enabling Services and Provisions	228
	4.8.4	Institutionalized Conflict Resolution Mechanisms	230
	4.8.5	Project Management Information Systems in PSNs	234
	4.8.6	Specific Business Interests of In-Between Contractors	235
	4.8.7	Work Flow Management Across a PSN	236

4.9	More Control on the Project Supply Network		238
	4.9.1	Naming, Nominating, or Approving Subcontractors?	238
	4.9.2	Coaching, Consulting, Mentoring	239
	4.9.3	The Other Stakeholders on the Customer Side	239
	4.9.4	Professional Integrity in the Project Supply Network	240
4.10	Chicken Races and How to Avoid Them		242
4.11	Closing Contracts and Projects		244
	4.10.1	Handovers and Acceptances	244
	4.10.2	Contract/Procurement Revisions in Project Supply Networks	245
4.12	Do We Need New Approaches to Contracting in Project Supply Networks?		246

Chapter 5 Project Business Management and Crisis Management — **251**

5.1	The Power of Documentation		251
5.2	Introductory Questions		253
5.3	The Dynamics of Success and Failure in Project Business Management		254
5.4	Causes for Conflicts in Project Supply Networks—A Survey		258
5.5	Benefit Engineering		260
	5.5.1	Benefit Engineering—The Process	262
	5.5.2	Negative Benefit Engineering	268
	5.5.3	From Cost Engineering to Benefit Engineering: A Deep Cultural Change	269
	5.5.4	Contractor-Side Benefit Engineering #1: Application by the Direct Contractor	270
	5.5.5	Contractor-Side Benefit Engineering #2: The Subcontractor's View	270
	5.5.6	Customer-Side Benefit Engineering	270
	5.5.7	The Caveats of Benefit Engineering	270
	5.5.8	Benefit Engineering as a Form of Change Request Management	271
	5.5.9	Project Managers as Benefit Engineers	271
	5.5.10	Does the World Need Professional Project Business Managers?	272
5.6	Turning a Customer Project Out of Crisis		276
5.7	Project Business Management—Is It an Open or Closed Skill Discipline?		280
5.8	Limitations of Tools for Project Business Management		282
	5.8.1	Market Window Theory	283
	5.8.2	A Method Cannot be Applied	284
5.9	Conclusion		285

Answers to the Introductory Questions — **289**

Glossary — **293**

Bibliography — **303**

Index — **311**

List of Illustrations

Figure 1.1   A majority of project managers believe in universally applicable practices.   5

Figure 1.2   Comparison of the two main station projects; the crisis in Stuttgart 21 refers to the years 2010 and 2011, and the project has since been led mostly back on a success path.   7

Figure 1.3   Responses from global group on how static and predictable their projects are.   12

Figure 1.4   The differences in lifecycle value creation of engineers' and gardeners' projects.   14

Figure 1.5   Multiple handovers deliver value in stages.   15

Figure 1.6   How project managers report deadlines.   16

Figure 1.7   Intercompany business and captive business.   19

Figure 1.8   The percentage of project managers in customer projects was roughly the same as in internal projects in a survey performed in September 2015.   21

Figure 1.9   The three options to get a work package done.   23

Figure 1.10 Three situations on the customer side that can lead to projects on the contractor side.   23

Figure 1.11 The complex flow of actions and documents in a customer project, assuming that Seller #1 wins in a competition against the other two sellers and becomes the contractor.   36

Figure 1.12 An example of a simple multi-tier contractor structure.   37

Figure 1.13 A stakeholder attitudes influence chart (SAIC) to be consulted when decisions are to be made, what influence they will have on stakeholder attitudes.   40

Figure 2.1   The word *cloud* created by the software from the frequency of words named.   46

Figure 2.2   The standard process flow, as it is often seen in project business with non-incumbents, begins on the side of the organization that is to become the buyer in the subsequent process, when a *make-or-buy* decision has been responded to with *buy*.   46

Figure 2.3   Investment and return expectations when work is delegated.   49

Figure 2.4   An example of the increasing speed of innovation and change of paradigms: telephones.   51

Figure 2.5   The different lifecycles around the project and the benefit realization for the performing organization.   58

Figure 2.6   The sliders used to answer questions on recent experience and future expectations on *make-or-buy* decisions.   59

Figure 2.7   The distribution of the responses on recent experience and future expectations regarding *make-or-buy* decisions using the seven-step sliders shown in Figure 2.6.   61

Figure 2.8   The past and future average trends for *make-or-buy* decisions grows with the workload assignment already outsourced.   62

Figure 2.9   The past and future average trends for *make-or-buy* decisions grows with the size of the project.   63

Figure 2.10   The average trends for the recent experience and future expectations on *make-or-buy* decisions are different over the world regions, but the trend is toward further growth in all of them.   64

Figure 2.11   Identifying sellers, approaching them, and requesting responses from them generally follows the *make-or-buy* decision.   85

Figure 2.12   The venue is now changing to the sellers from whom the buyers would like to receive an offer. The decision by the seller to make such an offer or not has many facets.   96

Figure 2.13   A survey among project managers on deadlines revealed that a three-quarter's majority of projects has time pressure, and that most of these projects have not only one.   97

Figure 2.14   TRAC subsumes many common influencing factors for the offer/no-offer decision.   98

Figure 2.15   TRAC used for a force-field analysis.   102

Figure 2.16   The competition for an offer is more than just the other offerer.   105

Figure 2.17   The top rankings of selling organizations' expectations of buyers' selection criteria are quite well aligned with the top-ranked criteria actually applied by buyers.   108

Figure 2.18   The cost calculation of the incumbent is commonly easier, because the investment in the development of the offer is paid back by the combined earnings from more customer projects.   109

Figure 2.19   Sending the offer to the buyer is a signal of preparedness and capability by the seller to do the intended project work, and also a commitment to satisfy the customer's needs, wants, and expectations.   112

Figure 2.20   An offer can be binding, which makes it an offer in both a commercial and a legal sense. The acceptance of the offer by the other party then makes the contract. An invitation to treat is an offer in a purely commercial sense. When the buyer accepts it, the contract is still not made in a legal sense— it is the acceptance of the seller that concludes the contract.   117

Figure 2.21   Different forms of teaming agreements for project work.   124

Figure 2.22   A typical build-operate-transfer (BOT) project has the intention that the contractor invests into the project (mostly for infrastructure) and later recovers the investment from the return made from operating the infrastructure. The example above has 35 years' total duration, including the build and the operation phase.   127

Figure 2.23   A swim-lane diagram showing acceptable price ranges by two parties. If these price ranges overlap, a contract is possible. A gap between the two price ranges makes it unlikely that the business can be developed.   131

Figure 2.24 The three types of limits on the buyer side relating to price and advance performance, and what response by the buyer is to be expected for the sectors that are delineated by these limits. (Numbers shown are examples.)   133

Figure 2.25 In a competitive setting, the "naked" price may be what matters most to the customer, but there are various options to package it and spruce it up with more beneficial combinations of numbers.   135

Figure 2.26 Quoting a naked price is the easiest task for the seller. Making the offer more relevant by adding further cost and benefit information is challenging, but may be the key to success.   137

Figure 2.27 Four quadrants describe whether a seller in a customer environment is expected to transform the environment or to act within it, and on what level this should happen.   144

Figure 2.28 Different forms of agreements.   153

Figure 2.29 The competition for the contract with the buyer and with the award and the signature of the contract.   157

Figure 3.1 The global distribution of legal systems.   163

Figure 3.2 At date *n* the project team and the vendors have to act in a competitive way in the procurement items that are in the acquisition phase, whereas they have to work in a collaborative style in the items that are already worked on.   174

Figure 3.3 The dynamics of stakeholder requirements on projects.   179

Figure 3.4 Differences in planning approaches in project management.   180

Figure 3.5 Assignment of cost risks depending on the project type.   184

Figure 3.6 A score sheet indicates the results achieved by the contractor and how these are computed from a rating and a weight.   190

Figure 3.7 Cost savings to the left of the target line are shared between customer and contractor. Sharing cost overruns ends at the point of total assumption, from which point on the contractor assumes the total cost risks.   193

Figure 4.1 Different structures of teaming agreements. The parties in the black boxes are in charge of managing the project. The arrows depict the general flow of money.   200

Figure 4.2 Development of Oil Price 2012–2017, Brent Crude, price development per March 13, 2017.   204

Figure 4.3 The number of interfaces in a simple PSN (left) with only three contractors is six, including the interfaces with the customer. The PSN on the right-hand side has twice the number of contractors, but the number of interfaces has grown by a factor of 3.5 to 21.   209

Figure 4.4 For a project with one customer and a larger number of contractors, managing the interfaces among the contractors can be among the most challenging tasks.   209

Figure 4.5 Privity of contracts: The customer has no contractual relationship with the sub-contractor.   218

Figure 4.6 Commonly found obligations in a simple two-tier PSN.   230

Figure 4.7 A project begins with high uncertainty. Then, stakeholders follow a learning curve.   231

Figure 4.8 The development of feasible decision options and the costs to implement them follow an opposite development to the learning curve. The response should be to accelerate the learning curve.   231

Figure 4.9   The workflow plan of the project in the case story before and after fast-tracking. The bar lengths are not proportional with the durations of the work items.   236

Figure 4.10   Preparation to pay or extort bribes in industry sectors from a Gallup survey in 2010.   241

Figure 4.11   In a project or a program consisting of (sub)projects with a common deadline, such as a project to enable timely start of production (SoP), one late project delays the SOP, the deadline of the program.   242

Figure 4.12   The common process flow in project procurement ends with formal contract close-out.   245

Figure 4.13   The procurement lifecycle includes the various contract lifecycles plus some time before and after the actual contracting period for preparatory work and final organization of documentation and other deliverables. Revision at the end of each lifecycle can help communicate a culture of "cleanliness".   246

Figure 4.14   The DIKA maturing process of decision making, beginning with (raw) data and ending with action, which in turn provides new data.   248

Figure 5.1   Many participants of the survey have collected experience in multiple roles, so that the numbers do not add up to 100 percent.   258

Figure 5.2   The most frequent cause of disruptions among the participants of the survey were conflicting business interests.   259

Figure 5.3   Repetition—for an internal project, benefits from the project are typically expected for the future. In a customer project, the contractor expects the benefits during the project lifecycle, beginning with the first payment and ending with the last.   261

Figure 5.4   Cost engineering addresses project costs, mostly on the side of the contractor. Benefit engineering addresses and increases the benefit.   262

Figure 5.5   Benefit engineering builds on a deep understanding of the customer as well as one's own organization.   263

Figure 5.6   A project business manager combines business acumen, which helps win attractive customer projects and expand existing ones, with project management competency to realize the potential from the business.   273

Figure 5.7   In old-style over-the-fence project management, a number of business units or independent contractors drove the project along a sequence of phases, with another unit responsible for another phase. Project managers integrated the project phases. Project business managers remove the fences and integrate the remaining phases at project beginning and end.   274

Figure 5.8   Project business managers oversee the entire process from the first contact relating to the project between buyer and seller to the final closeout and into utilization and servicing of the final deliverables.   275

Figure 5.9   The flow of invoices and payments between subcontractors and the customer. Invoice validation is the core process at the heart of the business.   277

Figure 5.10   Market window theory with six groups of participants. Note the development of initial outflow and consequential inflow over time.   284

List of Tables

Table 1.1 The Typological Dimensions That Were Identified During the Research 9

Table 1.2 Fundamental Differences Between Internal and Customer Projects 27

Table 1.3 Integrating and Fragmenting forces in PSNs 30

Table 2.1 Top Ten Countries with Participants in the Survey 60

Table 2.2 Answering Options for Workload Assignments in Projects and Distribution of Responses to Them 62

Table 2.3 Answering Options for Project Sizes and Distribution of Responses to Them 63

Table 2.4 World Regions Defined and Distribution of the Responses Across These Regions 64

Table 2.5 Example of a Set of Rules with Staged Requirements on the Procurement Processes to be Applied, Depending on the Value of the Procurement Item 94

Table 2.6 TRAC Used in a Weighting System to Develop a Total Score That Helps Make the *Offer/No-Offer* Decision 101

Table 2.7 A Sequence of Deadlines Imposed by the Buyer to Coordinate the Work of the Sellers with the Buyer's Schedule 122

Table 3.1 Some Typical Obligations Contract Parties Have Toward Each Other 161

Table 3.2 Motivational Price Adjustments Used to Motivate a Contractor to Deliver by a Certain Date Under Certain Fixed-Price Contracts 186

Table 3.3 Step 1: Cost Reimbursable Contract with Fixed Fee 187

Table 3.4 Step 2: Cost Reimbursable Contract with Fixed Fee and Cost/Benefit Sharing (Target Cost Contract) 187

Table 3.5 Step 3: Target Cost Contract with Price Ceiling 188

Table 3.6 Step 4: Calculating Costs at the Point of Total Assumption (PTA) 189

Table 3.7 The Distinction Between Refinement and Change 195

Table 4.1 Fortune 100 Top Ten Companies to Work For 205

Table 4.2 Objectives and Constraints are Separated by Reserves 228

Table 5.1 The Cost–Revenue Plan of Cicada for the 2013 Business Year, Projected at the End of the Previous Year 255

Table 5.2 The Cost–Revenue Plan of Cicada, Revised in July 2013 for the Entire Year 255

Table 5.3 The Cost–Revenue Plan of Cicada for the Given Business Year 256

Table 5.4   Requirements Commonly Placed on Managers Today, Compared with
             Those in the Past   265

Table 5.5   A Common Approach Generally Follows Three Steps   281

Foreword

I first met Oliver Lehmann in November 2016 at the Project Management Institute (PMI) "UK Synergy", the annual conference for its project management professionals, where I was giving a presentation. I had been developing relationships and collaborating with several global PMI Chapters (including New York City, Lévis-Québec, Montréal, South West Ontario, and the Philippines, as well as the UK's regional satellite branches). There were (are) other Chapters in the early stages of *kennenlernen* ("getting-to-know-you"), including Berlin, where I had met Chapter members in September 2016 after an introduction by Oliver via LinkedIn. Time constraints had prevented me from travelling on to Munich, where Oliver is based, after Berlin, but we had corresponded about collaborating and had resolved to meet when either of us was in the other's city.

And so it happened. A beaming man wearing colourful glasses came bounding over to me at Synergy, bearing a book. I had just given my presentation, *The Impact of Law on an Under-Rated Art,* to the assembled 750 delegates in the magnificent Westminster Central Hall in London. Readers who were there may recall the rather memorable start to my presentation of Taylor Swift's *Out of the Woods,* overlaid with my "new" lyrics about project management. I talked about the project management profession being a blend of the art and the science, and emphasised the importance of equipping project managers with basic legal skills so as to enable them to navigate their way through the straightjacket of law for the benefit of their projects, and not to their accidental detriment.

I also talked about "internal" and "external" projects, and the differences between when the project manager's customer is within one's own business ("internal") or is a paying customer ("external"). I warned of the need to take care with professional indemnity (errors and omissions) insurance, which every paying customer demands, and the mistakes which external project managers can easily fall foul of. Internal projects could be hazardous too, and open to political sabotage. Project managers had much to juggle and felt heavy responsibilities.

Oliver was keen to talk more about "internal" and "external" projects. "Let me show you my new book, *Situational Project Management—The Dynamics of Success and Failure,* which has just been published", he said; "I also talk about this differentiation from the project management perspective. I would like you to write the foreword for my next book". I agreed, and in return, asked him to autograph the book, which he obliged. By coincidence, we discovered that we sat at the same table for the rest of the day too. And so our professional connection was cemented.

I carefully read Oliver's book. It was a fascinating read and chimed with the way I consult and train on Law and Project Management. As a lawyer who specialises in infrastructure and commercial development, with over a decade of industry experience as a senior in-house counsel, and prior to that, six years of private practice experience, SitPM (Situational Project Management) makes obvious sense to me. The scales must balance for project success: on one side sits "flexibility" and the ability to adapt to change; on the other side sits "control" and the need for compliance and process, or protocol.

Oliver's book is simply very well written: one must thank the precision of the German language for having gifted Oliver with the ability to write clearly and meticulously. His explanations are comprehendible by experienced and aspiring project managers alike, and they have improved my (non-professional) understanding of SitPM—the technicalities of the profession, of methods of practice, and of the hazards, which prevent project managers from performing well. One hazard is being too rigid, or being fixed on the "plan" or the "process". Another is being too flexible, of performing without solid structure or robust plan. In my classes, I emphasise that one style of project management does not fit all projects, and despite so-called "best practices", none can be applied to every single project permutation. Above all, the contract will dictate the mechanisms and tools available to the parties, and the project manager. Oliver's book, and now his second book, *Business Project Management*, talk similarly.

Coming back to project strategy and structure, the focus on "internal" vs. "external" projects yields interesting "pull and push" concepts. This is the core of this book. "Internal" projects are ostensibly for the good of the organisation, but as a cost centre they must make impact and yet not unduly upset business as usual (BAU) operations. "External" projects make money (hopefully) for the organisation, and so are often seen as more important. In addition, the project management expertise may be "internal" or "external". All these permutations have different pros and cons. Oliver characterises the "internal" vs. "external" dilemma in his research as the "make or buy" scenario. Project managers will be interested to hear that his research shows that customers are increasingly tending to "buy" (i.e., go external) rather than "make" (i.e., stay internal). There is a sense of buying expertise, and cost-efficiency, whether in relation to tangible items or professional services.

Where the customer engages an "external" project manager to act on its behalf (whether for an internal project or an external project), the most important stage is contract formation. In my legal consultancy and training workshop organization, we have clients on both sides of the contracting table. One day, I might be working for a customer with a project "vision", building the contracting strategy and drafting the project documents, including the project manager's appointment. Another day, I might be working for a project management consultancy, supporting them to negotiate their services contract. Since they are at the "sell" end, they must make money from managing their customer's projects, making each one count to maintain profit level, reputation and market access.

A project manager's service contract may be as few as 10 pages or more than 100. The contract will consist of terms and conditions: the services to be delivered, the payment in return, and the attendant processes and procedures as well as caveats, applicable law and termination and dispute resolution procedures. There may be much more too, dependent on the subject matter (for example, intellectual property, health and safety, sector-specific terms, etc.). Fundamentally, this "buy" contract will be enforceable in law. By contrast, the "make" scenario involves no contract, since the project manager "belongs" to the customer (and even where there is a separate sister organisation for such purpose, which is usually tax-efficient, it would be usual to prevent the 2 organisations from suing each other). The "internal" project manager therefore considers only 1 contract (i.e., the customer-contractor contract). The "external" project manager must understand 2 contracts—the customer-contractor contract, and his/her own appointment. It is my job to ensure that such "buy" arrangements co-exist comfortably, without clash or conflict. The stakes for project managers between "make" or "buy" are thus different.

In addition to the growing volume of projects due to the "buy" scenario, projects seem to be getting bigger. This trend will call for more highly skilled project management professionals: for example, PMI estimates another 1.5m jobs will be created by 2020. It will also, inevitably, lead to more complex contract arrangements. In major infrastructure, development and technology projects, the stakes are particularly high, but in all projects, vendors want to win the business, and customers want to keep the price low and the delivery output high. This "pull and push" is normal in contract negotiations in a free market subject only to the straightjacket of the applicable law. But it also makes it incredibly important to get the contract provisions right.

Contract negotiations are often done by senior executives who may or may not be project managers themselves and understand what they are offering or demanding. In my experience, whatever the project size (value), it is common for the project manager, who "ends up" with the day-to-day project responsibility, to have no part in the engagement negotiations. In mature organisations, this is managed via a regular open news channel from the project director to the project manager team during negotiations to assist with planning resource and task allocation, review risk allocation/assumptions, seek input on proposed contract terms, etc. In less mature organisations, 'management' is lacking: news may only filter down to the project management team once the job is won and an individual project manager is chosen to service the customer's requirements. Most organisations I have worked with are somewhere in between. Thus, the person allocated to wear the project manager hat, whether "internal" or "external", is likely to be vulnerable from the start. It is incumbent on him or her to enquire about, and feel comfortable with, the scope of services to be delivered (although, if the scope has already been agreed, the chance to influence is almost nil). If it is unclear at the beginning, it will only get worse. Mismatched expectations are, in my experience, a sure-fire way to claims, disappointment and broken relationships, plus stress for the individual project manager. None of these consequences results in a happy customer nor brings money home for the vendor.

As such, I encourage my clients to include project managers in early contract discussions. It broadens their business acumen and gives early insight into the project, and enables the customer to get to know the project manager, both of whom can nurture a fledgling relationship before the project gets underway in earnest. On a practical level, if a project manager has sight of, and input into, the customer-contractor contract (plus the project management services appointment, if external), he/she is more likely to deliver success, and feel empowered in the role, since the act of inclusion in discussions promotes a feeling of value. Running the gauntlet between a contractor and a customer, whom the project manager barely knows, is hardly the right way to start a project! In such a scenario, I have facilitated a 1-day pre-works workshop, which is to be extremely helpful to project team *kennenlernen* and stress-testing the practical against the contract: "what could, and should, we do if X arises on this project?"

Charting such legal territory is necessary to protect the project, the customer, the contractor and the project manager. Losing contractual entitlements or inadvertently adding to a service scope are hazards to avoid. I have a challenging and exciting practice educating project management professionals about law, the core of which is: basing your arguments in fact and applying the correct law means you are rarely wrong. More easily said than done! At PMI UK Synergy 2016, I summarised the project manager role as one of "damage limitation". The other side of the same coin is to "maximise profit safely and make (keep?) the customer happy". With a similar voice, what *Business Project Management* (BPM) really does is to call for superlative, professional judgment, which is flexible within an appropriate legal and practical structure.

Ideally, a project manager's basket of skills should contain relevant technical discipline(s), plus experience in several business-oriented areas: leadership, law, governance and stakeholder management, people management, risk and financial management. In addition, if the project is one of a "bigger picture", there is scope for relationship solidification (or destruction). Individual and corporate reputations can be (and are) made or broken in such circumstances. However, Oliver's research shows, and this is also true of my experience, that project managers rarely have all these skills, and are often insufficiently trained to be able to execute all these functions well. There is much scope therefore for upskilling the profession, and this book will certainly help with this.

Throw in an international contracting dimension, and the project manager must be mindful of multiple jurisdictions and cultures. For example, a German project manager managing a Government-sponsored building project in UAE delivered by an American main contractor and British architect. Dig deeper, and the project manager finds the sub-contractor is Chinese, who is using labour from Pakistan. The supply network sending components emanates from a dozen other countries. Each of those 20+ contracts will be founded on a single applicable law agreed between the relevant parties. To some extent, this type of complex arrangement explains the popularity of international arbitration, which transcends geographical boundaries and provides a neutral dispute resolution forum. Nonetheless, a project of this sort is extremely challenging for a project manager. Whilst the project manager does not "rule" upon the legal position of any particular issue, he/she uses independent professional judgment to make decisions about project matters throughout the delivery phase, which are usually binding contractually unless challenged using the dispute resolution procedure. Binding is necessary to maintain project progress: building does not stop while everyone works out what their legal rights and obligations are. Projects are "live" and multi-layered. What affects one party may not flow down the contractual chain to another. The project manager must be alive to this as a principle.

Another important project manager skill is record-keeping. Contracts such as the construction industry NEC3 (as of June 2017, NEC4), commonly used in the UK and increasingly worldwide, require the project manager to be the custodian of the project record, which includes documents such as the early warning risk register and the programme, as well as plans, drawings, the specification, and communications (instructions, notices, submissions, etc.). In my experience, what may or may not be done with this data, including security and ownership, is poorly explained in contracts and equally poorly understood. The international rise of Building Information Modelling takes care of this to a limited extent. It is most definitely a contractual hazard for project managers.

In my opinion, contracts are fiendishly difficult and, although archaic language is (thankfully!) dying out, they can be long and laborious. Words, which look straightforward, can be deceptive. Taking time to read, understand and react to the contract provisions should be the right—and duty—of every project manager. I think Oliver would agree with me on this.

Over these last few months, a friendship has been born. Oliver is resolutely passionate about the value of project management to business, as I am. His humour is infectious. We share a love of languages too, which enriches both our lives. It seems appropriate that I wish him heart-felt congratulations, *herzlichen Glückwünsche, félicitations, felicitatiónes,* and *gratulatii* for *Project Business Management*!

— Sarah Schütte
Solicitor-Advocate, LLB, ACIArb, MPD, member PMI
Managing Director, Schutte Consulting Limited
"Making law work for the construction and engineering industry"
London, UK
July 2017

Preface

Whenever you find yourself on the side of the majority,
it is time to reform (or pause and reflect).

— Mark Twain*

Project management has turned from a technical discipline to a business. Today, many corporations and other forms of organizations not only attribute their business success to the advances that they achieve by doing effective projects; projects are their direct source of income as project contractors.

Others must learn as buyers to mature their capabilities in project procurement management from doing some purchases here and there to engineering and managing complex, dynamic, and often opaque *project supply networks* (PSNs). These PSNs often include large numbers of heterogeneous contractors with the potential of conflicts on various levels.

All these players on the customer and the contractor side have specific business interests and corporate cultures. Many are headed by big egos, and when these supply networks become international, differences in time zones, legal systems, and business principles will further impact the projects. Working under contract in their projects, these corporations have to act together as PSNs to achieve mission success while their differences permanently threaten to impact trust, communications, and collaboration. There is a large number of examples of failed projects that would have been successful if the PSNs had been managed more proficiently.

From time to time, it seems advisable for the project management profession to examine whether its self-perception of what it is and what it does is still lined up with the actual practices of its practitioners. It is to some degree satisfying to see our representatives invited to speak at congresses that focus on economy and international politics, and when new societal concepts turn up—such as "Industry 4.0" or "Smart cities"—experts from project management are now commonly among the first people asked by media what they think about it.

While striving toward new devotions, we have to take care that we do not lose the firm ground of practicality and realism under our feet, The tendency is going towards more projects performed for customers in PSNs, and while it is taken for granted that qualified project managers are all also capable of managing these supply networks, many of them cannot. In

---

* http://www.twainquotes.com/Majority.html

addition to project management skills such as planning, scheduling, assigning people to tasks, and managing risks, they also need business acumen, something not taught thus far in project management education.

The same is true in the literature. Many publications on project management, and also on the related disciplines of program and portfolio management, emphasize the strategic aspects of projects: They say that projects are temporary endeavors that must be aligned with the corporate strategy, support the corporate strategy, and help lead the organization in tumultuous times to develop a future as intended by management. No doubt, this statement is true for many projects. For all projects? Probably not.

Project management is about coping with uniqueness. Some projects, as I will show, have a setup that is different to this description. A good example of a type of project that does not necessarily follow strategic considerations are mandatory projects, performed by organizations to meet requirements imposed by law, investors, business partners, or other entities that are powerful enough to enforce these projects on the organizations without asking first whether they comply with its strategic goals.

Customer projects are also different. These projects are performed by one or more contractor organizations for paying customers that are separate economic entities to them, and these contractors may then use subcontractors, who in turn may hire sub-subcontractors, and so on. There are exceptions, but as a rule, it is safe to assume that the contractors on these varying levels generally work for the customer to bring money home. This book focuses on projects that are performed in such a cross-organizational fashion, which comes not only with new opportunities, but also with substantial risks for all parties involved. Project business management with paying customers and earning contractors generally means high-risk business.

A customer project can be high-profit business for a contractor and can provide the company with a lot of satisfaction, knowing that one had contributed to a great project and helped a customer achieve things that would otherwise not be achievable. It can also turn into an economic nightmare and destroy companies.

The audience for this book are practitioners in "Project Business Management" (PBM), who manage projects in complex and highly dynamic project supply networks (PSNs). The number of these networks and their complexity are increasing, and project managers are facing this development mostly unprepared. I see an urgent need for a book that helps project managers understand and master the challenges that come with this development.

In these networks, there are project managers who are managing projects under contract for external, paying customers. Then, there are commonly project managers on the customer side, who are running internal projects with major procurement activities, who have to work together with project managers from the first group, and who need an understanding of the businesses of the vendors to be effective as customers.

A further audience group is both at the same time—contractors to paying customers and customers to their subcontractors in complex and dynamic PSNs. These prime contractors and other intermediates have to consider a multitude of interests: those of their own organization, then those of the direct customer, and finally the interests of the project as a whole, whose overall mission transcends the companies involved, and whose dynamics of success and failure depend on the ability of the organizations involved to act together as a well-aligned system, resisting the temptation of the particular and short-term benefit that at any given moment threatens the joint success of the parties involved.

A further group addressed in this book are the business development managers on both sides of a contract—the vendor-side business developers and the customer-side purchasing executives. Although each of these audiences looks at project business from a different perspective, the businesses looked at are in essence the same.

This is my second book on Situational Project Management. In the first book, *Situational Project Management: The Dynamics of Success and Failure*, I gave an introduction to the basic concept of Situational Project Management—in short, SitPM—and how project managers should adjust their practices to ever-changing situations. The word *situational* reflects the simple observation that the same practices that have been successful in certain situations in the past may fail in other ones in the future. Practices should be understood here as the collection of approaches, behaviors, methods, tools, and techniques applied by project managers and their teams on organizational and interpersonal levels to their projects in order to avoid failures and create successes.

In this first book, I described the development of a typology for projects and project situations that should help better understand the differences between them and select the practices that are appropriate. The typology is an open one; I described 14 dimensions, but there are probably more. Adjusting practices to different projects and changing project situations is a difficult task, not an easy one. A situational project manager no longer follows a simple "best practice" model, but needs to master a variety of practices and select the ones for a given situation that are most favorable and likely to lead to success for both the project and the person.

Alternatively, a project manager may have access to another person who can support him or her in this role with behaviors that the project manager has not mastered so far. Situational project management is based on the understanding that project management is not a closed-skill discipline, in which the project manager simply focuses on his or her own performance and that of the team and executes a predefined plan without regard to changes that happen inside the project and around it. Project management is an open-skill discipline, which requires a proactive and responsive attitude and a lot of situational intelligence to master the dynamics of success and failure. This is not easy but comes with two valuable benefits: reduction of team stress and better projects.

One of the typological dimensions described was the distinction between internal projects and customer projects. An internal project is a cost center for the performing organization. It does not bring money home, but generates costs, often with the intention of a future business benefit to be generated in another workstream outside the project—for example, in ongoing and repetitive operations. The benefits may be additional income or cost reductions, but these would then be generated by this other workstream. Some internal projects do not create economic benefits, but are performed based on strategic considerations to make the organization future proof. Internal projects may serve other purposes, which could be curiosity or a social good. Sometimes, none of these business justifications apply, as many internal projects are simply mandated by law or other undeniable obligations, the mandatory projects mentioned above.

Customer projects, in contrast, are profit centers performed under a written or verbal contract. Project managers in organizations whose business is customer projects typically provide income for the performing organization. Some organizations performing customer projects may rely on this income as their only income, others have it as a more-or-less valuable add-on to their operational business. In both cases, the environmental factors and requirements on professionalism for project managers are different for customer projects and internal projects—a distinction rarely mentioned in the literature and, as far as I have seen, not elaborated in detail anywhere else.

One focus of the current book will lie on the second type of projects. The first intended audience of this book are therefore project managers whose projects are acting as profit centers. Many customer projects are embedded within complex PSNs that extend organizational, national, cultural, time-zone, etc. borders and come with legal, technical, organizational, and interpersonal interfaces that develop complex dynamics, which often grow during the course of a project to a degree that was not predicted at the onset of the project.

Understanding these project supply networks (PSNs) and their interfaces takes a lot of courage and empathy for all parties involved. They bring new opportunities to corporations and other organizations and increase their adaptiveness.

The basic desire of parties involved is to tap assets of other organizations and turn them into project resources. For buyers, these assets may be skilled people, machinery, patents and licenses, and others. The most important asset that buying organizations commonly want to tap is management attention—finding someone who will take care for the correct performance of the outsourced work.

For a seller, the most obvious asset to gain access to is the money of the buyer.

Project Business Management comes with high financial risks that can jeopardize the existence of an organization, and the troubles or even default of one organization may impair the business success of other organizations in the network and, in a worst case, endanger the entire mission. Customer project management is a business task with maximum stakes, and project managers on all tiers in a complex PSN act at the limits of what is actually manageable. The same is true for the buying organization, which can massively suffer from troubles on the side of its contractors.

Readers who know my first book, *Situational Project Management—The Dynamics of Success and Failure,* will find some overlap and repetitions in this book. This is natural and to some degree intended. The two books do not describe different topics but reflect on basically the same subject—Situational Project Management (SitPM)—from different viewpoints. The first book describes the basic approach of SitPM and typifies projects and project situations to allow for deeper analysis and focused problem solutions. This book, *Project Business Management,* now puts the spotlight on one type of projects—customer projects—and shows what new knowledge can be gained from this narrowed focus of research and contemplation.

Another reason for possible repetitions is my wish to make this book sufficiently self-contained, so that a reader who does not know SitPM in depth can still use it and benefit from reading and from applying its concepts, tools, techniques, and approaches. A third aspect is of didactic nature: *Repetitio mater est studiorum;* repetition is the mother of learning.[1]

---

[1] Marketers know this principle as the "Rule of Seven": The target audience of an advertisement or a commercial must see a message at least seven times before it is grasped and taken in (Eaton, 2014, p. 48). The same principle applies in education, where repetition can help students learn. It can also make studies, and commercials, very boring, of course.

My intention with this book is twofold: raising awareness of risks and specific problem fields and develop solutions that are easy to implement in practice. The audience for which I wrote the book are seasoned practitioners, who will often recognize the situations described in my case stories and will value the warnings and advices. I also intend to address educators and other groups of rarely practitioning experts.

Project management offices (PMOs) in organizations doing customer projects and using them as their source of income should consider developing themselves into *project business management offices*, to not only streamline methodologies, terminology, and basic approaches, but also to monitor profitability and liquidity of portfolios with customer projects to ensure early response when these portfolios run into crises. I hope their staff will find the book helpful.

An audience from companies with increased exposure to liabilities are people working for prime contractors and other forms of "in-betweeners". These companies act as commercial intermediaries between customers and subcontractors, being contractors to the former and customers of the latter. The position may be a profitable one, which engages subcontractors' skills, knowledge, experience, etc. to earn good money from the customer without overburdening their own resources.

In other moments, diminutive margins, earned when the costs that are invoiced by subcontractors have been deducted from the price paid by a customer, may make this "ham in the sandwich" position deeply unappealing, particularly given the risks on financial, technical, and legal levels that the prime contractor has to assume. When the book looks at both sides of project business, buyer and seller, staff members of prime contractors actually do business on both sides.

There has not been much literature written on project business management, and not much research has been done in the field.[2] This in turn is an opportunity for scientists and also for students to build new knowledge. During my time writing this book, I did some research on a scale that is compatible with my profession as a trainer and my additional work as an association volunteer and author of books and papers. Whenever this research answered a question I had, two or three new ones turned up. If a reader is interested in *Project Business Management* as a topic for academic research, possibly for a thesis or a dissertation, I will be happy to communicate these questions and give advice on how to design studies to find answers and publish them.

Finally, if the book helps practitioners to do better projects in customer/contractor supply relationships, turning contract parties into project partners, raising awareness of the risks in the business and support solutions to manage these risks, it satisfies my intention of giving something back to the profession that has given me income and satisfaction for over three decades and whose staggering growth I had the joy and honor to observe over that time.

— Oliver F. Lehmann, MSc, PMP®
Munich, 21 April 2018

---

[2] An exception is the book by Robin Hornby, *Commercial Project Management* (Hornby, 2017).

Acknowledgments

I am most grateful to my wife, Silvia; my children, Antje, Daniel, Sandrine, and Tizian; and my grandchildren, Amelie and Helena, for their patience with me, when I was busy with this book and could not give them the time and attention that they deserved from me.

This book is to a great degree a result of deep and lasting friendships. Friendship can be built on common interests, common enemies, and common values. Interests and enemies are changing over time. Values last forever.

My special thanks go to Dr. Ginger Levin, my editor, and to John Wyzalek from Taylor & Francis. I would also like to thank the production staff at DerryField Publishing Services— Theron Shreve and, of course, Susan Culligan, with whom I worked in tight cooperation, and who was most critical and helpful to ensure that my English was not too much tainted by Germanisms.

In addition, I am thankful to the many educators and experts whom I had the honor to meet in person, and whose thoughts and beliefs helped shape mine, including James R. Snyder, one of the founders of PMI; William R. Duncan, primary author of the original *PMBOK® Guide* 1996 and the trainer for my PMP® preparations; Patrick Weaver, who is one of the few experts I have met so far who shares my interest in the history of project management; and Cornelius Fichtner and Jonathan Hebert from OSP International, who challenged me on important questions in the field and raised my awareness that such a book must not only suit experts but also give help to practitioners. Thanks also go to Deanna Landers, Robert Monkhouse, and Kris Troukens, who together with other idealistic friends and colleagues founded Project Managers Without Borders and opened the profession to a new degree of altruism, benevolence, and magnanimity.

Antje Lehmann-Benz, my daughter and colleague, challenged me repeatedly on the compatibility of Project Business Management with agile practices, a field that definitely deserves more research.

My gratefulness also belongs to Goran Krstulovic, who as a training provider immediately identified the new opportunities that come with Project Business Management and who put a great deal of effort into making customers aware of the educational gaps present in this field.

Very helpful was also David Pells, who gave me the opportunity to write a series of papers for his monthly magazine *Project Management World Journal,* where I could address many topics that also found their way into this book and test some of them for acceptance by an international readership. One of the papers, named "Crisis in Your Customer Project? Try Benefit Engineering", received the magazine's Editor's Choice Award 2017, and benefit engineering is indeed at the heart of my recommendations for solutions to improve the profitability of a customer project and the happiness of the customer.

In discussions with several women—particularly Vivian Isaak (Magnum Group, Inc., Philadelphia, USA), Barbara Wichmann (Artemia Communications, Inc., San Francisco, USA), and Sarah Schütte (Schutte Consulting Ltd., London, UK)—I became aware of the importance of woman-owned business enterprises in Project Business Management. This gave me the sensitivity to be careful with language and to ensure that the wording used for the book is fully inclusive to women and the organizations they own.

In the past, I had the opportunity not only to follow a project management path of qualification, but also to get certified in the field of business development and proposal management, which led me to becoming an APMP Certified Trainer in this discipline. I am thankful to Tony Birch and Cathy Day (Shipley UK Ltd., Yeovil, UK) for making this possible. The lessons from this qualification influenced this book as much as my personal experience in the field. Kudos also to Wolfram Seyring (Shipley GmbH, Hilden, Germany), who spent many hours with me critically discussing the concepts of the book and refining my understanding on the business development side.

Another deeply influencing person for me and this book is Peter Eigen, the founder of Transparency International (TI), the world association against corruption. During a joint car trip in 2005, he explained to me the rationale of the foundation of TI and toughened my position against corruption, unfortunately something that impacts Project Business Management in some countries and industries on a massive scale.

My last "Thank you" goes to my customers and students over a time of more than two decades in the professional training business. They believed in me but also challenged my understanding with often tough questions. They helped me question and validate the models and practices that I trained against their individual realities of day-by-day project management and sharpened my understanding of the field of business that deals with the collaboration of companies in contractual situations. Mission Success First!

About the Author

 Oliver F. Lehmann was born in 1957 in Stuttgart, Germany. He has studied linguistics, literature, and history at the University of Stuttgart and project management at the University of Liverpool, UK, where he holds a Master of Science degree. He had practiced project management for more than 12 years, mostly for the automotive industry and related trades, when, in 1995, he decided to make the change and become a trainer, speaker, and author. He has performed seminars for employees from international companies such as Airbus, DB Schenker Logistics, Microsoft, Olympus, SAP, and Deutsche Telekom, and he has had assignments as a trainer in Asia, Europe, and the U.S. He works for several training providers and is also Visiting Lecturer at the Technical University of Munich. Lehmann is frequently invited to speak at congresses and other events, where he motivates his audience to welcome the incredible diversity found in project management around the world, and for the new experiences that wait for discovery by project managers.

Among his diverse certifications are the PMP® certification of PMI, the Project Management Institute from Newtown Square, PA, USA, which he obtained in the year 2001. In 2009, he acquired the prestigious CLI-CA (Certified Associate) certification issued by CLI, the Connective Leadership Institute from Pasadena, CA, USA, which is only held by a small number of people. He is also a Professional Scrum Master™ (PSM 1), certified by Scrum.org.

Lehmann has been a member of and volunteer at PMI, the Project Management Institute, since 1998, and serves currently as the President of the PMI Southern Germany Chapter. From 2004 to 2006, he contributed as an analyst for troubled projects to PMI's *PM Network*® magazine, where he provided the monthly editorial on page 1 called "Launch", analyzing troubled projects all around the world.

Lehmann believes in three driving forces for personal improvement in project management: formal learning, experience, and observation. He considers the last one often overlooked, but important, because attentive project managers do not need to make all errors by themselves, they also learn from the successes and errors of their colleagues.

Image courtesy of Andreas Madjari

Chapter 1

Laying Out the Scenery: The Business Side of Project Management

1.1 The Importance of Project Business Management

The project managers at Bumble Bee Inc.,[1] the prime contractor organization in the software development project performed for Woodpecker Ltd., their customer, hoped to have a party after final handover of the project's deliverables to and acceptance by the customer. They were certain to get the handover finished that day and to receive the final payment from the customer by the end of the month. The project so far had been successful, and the revenue–cost calculations promised a sound profit that would please management and shareholders. The expectation was that this would be like a great performance in a car race, a pole to flag victory, with the driver finally standing on the top step of the winners' podium.

Everything was going great . . . until the customer's representatives showed up, began their acceptance testing following the functional and technical requirements that had been laid out, and finally said "No" to the deliverables. Too many requirements had not been met, some of them described in the contract, others in e-mails and communicated in meetings. While the contractor at one point in time had lost track of them, the customer had not. The condemnation by the customer on what was developed so far was immense, and Bumble Bee had to accept that in order to meet contractual obligations, a lot of additional work needed to be performed. They also had to accept that they would not be able to bill the customer for the additional work;

---

[1] The names of the organizations are changed but the case story is real.

instead, the unplanned rework would cause delays that would lead to price deductions, and it was not clear when the resources would be available again to do this work. The Bumble Bee project managers also understood that their subcontractors would charge full rates for the additional work, and that the margin between their process to the customer and the costs incurred with their contractors would become very small and could possibly turn into red numbers.

In addition, instead of utilizing the opportunity to shine in front of the customer, Bumble Bee made the impression of being badly managed and disinterested in customer satisfaction, which they had never been. From a promising business, the project turned into a financial nightmare just in some hours, and at this time, there was no way out of the entrapment in which the contractor found itself.

Although modern project management originated as an engineering discipline, it has always had a business aspect attached. Funding and costing of projects are central tasks that performing organizations need to carry out, and there are many other aspects of projects that are more linked with the business character than with the technical challenges of the endeavor. A very central one is the mutual dependency that the parties (or partners?) in a project performed under contract develop once the contract has been closed. The contract comes with obligations, but also with high risks for both parties involved.

Project Business Management, or PBM, to simplify addressing to it, is definitively *not easy money.*

The following questions are written in the style of a certification test. They are intended to give you an understanding of the contents of the following text section and the questions that will be discussed in it.

1.2 Introductory Questions

It may be interesting for you to answer these questions before you read the section, and then again once you have finished it. There will be such a section with questions at the beginning of each chapter.

The answers are found at the end of the book.

1. A project may have no deadline, a single deadline, or a sequence of deadlines. These deadlines may be internal, may be defined in a contract with a customer, or may be otherwise mandated. Later in the project, deadlines are often found to be unrealistic, detrimental to the project, or just impossible to meet.
   What should a project manager do in such a situation?

   a) The project manager should try to get replaced by a colleague, before the difficulties from the deadline become visible to management.
   b) The project should be immediately terminated and the causes for termination updated to the corporate lessons-learned database.
   c) The problems that arise from the deadline should be communicated early to appropriate stakeholders to allow for timely resolution.
   d) The project manager should make a note and go on with the project—many problems are self-resolving over time.

2. The organization for which you are working has a new CIO who has recently sent out a message to all employees in her department, stating that the company IT wants to "go agile". Today, you were told that the software implementation project you and your team are currently managing has to be adapted to agile methodologies.
   How do you react?
   a) You immediately transform the methodology applied in your project to Scrum, appoint yourself as Scrum Master, select a product owner, and reduce the team to nine people.
   b) You make yourself familiar with agile methods in order to understand if these could be helpful for your specific project situation or are detrimental. Then you report your findings to management.
   c) You tell management that agile is not for you, that you think it never works, and that when an organization says something like that, it can be nothing but hot air. Therefore, you prefer to hand in your resignation.
   d) You are not sure what you are supposed to do. Therefore, you ask to be transferred to another project. You know that you want to avoid having to switch methodologies in a running project.

3. You work for a company that has contracted you out to a big corporation to manage a project for the financial sector. The head of that organization's PMO approached you today and suggested that her team could give you some proven best practices for the management of projects for the insurance industry.
   What is your reaction?
   a) You assess whether these "best practices" are rather favorable or detrimental in the given situation. If they are not, you reject using them.
   b) You generally reject using them, as you have not developed them by yourself.
   c) You ask her to immediately pass on all relevant information to you, to allow you to copy the past success in your project.
   d) You approach other project managers in the financial sectors about their best practices and apply these instead.

4. A number of companies have founded a consortium. In the context of project management, what is their most probable intention?
   a) To organize the lessons-learned documents in a way that will facilitate decision making in future projects
   b) To share organizational process assets (templates, forms, procedures, etc.) with the other venturing companies
   c) To reduce the strain on the credit line from acting for the project by "hiding" inside the consortium
   d) To build a temporary joint venture to perform a project together that each company alone would not be able or willing to do

5. An internal project is different from a customer project in what form?
   a) An internal project has a project manager, a customer project a project engineer.
   b) An internal project is a cost center; a customer project is a profit center.
   c) An internal project is a profit center; a customer project is a cost center.
   d) An internal project is more operational than a customer project.

6. As the project manager for a customer project, you find out that on the buyer side, no internal project has been defined, and that there is therefore no buyer-side project manager and project sponsor formally assigned and documented in a project charter. What should you be acutely aware of?

a) Omissions and errors on the customer side can impact the contractor's success. You should recommend the customer to also build a project structure, and if this fails, to find ways to protect the project.

b) It is of utmost importance to immediately define the role of each team member on the contractor and customer sides in a way that helps ensure that everyone knows you are the only person in charge.

c) The customer is contractually and legally obliged to also build an internal project when work is outsourced. You make the customer aware of this and prepare to sue the company in case of noncompliance.

d) It does not matter if the customer is organized for the project. The full responsibility for the success of the project lies with the contractor and its dedicated project team.

1.3 The Story So Far

The following paragraphs provide a brief summary of my efforts to develop a typology of projects[2] between 2014 and 2016. If you have read my book *Situational Project Management: The Dynamics of Success and Failure*,[3] you may skip this section and go to the next. If you have not, it may be helpful to understand the starting point—situational project management, or SitPM—better, of which Project Business Management (PBM) is a practical implementation.

Could it be that in its current self-conception, project management is much more similar to ancient alchemy than to a modern science or an art? Alchemists were driven by the desire to find the philosopher's stone that could turn lead and other cheap metals into gold. They searched for panaceas, cures for all diseases, and while they developed various laboratory methods—some of them still in use today—their activities were mostly performed against a background of mysticism and magic.

There were several steps that took practitioners and scientists from old alchemists' approaches to those of modern chemists. A central one was the publication of the periodic table (Mendeleev, 1869), which allowed chemists to classify and typify chemical elements and improve their understanding of chemistry through the identification of an inner order in the diversity of elements. A similar step was achieved in biology with the development of the Linnaean taxonomy, which allowed scientists to classify species and understand their relationships and also their differences.

Typologies and with them classifications allow us to better manage diversity. Another example is provided by burns. Burns happen on a continuum between a minor injury and the most

---

[2] The following paragraphs were previously published in *Project Management World Journal* (Lehmann, 2016a).

[3] (Lehmann 2016b)

dreadful damage to tissue that can happen to humans. Each burn is different, but a typology in the form of a system of degrees helps medical practitioners respond appropriately to them. Burns of a first degree are mostly treated by applying outpatient care and superficial methods. Burns of the third or fourth degree (depending on the system) will be treated in intensive care within a hospital. Despite the uniqueness of burns, the typology helps to better select the most suitable response.

One should note that the classification systems in chemistry and biology are open classifications, which can be expanded when new knowledge has been explored and new elements or species, genera, and so on should be added to the existing ones. This is different to the closed classification of burns; this classification is generally considered to be complete.

1.3.1 "Best Practices" or Uniqueness?

In project management, the common belief in the existence of a "best practice" approach is a concept comparable to alchemy, and it is widely held. Many project managers believe that there must be a practice that is applicable to all projects that will generally ensure success in all of them. Interested in the question of how popular this concept is, the author asked project managers between April and August 2015 whether they believed in universal best practices. He received 189 responses, and the majority confirmed that they believed in best practices within the discipline. Figure 1.1 shows the results of the survey.

In scientific papers and articles, any differences between project types are also commonly ignored. Searching websites that provide links to published work on project management gives many results of research in project management generally, but the vast majority is not linked to specific types of projects. The questions that they raise would be similar to scientific papers and articles in chemistry asking, "What is the boiling point of an element?", or in zoology, "How

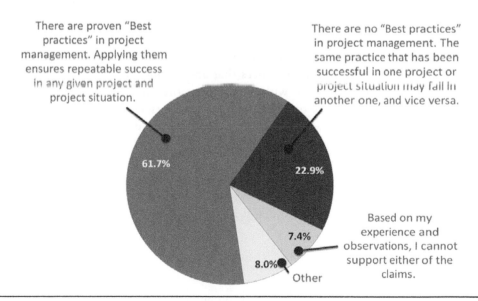

Figure 1.1   A majority of project managers believe in universally applicable practices.

do animals survive?", ignoring the fact that boiling points are different from element to element and also depend on environmental conditions, and that the latter is also true for the survival strategies of animals.

There are also promotion campaigns for "proven best practice methodologies" that can be "applied to all types of projects", which would be comparable to a description of the best treatment practice for all burns, ignoring their degree.

This last point is very common in project management in organizations. When one talks with managers from the project governance functions, statements such as, "We are moving all projects to agile methods" (or any other methods that are considered cooking recipes for project management) is possibly good news for some of their projects, but may be bad news for others.

The claim of "best practices" contrasts with the definitions of the term "project" used in the various international standards. Here are some examples:

- "A project is a temporary endeavor undertaken to create a unique product, service, or [other kind of] result".[4]

- "Project: [A] unique process, consisting of a set of coordinated and controlled activities with start and finish dates, undertaken to achieve an objective conforming to specific requirements including constraints of cost, time, and resources".[5]

- "*Projekt: [Ein] Vorhaben, das im Wesentlichen durch Einmaligkeit der Bedingungen in ihrer Gesamtheit gekennzeichnet ist*".[6]

The International Organization for Standardization (ISO®) standard 21500 (ISO, 2013) also emphasizes uniqueness as a main characteristic of projects and explains how differences between seemingly similar projects can arise due to the specific processes applied or how they may be affected by the unalike environments in which the projects are performed.

In reality, the same practice that has led to success in past projects may lead to failure in others, and vice versa. The principle may be relevant not only to entire projects and their lifecycles, but to situations during these times. Approaches that have led to success in certain situations may cause troubles in other situations still within the same project. Accordingly, Situational Project Management (SitPM) is the application of those practices that are favorable in given project situations while avoiding other practices that are considered detrimental. Situational project managers are not just confident with the practices that they master but go through a lifelong learning process, adding new tools, techniques, behaviors, etc. to their existing capabilities, much as a craftsman or craftswoman adds new equipment and tools to their job shops to help them meet varying demands and requirements.

What happens if a project manager, believing in the comforting certainty of a best practice, avoids this continuous learning process? If a person only has a hammer to make a living, the person must convince the world that it is made of nails. Another person may only have a screwdriver, and this person must tell the world that it is screwed.

---

[4] (PMI 2014, p. 4)

[5] (BSI 2000, p. 10)

[6] An undertaking that is chiefly characterized by the uniqueness of the conditions in their entirety. Own translation (DIN 2009, p. 11).

An example of how a practice that has been successful in one place may lead another project into trouble is illustrated through the two central station projects in Germany—Berlin and Stuttgart. They can highlight the insufficiency of "one-size-fits-all" approaches that come with the postulation of universal best practices:

- **Berlin Hauptbahnhof.** Berlin Central Station was a new construction opened in 2006. In its overall appearance, it is mostly considered a successful project, a piece of modern traffic infrastructure, which meets functional purposes and is aesthetically impressive.

- **Stuttgart 21.** Stuttgart Central Station is a reconstruction of an existing station, turning the tracks by 90 degrees to convert a 16-track dead-end station into an 8-track through station. The project began in 2010, and it was planned to have the new station operational by 2016. Still in the first year, the project was facing massive resistance from local citizens who took their rejection of it to the streets and demanded complete project termination. Their protests caused a delay in the project schedule and led to the resignation of the project manager in May 2011.

It is interesting to compare the two projects, because at first glance, they are almost identical (see Figure 1.2).

	Success → **Berlin Hauptbahnhof**	**Crisis** → **Stuttgart 21**
Industry	Railway	Railway
Application area	Construction	Construction
Deliverable	Main station	Main station
Organization	Deutsche Bahn, local and national government	Deutsche Bahn, local and national government

Figure 1.2 Comparison of the two main station projects; the crisis in Stuttgart 21 refers to the years 2010 and 2011, and the project has since been led mostly back on a success path.

To make the difference even harder to understand, both projects were performed by the same project manager (in the Stuttgart 21 project until May 2011), using the same approach, which was obviously beneficial for the Berlin project but detrimental in Stuttgart.

On the surface, it is hard to understand why the project in Berlin was successful while the project in Stuttgart was not. The typology described below should help in making sense of the differences and lead to adaptation of approaches by project managers and other supervisors.

1.3.2 A Research Project

To better understand how projects have commonalities and differences that can influence the dynamics of success and failure, the author performed a research project with the intention to tap knowledge from experts. He brought together a group of 17 project management experts, who have been practitioners in the past, but used their experience in the field in their new roles as instructors; authors of articles, books, and blogs; in project governance or in volunteering in professional associations. The experts had between seven and 36 years of experience each, which accumulated to 393 years of project management experience as practitioners and in the later roles. Twelve of the experts held the PMP® certification[7] from the Project Management Institute (PMI), one of them also held a Level B® certification[8] from the International Project Management Association (IPMA).

In interviews, the experts were asked to answer two questions:

- Dysfunctional question: "During your time as a practitioner or as an expert, do you remember a moment when a practice, a method, a behavior, or a tool for project management that had led to success before, then led to failure?"

- Functional question: "During your time as a practitioner or as an expert, do you remember a moment when a practice, a method, a behavior, or a tool for project management that had led to failure before, then led to success?"

All experts had examples that answered the dysfunctional question, sometimes more than one, but not all had an example that answered the functional question. During these interviews, the examples that the experts remembered were recorded and, in a next step, further investigated by applying the "Five whys" technique, a root-cause analysis method used to dig deep into the underlying origins of problems in management and production.[9] Another applied technique was "Affinity diagramming", which allowed the team to consolidate the thus far anecdotal stories and to identify underlying principles, finally resulting in the definition of project types that are relevant for the selection of project practices.

1.3.3 A Typology of Projects—Project Types

The project typology is considered open, which means that the following dimensions are not considered a complete description and that further exploration could identify many more. The dimensions that are described below were just those which turned up during the research. Table 1.1 gives an overview of these dimensions. The column "Occurrences" describes how often the dimension appeared among the answers. The column "Mode" was introduced based on a discovery by one of the experts that some dimensions describe dichotomies ("B/W" for black and white), while others describe continua ("Greyshades").

1.3.4 Mark 1 vs. Mark *n* Projects

The terms are borrowed from British engineering, sports cars, and Japanese cameras.

A Mark 1 project is the first of its kind, at least for the people involved in it. As a breakthrough project, it has a high degree of novelty and cannot rely on existing processes and

---

[7] Project Management Professional®.
[8] Certified Senior Project Manager.
[9] (Adams 2008)

Table 1.1 The Typological Dimensions That Were Identified During the Research

	Types of projects and project situations identified		Occurrences	Mode
	Typological dimension		Occurrences	Mode
1	Mark 1 project	Mark n project	|||||||	B/W
2	Greenfield project	Brownfield project	|||	B/W
3	Siloed project	Solid project	|||	Greyshades
4	Blurred project	Focused project	||	Greyshades
5	High impact project	Low impact project	||	Greyshades
6	Customer project	Internal project			B/W
7	Stand-alone project	Satellite project			B/W
8	Predictable project	Exploratory project			Greyshades
9	Composed project	Decomposed project			B/W

solutions; they must be developed during the course of the project. Mark *n* projects, in contrast, are similar to former projects, and the teams involved have a lot of experience with this kind of project. They often have processes and readily developed solutions to rely on.

This dimension has been identified and described above by Aaron Shenhar and Dov Dvir,[10] who stated that a Mark 1 project has higher risk than a Mark *n* project, which seems generally plausible. However, the result of the author's research gave a different picture: In two out of seven cases in which the dimension was influencing success and failure, it was the novelty of the project that caused the problems. In five out of seven cases, troubles came from the Mark *n* character of projects, from complacency and from the lack of attention to seemingly small issues that grew and became major crises later in the project.

Seven cases for an analysis may be too small a sample size to make a final statement on the risk exposure of Mark 1 projects versus Mark *n* projects, and the topic would be an interesting one for further in-depth research. The example shows how a typology can open doors for future exploration and discovery that is more focused and more tightly connected to the realities of projects in this world.

1.3.5 Greenfield vs. Brownfield Projects

These two terms are quite popular in construction and infrastructure projects. Greenfield projects are built on virgin ground, literally or metaphorically. Their managers do not have to take too much care of legacies, and the number of stakeholders involved is mostly small, allowing project managers to focus on the project. In a brownfield project, there may be a lot of legacies and stakeholders impacting the project, often massively, and expectations, hopes, and fears raise high. Organizational and interpersonal issues add to that.

Berlin Central Station is an example of a greenfield project in a literal sense. The station was built on a green strip, which had been left over from the former "death strip" between East and

---

[10] (Shenhar & Dvir 2007)

West Berlin. The wall was dismantled, the barbwire and spring guns were removed, and a strip was left crossing the city from its north to the south, wide enough to make space for a new and modern traffic infrastructure. The project team did not have to give much consideration to nearby residents; instead, they focused on keeping too much political influence at a distance to the project.

Stuttgart 21 is a brownfield project. Due to the hilly surroundings, it consists of vast tunnel drilling activities, which need to be undertaken inside difficult geological layers (anhydrite) that can swell in contact with water, with a chance to cause massive damage to the houses atop it. The owners of these houses asked the project manager to meet them and talk about their concerns, something he rejected.

Repelled stakeholders often come back bringing friends, lawyers, and the press with them. The approach of isolating the project from its stakeholder environment—something that had been successful in the greenfield project in Berlin—drove the seemingly similar project in Stuttgart into crisis.

Meanwhile, the project managers in Stuttgart have learned their lesson the hard way and implemented a system to improve stakeholder involvement and engagement called the Bürgerforum 21 (Landeshauptstadt Stuttgart, 2011), in which the concerns and worries of citizens are discussed in a transparent and open fashion.

1.3.6 Siloed vs. Solid Projects

Projects can be siloed in various ways. There may be different organizations that work together as partners or are distributed over complex and often highly dynamic *project supply networks*, each of these organizations with specific business interests and different ways of how they want to perform the project. Teams may be distributed over various countries, cultures, legal systems, etc. "Siloing" may relate to age groups, genders, and many more aspects.

Siloing may also be a result of phase models, when a sub-team has the responsibility for a project phase and, when this has been finished, throws the result "over the fence" in a figurative sense to another team that will be responsible for the next phase. Siloing can make project management difficult, but it is often unavoidable.

Solid projects instead are like "bands of brothers"[11] and sisters that act tightly together, understanding that then the result will be more than just the sum of its parts. Solidifying projects can include measures such as colocation and concurrent engineering, overlapping of phases with the intention to improve communications. There are limitations for the application of these measures, so project managers should be able to manage siloed projects as well as solid ones.

1.3.7 Blurred vs. Focused Projects

According to a major part of the literature, each project has a clear start date and a clear end date, desired deliverables have also been specified and agreed upon, and the team is assigned to the project, so that people know if they belong to the project or not.

---

[11] (Shakespeare 1599)

This organizational and interpersonal separation of the project from the performing organization(s) is often not more than wishful thinking. The internal requestor or paying customer has no clear understanding of the deliverables that would allow for specification, and if they have, this understanding is often open to change. The performing organization too often does not have the resources at hand that it can dedicate to the project, so there is a continuous coming and leaving of human and other resources.

And as much as the project has slowly grown from some kind of limbo into existence, there is also no clear, definable point at which it can be said that the project is over and closed. While focusing is desirable, project managers must also be able to manage the ambiguities and uncertainties of blurred projects.

1.3.8 High-Impact vs. Low-Impact Projects

High-impact projects commonly have more management attention than low-impact projects. The impact comes with opportunities, but also threats, and the higher these are, the more management attention that can be expected. Management attention is often the scarcest, but most valuable, resource in a project. While its presence does not guarantee the availability of other resources such as funding, people, or equipment, its absence is a sure reason that these other resources will also not be available for the project.

1.3.9 Customer Projects vs. Internal Projects

This is the most obvious typological dimension in SitPM and is the focus of this book.

An internal project, performed for an internal requestor, often called an "internal customer", is a cost center. There may be future expectations that the deliverables of the project will give the organization monetary benefits, but the project as such costs money and does not earn it. Projects can be performed for a variety of future goals, including new income, cost savings, or strategic benefits. Some are made to build a monument to an influential person or just for fun. Internal projects may have complex business cases or are initiated in an ad-hoc decision.

Customer projects are mostly profit centers. The organizations involved perform these projects for paying customers, and it is the job of the project managers to bring money home. Initiating these projects is far more complicated, as it involves a business development process jointly performed by a buyer and a seller, who will later become the customer and the contractor.

The following chapters will dig deeper into this differentiation. The book's focus will then be on:

- Customer projects, seen from the perspective of the contractor.
- Internal projects that include procurement, especially when this procurement leads to complex and dynamic *project support networks* (PSNs).

Project Business Management is a discipline that I recommend to develop in economic theory and business practice—one which describes the complex interfaces among the organizations inside these projects.

1.3.10 Standalone Projects vs. Satellite Projects

Many projects do not stand alone, as is normally assumed in the literature. Their project managers and teams perform projects "in the wake" of other projects (which the experts call "principal projects"), and the success of the satellite projects relies on the work of the project's own team, as well as on that of the principal project's team. A crisis in the principal project may swiftly translate into a crisis in the satellite project. The dynamics of success and failure can become very complex, especially when there are more than two projects involved.

1.3.11 Predictable Projects vs. Exploratory Projects

The most significant discussion in project management during the last couple of years explores "agile methods" versus "waterfall methods". Agile was a hype in recent years, but looking at presentations in congresses and also observing publications, it seems that a renaissance of the predictive approaches is occurring at the moment. The author also made a survey on this topic in 2012, asking project managers how static and predictable requirements in their projects are. The responses from 140 respondents from a global group are shown in Figure 1.3.

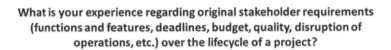

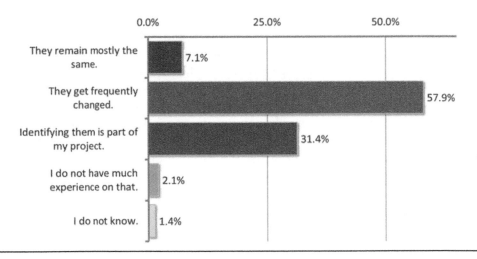

Figure 1.3   Responses from global group on how static and predictable their projects are.

The first group of 7.1 percent is the group of projects that require predictive approaches with long-term forecasts and planning. The third group are projects in which "the way is made by walking"—projects for which agile methods have been developed.

Between these two groups that are best managed using waterfall or agile methods is another one in which requirements have been defined but are open to frequent change, and for which the teams should apply an approach variously called "Progressive elaboration", "Iterative incremental", or "Rolling wave". A decision to perform all projects in a portfolio using highly predictive methods, or performing them all applying agile methods, is probably a decision that will benefit a minority of projects but will be detrimental to others.

1.3.12 Composed Projects vs. Decomposed Projects

A traditional approach to project management responds to a challenge that comes with probably the oldest definition of a project, written in the late 17th century by Daniel Defoe, who stated that, "The true definition of a project, according to modern acceptation, is a vast undertaking, too big to be managed, and therefore likely enough to come to nothing".[12] In order to manage this undertaking, it is commonly decomposed into smaller, better manageable items along a tree structure called the Work Breakdown Structure (WBS), hoping that the re-integration of these items will lead to the complete set of results that the project is required to deliver. The last pieces of wood in this tree are commonly called *work packages,* and these work packages can then be performed either by the project's own team, by other business units as internal providers inside the performing organization, or by external vendors.

Some projects are developed using the opposite approach. Friends come together, or organizations that have a more or less vested interest in the project join together, or a customer organization requires contractors to work together. These organizations then come together, each offering a contribution to the project, and the WBS is then not developed by decomposing the project but by composing it from individual contributions. If the contributions are able to bring about a complete set of deliverables, and if the parties involved adhere to their commitments, such projects can become very powerful. They are also vulnerable to changes in the business situations of the contributors, and project managers must rather have great moderating skills than be traditional managers.

1.4 Further Types of Projects

I found further types of projects in my own work or in discussions with and observations of my students, types that did not turn up in the research project, as follows.

1.4.1 Engineers' Projects and Gardeners' Projects

See Figure 1.4 for the different value streams in the two project types.

Project management has its origins in engineering, and for many project managers, there is only one value development curve that they know: The value of their deliverables is highest during handover and will then depreciate because of wear and tear, aging, outdating, and other causes. People who are responsible for the deliverables may slow down the process by careful treatment, regular maintenance, and updates from time to time, but they will not be able to stop

---

[12] Abridged quote (Defoe 1697)

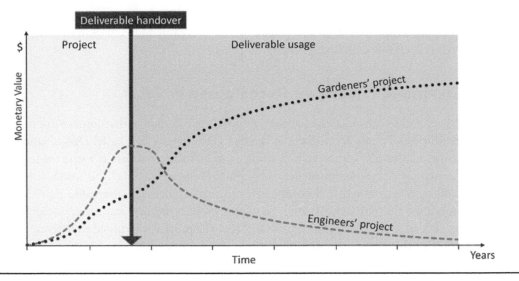

Figure 1.4   The differences in lifecycle value creation of engineers' and gardeners' projects.

the process. Gardening projects are different, because the value of the deliverables is expected to grow after handover, and it may take years for the deliverables to develop their full value.

There are projects that deliver benefits that are more similar to gardening: At the moment of handover, the deliverables are still immature and their value—business or otherwise—is rather low but is expected to grow over time. The maximum value may be achieved years after the transfer of the deliverables, comparable to a garden, in which the plants then have grown to a size that they can yield a valuable benefit.

1.4.2 Discretionary Projects and Mandatory Projects

Discretionary projects have a business case or another aspect that make them attractive for the performing organization. The business case is often a tangible benefit that the organization wishes to obtain by the project, such as additional earnings, cost savings in operations, or improved strategic position. Some readers will be familiar with project selection methods such as strategic scoring, net present value, or internal rate of return; discretionary projects are those to which these methods would be applied.

Mandatory projects, in contrast, are required by law or are necessary to avoid an emerging business crisis. Their purpose may also be to master such a crisis if it has already occurred. Mandatory projects may be deliberate responses to compliance rules, but they also may be done in sheer panic, especially if they have been procrastinated and started too late, and law has set firm deadlines that are enforced with severe penalties but are difficult to meet.

Although internal projects are commonly performed in a weak matrix, which means project managers have difficulties to obtain resources for their projects, these projects generally enjoy a strong matrix, and the project manager is generally expected to obtain tangible support from the functional organization.

1.4.3 Single-Handover Projects and Multiple-Handover Projects

A project may have one moment of handover, which mostly finishes the work in the project and starts the usage of the deliverables of the project. Depending on the industry and on the deliverables of the project, the handover may be given a different name; common other names are *delivery, start of production* (SoP), *start of service, goLive, grand opening,* and more. The left part of Figure 1.5 describes how the value of the deliverables from the recipients' perspective develops in a project of this type.

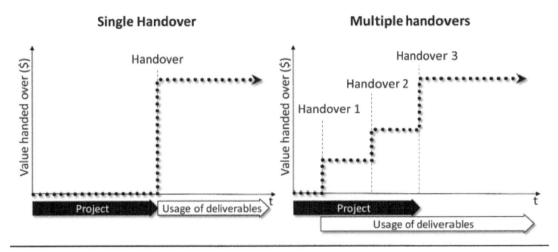

Figure 1.5   Multiple handovers deliver value in stages.

Projects with multiple handovers are different: The deliverables from these projects are not transferred as one massive piece but in portions during the course of the project. Multiple handovers can be a solution in a situation of time pressure: A project will not be able to finish all work by an imposed date, so prioritization is done. What is most important? What is mature enough to be finished on time and then delivered? Multiple handovers may also be part of a proactive organizational project management strategy of providing quick wins to the requester or customer and allowing the use of a partially finished product in order to gain the first benefits early, build trust in the project team, observe how users come to work with the deliverables and how these deliverables perform, and to make adjustments for further development where necessary.

Staged deliveries in this strategic understanding are also called *evolutionary deliveries,* and there is even an extreme form called *continuous delivery.* This latter approach is mostly used for software development with the intention to have frequent handovers of small increments of the software to its users. Some users will find the approach fantastic, as it ensures tight interaction between them and the developers; others may be terrified by the expectation of the need to learn new software functions every other week and the outlook that functions one has understood may be changed again. One obviously has to use these approaches with a sense of situational proportion and empathy for people involved.

1.4.4 No-Deadline Projects, Single-Deadline Projects, and Multiple-Deadline Projects

In the literature on project management, but also in training, a common presumption is that a project has a deadline that it must meet. This presumption often comes together with the previous one of the single handover or delivery. Again, I wondered to what degree this presumption correctly describes the situation that project managers are experiencing, and I asked them in an ad hoc survey in 2015. I received 402 responses by individuals who are managing a project, with results shown in Figure 1.6.

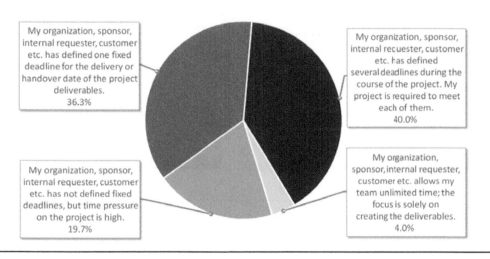

Regarding time pressure, which statement is true for your current project?

My organization, sponsor, internal requester, customer etc. has defined one fixed deadline for the delivery or handover date of the project deliverables.
36.3%

My organization, sponsor, internal recuester, customer etc. has defined several deadlines during the course of the project. My project is required to meet each of them.
40.0%

My organization, sponsor, internal requester, customer etc. has not defined fixed deadlines, but time pressure on the project is high.
19.7%

My organization, sponsor, internal requester, customer etc. allows my team unlimited time; the focus is solely on creating the deliverables.
4.0%

Figure 1.6   How project managers report deadlines.

Forty percent, the largest group of respondents, said that their projects have not just one deadline but several of them. The second group of a bit more than one-third of the respondents met the presumption of one deadline. Almost a fourth of the respondents stated that their project does not have a deadline, which does not necessarily mean that the project is not performed under time pressure.

1.4.5 One-Shot Projects vs. Multi-Shot Projects

In many projects, the team has only "one shot" to deliver results successfully. The situation is indeed similar to shooting: If you have only one shot, you have to aim more diligently—if you miss the target, you will not get a second chance. In one-shot projects, the project manager needs a realistic and viable plan.

One has to create predictability and must plan for the things one knows as well as for the foreseeable uncertainties. And one has to anticipate the relevant bottlenecks and develop the entire project around them.

There are other projects in which the project manager and his or her team have several shots. They can use staged deliveries and handovers of partially completed products to develop their final product in a step-by-step approach and fix errors as they occur. They can easily allow for change requests, as long as they are given the resources and time to manage them in a coordinated and controlled fashion.

1.4.6 Open Typologies and Closed Typologies

The typology above is not a closed one, postulating "the 32 types of projects, structured along the 16 typological dimensions", but instead assumes that there are many more criteria that can be used and that will lead to identification of further types. There are probably many more dimensions to typify projects that are relevant and helpful for the day-to-day practice of project managers. These are just the types that I identified from my own practice, from talking with experts and practitioners in project management, and from observing others, always wondering what the root causes are that practices that have been successful in one project may fail in another one.

It may also happen that a project changes its type, and some projects do this repeatedly during their lifecycles. To give an example: A project to build a bridge may go through a first phase, which is mostly driven by creativity, asking, "What should the bridge look like?", and by exploration of the underlying geology, of future traffic requirements, and of other considerations. It may be impossible to predict and plan the details of this phase—if these details were known, the phase would become unnecessary.

Later, when the bridge is planned and construction is on its way, long-term planning on a granular level will be necessary to ensure the availability of funding, people, equipment, and materials at the time when they are needed and to have all approvals granted before their absence leads to delays. A street-art project may have the opposite lifecycle: First, resources must be planned, approvals must be obtained, and the project must be publicized—activities that depend on a long-term approach to develop the free space for creativity that will follow on the days of the event.

The typology relates to entire projects, but also to situations inside a project. A challenge for a project manager adopting Situational Project Management (SitPM) is to master different practices and to be able to choose the one that is most promising in a given situation.

Another challenge is to make such changes without giving the impression of being erratic and unreliable. Situational project managers understand that their discipline is built on lifelong learning. They are good communicators who can explain to others the details of their approaches, helping others to develop a deeper understanding of the needs of the project and how it interacts with the organization's operations and with the often conflicting needs of stakeholders.

A good friend of mine is a project manager in the information technology department of a major automotive corporation. He often boasts of his "agilism", something he is a true expert in. Recently, he explained to me that for a certain project phase, he had to adopt another approach, as he believed that his favored agile methods would not be appropriate. This is a perfect example of a situational project manager.

1.5 An Introduction to Project Business Management

There are internal projects and customer projects. "Customer" in the context of this book means a real customer—an organization that is external to the organization that performs the project and pays this performing organization, called the "contractor", for this performance. "Customer project" is understood here as a project in which deliverables travel across organizational borders—often also across national borders—and so does the money involved. It is also understood here as a general rule that the contractor, the contract party whose job it is to deliver, creates financial income to secure its own existence. There are exceptions to this rule that I will discuss later—customer projects not performed to generate income but to support other business goals, sometimes called "strategic projects", and even "pet projects", wherein a manager wishes to have a customer among the company's customers because he or she likes the company or its products so much.[13]

I am using the term "customer" here in this very narrow sense and separate from its use in a looser habit: In many organizations, a model called the "internal customer" is used, often mirrored by an "internal vendor". In these organizations, the flow of deliverables and money inside the organization is modeled in a manner similar to an external supply channel, with internal charges mimicking billing systems between organizations. These models are mostly connected with operational "profit center" concepts, in which each business unit acts as an almost independent little company inside the organization. Promoters of such systems commonly state that internal and external customers should be regarded as equal entities, which may be a strong approach in certain situations but can lead to troubles and inefficiencies in others, when work and, with it, money that could circulate inside the organization leaves it instead to pay bills. This happens when the external contractors' bills would be less costly than the charges of the internal provider of the same products or services. From the perspective of the project, this decision looks like a cost reduction, while from a costing perspective of the entire organization, the project becomes expensive, especially when it leaves the resources of the rejected internal contractor idle. Profit center models can lead to disintegration of organizations.

Peter F. Drucker, who coined the term "profit center" in 1945, later changed his mind on the concept ". . . because inside a business there are no profit centers, just cost centers. Profit comes only from outside. When a customer returns with a repeat order and his check doesn't bounce, you have a profit center".[14]

There are two criteria that separate the "internal customer" from a true and real paying customer: The risk of (1) lawsuits and (2) public embarrassment when things go massively wrong.

When a project under contract between two separate legal entities gets into trouble, the worst case is that it will end at court. Parties must be acutely aware of these risks at any given

---

[13] It may sound ridiculous, but I had a boss in my younger years who wanted me, the General Manager of the German operations, to do business with Mercedes Benz and Porsche, because he drove their cars and wanted to tell his friends during golf matches that his company had them as customers. Developing business with these companies in our field of business meant stepping into already existing high competition, at a time when it would have been much easier and more profitable to attract business from other manufacturers who were currently receiving shoddier treatment than these giants.

[14] (Drucker 2013, p. 34)

moment in the project, and all actions must be documented in a way that these documents could support their own case in such a situation. The threat of legal action also means that poor performance of the contractor may get scrutinized in public, at least inside the industries to which customer and contractor belong.

A project performed by an internal vendor for a customer in trouble will not be brought to court but will be resolved inside the organization, commonly by high-level management. It is often "swept under the rug" to avoid the public embarrassment that comes with such fiascos.

In this book, whenever I use terms such as "customer" and "client" for one side and "seller", "vendor", or "contractor" for the other, I mean two or more organizations that are organizationally and financially independent from each other but have made a voluntary and more-or-less well-prepared and -contemplated decision to temporarily act together—one providing goods and services, the other one in most cases paying for them. Through the contract, they finally step into a voluntary dependency on one another, a dependency that is often hard to escape.

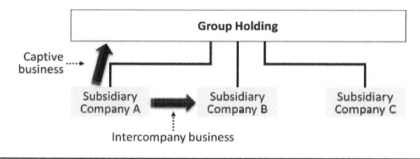

Figure 1.7   Intercompany business and captive business.

What about "captive outsourcing"? The term describes a business situation in which the "internal vendor" is a wholly-owned subsidiary company of the "internal customer". Figure 1.7 shows the difference for a group of companies under a holding. When Company A does a project for the holding of the group, the internal customer and vendor are not on the same level, as the group holding is also the governing body for the company. Based on the two criteria above, this is also an internal situation—a group internal one, to be more accurate: Company A and the group holding company would never sue each other at court, and failures are commonly kept private.

A subsidiary in a captive business is often in an ambiguous position: The customer is at the same time the owner and therefore dictates what this company must do and not do, hence the name "captive", which sounds a bit like slavery. At the same time, this subsidiary company is measured for its business success as if it were an independent company. The owner company may allow them to win external customers from outside the group or not, but in both cases expects full availability of the subsidiary's resources when it needs them. This expectation generally does not come with a guarantee of the utilization of these resources, and often, the subsidiary must quote for the business of the owner against competition from external companies.

Another expectation by the owner/customer company is that captive providers are better at protecting business secrets and intellectual property than actual third-party vendors. The group may further benefit from a different location of the subsidiary company with better infrastructure, easier legislation, or less expensive workers. Employees of the subsidiary company often have a deep feeling of uncertainty as to whom they should dedicate their loyalty, the group or the company. They may feel that their colleagues on the group level treat them as second-class employees, and the payment schemes inside the group often confirm this perception.

Business models have become highly dynamic. At one point in time, the group may find someone to sell the subsidiary to, and then things may change completely. The customer is no longer also the owner but just a customer, and the subsidiary company will become free to find other customers from outside the group. The guarantee of availability of resources for the group, as weak as this assurance often is, will finally vanish, and the business relationship could undergo legal remedy when it sours beyond a degree that can be handled in negotiations. The former subsidiary will then get into the same situation as other external companies that are financially and management-wise independent but perform a project under contract for a customer.

Sometimes the opposite happens: A formerly independent company is bought by another one, possibly an existing customer, and becomes its dependent subsidiary. Such a case was Dragonfly Inc., a company developing mission-critical software solutions for project management in large-scale development environments. They had a strong focus on building systems that combined cutting-edge development methods with high reliability of their products and applicability in highly complex environments. They had a tight focus on one market segment—in which they were the global market leader, while they were almost unknown elsewhere—and they paid their staff better than their competitors did, thus ensuring technological excellence and perfect responsiveness to customer needs.

They had an all-around perfect business approach with only one exception: They were a "JAM" company, an acronym sometimes used for families that are "just about managing",[15] but also suitable for many companies, especially from the field of customer project business. JAM companies bring enough money home to avoid going bankrupt, but not enough to allow for major investments into the organization's future and to build sound contingencies for times of crisis.

Dragonfly Inc. had a profitability that allowed the company to run the daily development business as long as it followed the expectations by management. In 2012, one customer required massive investments in network diagramming software functionality, something that many competitive products had, but not Dragonfly's. Their CEO had always insisted that customers would not use network diagramming anyway, so that the company could save the investment.

This had been true for a while, until their largest customer made a decision to implement network diagramming in their product development. They were tired of late projects and even more of late communication of project delays, a normal consequence of not using network diagramming—without application of this technique, it is hard to make long-term predictions.

Developing the functionality (together with some other financial burdens) soon overstretched Dragonfly's financial strength, and it became illiquid and had to apply for insolvency and creditor protection.

They were finally taken over by their largest customer, who was interested both in keeping alive the service for the existing solution, which was used to manage several hundred development projects and could not be easily replaced, and also in getting new functionality developed when the company identified a need for that.

---

[15] (Tetlow 2016)

Dragonfly's project managers originally had the job to bring money home; after the take-over, they had to guarantee to the new owners that resources would be available for them when needed. After the formal end of Dragonfly's insolvency phase, shares of the company were sold to other investors, and the company again became an (almost) independent software vendor.

The dynamics of success and failure in *project supply networks* (PSNs) are complex, and it is often hard to predict what the future of a company may be.

1.5.1 The Significance of Customer Projects

I have asked project managers in my seminars whether they are in customer projects or in internal projects, and the distribution was roughly 50/50. For my previous book, *Situational Project Management: The Dynamics of Success and Failure*,[16] I wanted to replace my observations with numbers, and therefore, in September 2015, I asked project managers in a dedicated group on the social network LinkedIn in what type of project they are active. I received 246 responses, and the result quite well confirmed my experience from the classroom (see Figure 1.8).

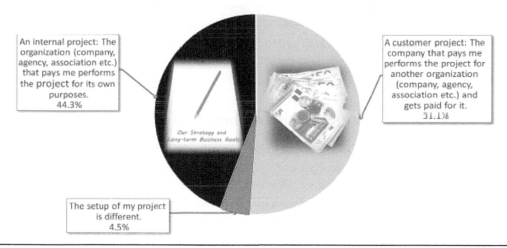

What Type of Project are You Currently Managing?

An internal project: The organization (company, agency, association etc.) that pays me performs the project for its own purposes.
44.3%

A customer project: The company that pays me performs the project for another organization (company, agency, association etc.) and gets paid for it.
31.1%

Our Strategy and Long-term Business Goals

The setup of my project is different.
4.5%

Figure 1.8   The percentage of project managers in customer projects was roughly the same as in internal projects in a survey performed in September 2015.

This is an interesting point: While I am writing this book, the project management literature generally focuses on strategic alignment of project management and the related multi-project disciplines of the following:

- **Program management.** Managing a number of projects that contribute to achieving a common goal.

- **Project portfolio management.** Selecting projects, prioritizing, load balancing across the performing organization, and some more tasks.

---

[16] (Lehmann 2016b)

The Project Management Institute (PMI), for instance, requires that the collective portfolio of projects and programs advances the organization.[17] The competing International Project Management Association (IPMA) defines projects as means for implementing organizational strategies.[18] Both definitions assume that an organization does projects for its own future development and to meet strategic goals. Projects in this understanding are investments done today to secure benefits for the future. This is generally true for internal projects.[19]

There are probably exceptions to all rules in project management, including the following, but customer projects are generally performed to bring money home. In the survey cited above, more than 50 percent of the responding project managers had a main objective different to the great strategic goals of so many internal projects: managing a project so that it is profitable for the performing organization as a contractor. A second objective is often to not overstretch the liquidity of this organization. In short: bringing money home with projects.

Active empathy for this group of project managers, possibly the majority among them, is also important for project managers doing internal projects: When they procure work items for the project from outside the organization, they have to understand the business situation of their contractors and the people working there. The contractors may in turn have subcontractors, who can further employ sub-subcontractors, and so on. Understanding, managing, and integrating such multi-tier PSNs can become complex. They often develop their own inner dynamics, and project managers are rarely educated for the challenges in technological, legal, social, and interpersonal dimensions that they encompass. While integration management is the pinnacle of project managers' competencies anyway, integrating across complex PSNs is particularly exigent.

1.5.2 Procurement as Part of the Architecture of a Project

It is helpful to first look at a Work Breakdown Structure (WBS) to understand whether contracting can be an essential part of a project, act as a rather marginal activity, or may be not done at all in other projects. I already quoted above the oldest definition of a project that I am aware of, made in the late 17th century by Daniel Defoe: "The true definition of a project, according to modern acceptation, is a vast undertaking, too big to be managed, and therefore likely enough to come to nothing". A commonly used approach to manage such a vast undertaking is to decompose it into smaller, more manageable components over several hierarchical outline levels. The last level of this decomposition is commonly called the *work package,* which is the lowest level of management supervision. Beyond this level, another planning element called "Activities" would be defined, which relates to people performing the actual work.

Every complex system needs a kind of architecture, a fundamental design structure that provides inner firmness and gives each of its constituting elements a place where it belongs and to which it is attached. Complex software has an architecture; cars, aircraft, spacecraft, and buildings have it. Architectures are commonly made in a combination of stiff and flexible elements

---

[17] (Langley 2015, p. 2)

[18] (IPMA 2016, p. 33)

[19] There are also exceptions here—for instance, mandatory projects, which do not enhance an organization's future but ensure compliance with laws or other kinds of binding rules.

to make the systems both resilient and flexible to the degree needed for the task they have been created for. The WBS is the architecture of a project.

There are three options for how the activities inside a work package can be performed (see Figures 1.9 and 1.10):

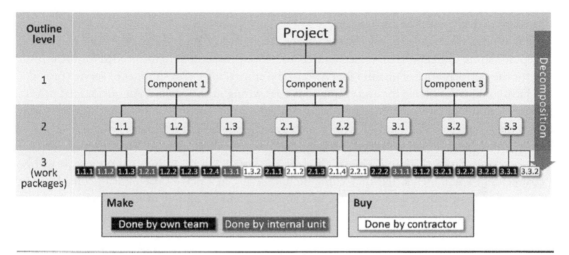

Figure 1.9   The three options to get a work package done.

- A work package can be given to a contractor—a legally separate entity. From the point of view of this contractor, the work package is often considered a project on its own—a customer project.

- Another business situation that can lead to a customer project is a decision to outsource an entire internal project. The customer in this case gives away the management function of its own project, not just of parts of it, and generally expects a turnkey solution at the end.

- A third situation is a customer project with no internal project on the customer side. The customer has some kind of business goal and asks a contractor to make all arrangements necessary to meet it, and the contractor then makes a project from that.

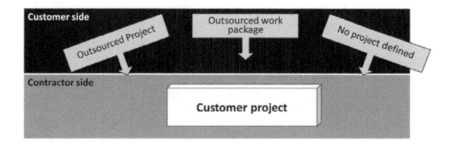

Figure 1.10   Three situations on the customer side that can lead to projects on the contractor side.

Each of these common project situations comes with specific challenges to the contractor; among them are the following.

Outsourced Projects

The responsibility and with it the accountability for success and failure of the entire project lies almost exclusively with the contractor.

This model takes most of the burden from the customer and places it on the contractor. Except for some obligations on the customer side—for provision of items needed for the project and for the performance of enabling services to the contractor—this supplier or service provider will have to take the blame for everything that goes wrong in the project. An outsourced project consists actually of two projects: an internal project on the customer side and a customer project on the contractor side, and the two projects are tightly linked together.

One can take as an easy example a family that has a turnkey house ordered from a construction company. From the family's point of view, the house will be an internal project. It will be a cost center: The benefits that may come include living in a new house built in consideration of their own requirements, saving rents to a landlord, and being free to make decisions on the house that one could not make in rented space. These benefits will be enjoyed only after the family has moved into the house. From the perspective of the construction company, the house is a customer project, undertaken to bring money home, and the company has to consider and reconcile two spheres of interests—its own and those of the family. Sometimes, the interests will be shared; often, they are in major conflict.

Many customer projects are performed physically inside the customer's premises and can even be performed organizationally inside the customer's functional structure. Then, situations can develop wherein the employees of the contractor feel more familiar and "comfy" inside the customer's than in their own firm, losing the natural and often necessary distance that employees of a seller should have to those of a buyer, and vice versa. In such situations, it is then often difficult to separate the customer's employees from those of the contractor, and sometimes, the latter tend to forget who sends them their pay checks and to whom they owe their final loyalty in moments of conflicts.

Outsourced Work Packages

In these projects, the contractor does not have the responsibility for the customer's full project but for a component of it, which may be anything from a critical element to a fringe task. The customer has an internal project defined and develops a more-or-less formalized WBS, a tree hierarchy of tasks and their deliverables that the project has to achieve. When a WBS component is then procured by an external provider, this company develops a customer project out of it.

This approach leaves more responsibility with the customer: In addition to obligations on the customer in the form of the already mentioned provisions and enabling services, the customer must coordinate the contractor's work package with the other actions inside the project, some or all of them possibly done by other contractors, and must protect them from too much disruption caused by the concurrent work on the same project.

The customer will also have to manage a number of dependencies between these work packages—for example, when one work package creates a result that another work package has to use as input for its work, or when two work packages occupy the same space and must therefore be done in a sequence so as not to clash.

In some industries, the procuring projects on the customer side are called programs, particularly when the scale of procurement is high; in others, the name project is still used for them, and the outsourced work packages may then be called external subprojects or something similar. There is no uniform language for these structures, but the structures are very common.

No Project Defined on the Customer Side

This can become uncomfortable for the contractor, as on the customer side there is no project manager as a single point of contact. Instead, a multitude of communication channels between contractor and customer can make managing the project difficult. Organizational conundrums of the customer can turn into financial problems for the contractor, and understanding what needs to be delivered gets confusing, especially when different stakeholders on the customer side communicate contradicting requirements on the work and its results.

Another aspect that can bring projects into difficulties is chemistry among people—each communication channel can become a source of misunderstandings and conflicts. While chemistry can be a problem in all contracting environments, without a project manager on the customer side, no one has the formal responsibility to deal with that. Small misunderstandings, disagreements, and personal wrangles can then grow and become major disturbances that threaten the entire project.

The External Project Manager

There is a fourth form of project work under contract that is not quite a customer project but often also called so: The external project manager assigned to an internal project. It is still an internal project, as the performing and the requesting organizations are identical and the resources are provided by the requesting organization itself, including the externally hired project manager. This person is often self-employed or may be leased to the customer by the person's employing company. It is also not untypical that the project manager brings a team with himself or herself, but this team is not performing the project for the customer in its own responsibility—the decisions in the project are finally made on the customer side. This model is common for consulting jobs and for professional development projects, including major investments in training and coaching.

1.5.3 The Basic Nature of Customer Projects

Internal projects and customer projects have many things in common—for example, their temporary nature and their fundamental uniqueness. They also have differences. As mentioned above, the most fundamental difference between an internal project and a customer project is the legal nature of the contract in the second.

Think of a project in dire straits. An internal project in crisis needs to be remedied by management, often in a confidential setting, as the failures are too embarrassing and could damage the reputation and trust of shareholders, customers, and other important stakeholders in the organization.

Customer projects in crisis frequently culminate at court, as mentioned earlier. This permanent threat is the strongest characteristic of a customer project, and many experienced project managers have already had more than one lawsuit related to their projects in their professional life. There are also lawyers who have specialized in lawsuits relating to projects, and many of them have a natural preference for projects with high monetary value, which promise attractive fees for them. Court cases are among the things that most project managers and their companies try to avoid at almost any price. The old Romans had a saying that *"Coram iudice et in alto mare in manibus deorum soli sumus"*,[20] pointing to the unpredictable nature and therefore high risks that come with legal proceedings. Judges often have difficulty understanding what the contested project is about and why people behaved as they did, and must then render their verdicts based on incomplete and not fully understood facts. Another impact of a lawsuit for both parties, customer and contractor, is that it massively binds management attention—the scarcest and most valuable resource in a company. The desire to avoid the threat of legal action is a strong disciplining factor. Project managers in customer projects are mostly in a much stronger position than their colleagues in internal projects, and the desire to avoid lawsuits is a strong factor that empowers them: Most companies do not want to be in a breach of contract situation.[21]

Table 1.2 describes the most fundamental differences between internal projects and customer projects that can often be found.

The risk of legal action is indeed a permanent Damocles' sword hanging over the engaged teams. They must always be acutely aware that performing customer projects is a high-risk business for themselves and for all other parties involved, including customers, subcontractors, business partners, and so on. This does not necessarily mean that project managers find themselves frequently at court. The desire to avoid legal action, which comes with uncertainty and possibly public embarrassment, is in most cases sufficient to drive parties to compromises, but the party that could develop a stronger position at court would also be the one that gets the more favorable arrangement.

Sometimes, conflicting parties are not able to find an out-of-court settlement, and then a customer project may finally end at court.

In a project under contract, two or more organizations are entangled with each other in a tightly woven and complex arrangement, and minor hiccups in one organization can easily translate into massive problems for the other one(s). Managers of these organizations are basically aware of these risks and therefore desire to protect themselves and their companies from such influence. Then, these organizations tend to develop protective mechanisms for themselves to avoid being damaged or held liable for things going wrong under their own domain or that of the other party, and the more corporations apply these protective mechanisms, the more their projects will suffer from organizational fragmentation and disintegration. A central element of any business relationship is the set of legal risks involved, and legal remedy of conflicts is highly competitive by nature.

---

[20] "Before a judge and on the high seas, we are left alone in the hands of the gods". Own translation.

[21] One should also be aware that the concept of *"Pacta sunt servanda"*—of the strong binding character of contracts—has a cultural perspective: Different traditions exist as to whether a contract should be considered "sanctimonious" or is just seen as a sketchy and non-binding memorandum.

Table 1.2 Fundamental Differences Between Internal and Customer Projects

	Commonly observable differences	
	Internal projects	**Customer projects**
Are . . . for the performing organization	Cost centers	Profit centers
The project requester is . . .	Located inside the own organization	A legally separate entity
The project team has to consider . . .	The interests of the performing organization	The interests of the performing organization and the customer
Project approval mostly follows . . .	A project request/approval process or no process at all	An offer/acceptance process leading to a contract
Disputes are to be finally resolved . . .	By management	At court
The performing organization does the project to attain . . .	Deliverables and change	Income
Project selection is made as . . .	A sequence of internal decisions	A bid/no-bid decision (contractor side), contract award (customer side)
Project work for the requester is based on . . .	Internal requests and agreements	Legally binding contracts
Team's familiarity with the target environment at project start is generally . . .	High	Low
A project budget is developed through . . .	A more-or-less informed management decision, or not at all	Deducting a margin from the price to the customer
A project budget is usually managed by . . .	The project sponsor or a supervisory board or may be nonexistent	The project manager
Inside matrix organizations, most project managers are . . .	Rather weak	Rather powerful
Obtaining internal and external resources is generally . . .	Rather difficult	Rather easy
Availability of booked resources is rather . . .	Unreliable	Reliable
Management attention for the project is mostly . . .	Rather low	Rather high
Project managers must consider . . .	The interests of the own organization	The interests of both the customer and the contractor
Staffing and procurement is mostly managed by . . .	Functional units	Project manager and project management team
Reputation of project managers inside the performing organization is mostly . . .	Rather low	Rather high

In opposition to this legal perspective stands the relational perspective. Melvin Conway observed that the functioning of complex systems necessitates effective communication structures of the teams involved in developing them.[22] I will discuss Conway's law in more detail in the second chapter. In projects with complex and often dynamic team structures, mutual empathy, communications, and common understanding are the key skills necessary to build functioning systems. There are many examples of failures due to the lack of such skills:

In 1999, NASA lost its Mars Climate Orbiter probe when it maneuvered to insert itself into an orbit around Mars. The cause was a confusion of non-metric and metric units in communications between teams, which led to a trajectory that brought the probe into the Mars atmosphere, where it disintegrated.[23]

Just a month later, still in 1999, NASA lost its Mars Polar Lander, probably due to miscommunications between the landing gear software and that of the main descent engine, which assumed that vibrations from the operations of the systems were a signal that the probe had touched down on Mars' surface, which led to a cut-off of the engine while the lander was still 40 meters above.[24]

In Taiwan, the Taiwan High Speed Railway Consortium, which was tasked with the construction of a high-speed railway link—a "bullet train" designed for a speed of 300 km/h (186 mph)—from Taipei in the north to Kaohsiung on the southern end of the island, over a distance of 345 km (214 miles), became a victim of the dynamics and complexity of PSNs. In 2009, after they had switched contractors during the course of the project from German–French Siemens–Alsthom to Japanese Shinkansen, mixing two independently developed technologies and cultures without any experience in such integration, thus frustrating their vendors—both the rejected ones and the new ones, who had to work according to standards set by the original vendors. Being a private venture, they had to be bailed out by the state before they went bankrupt and left the island with one of the largest industrial ruins in the world.

Boeing's 787 Dreamliner passenger aircraft had cases of burning batteries in 2013. These were purchased from French Thales Group, which made the electronic control systems and sub-contracted the batteries to the Japanese manufacturer GS Yuasa. The burning batteries led to the grounding of the entire fleet of aircraft for three months.[25]

In 2015, German car manufacturer Volkswagen found itself confronted with a time bomb—hidden in the engine control software of its most widely distributed motors—that changed the parameters of the engine's exhaust treatment system when it identified that the motor was on a test bed and not on the street. Volkswagen management said that they were unaware of the deception.[26]

While all these failures seem to be first of all due to technical causes, they have root causes underlying organizational errors and confusion. Effective communications among sub-teams in a solid project may not guarantee success, but fragmentation among these teams and siloing of

---

[22] (Conway 1968)
[23] (Stephenson et al. 1999)
[24] (Casani et al. 2000)
[25] (NTSB 2014)
[26] (VW 2015)

the project by fragmenting its inner organization beyond the unavoidable almost always guarantees failure. Projects with different organizations working together are at higher risk of siloing. In addition to the many aspects of fragmentation that one can find in all major projects, different business interests among contract parties are added as another strong disintegrating factor.

The problem of siloing can occur in any project. In customer projects and in complex PSNs, the dimension of the problem increases, when in addition to the technical, organizational, and interpersonal aspects that require attention, a fourth is added: the legal aspect.

To complicate the matter further, modern PSNs are developed crossing national borders as well as spreading over business cultures and different legal systems. Having a contractual relationship over two countries means that at least one party must act in a cultural and legal environment that is not its familiar environment and may often be largely unknown. This is a strong handicap: What is perfectly legal and ethical inside its own country may be unacceptable or even illegal in the other country. From the standpoint of a lawyer educated for a generally competitive world,[27] the answer to such a challenge is simple: Make sure that the law of your own country applies, so the other party has the handicap when the relation is getting sour and the project is turning to a bumpier road. From a project manager's point of view—someone who is much less interested in the outcome of a court case than in the forming of a mutual success culture, ideally based on reciprocated empathy and cooperation with the intention to have a great project—the focus is more on avoiding the misunderstandings and mishaps that can damage the project, and competitive behavior is a strong driver for these misunderstandings. A major goal of this book is to help project managers in complex PSNs develop a situational "Mission Success First" culture, in which all parties involved are prepared to invest in project success, placing their particular interests on second priority and upholding the team spirit necessary to go beyond not failing to developing true successes.

There is a multitude of forces active in contract-intensive projects, in which project managers on the vendor side have the task to bring money home. Table 1.3 shows some of them. The larger the project and the more parties involved, the stronger the fragmenting forces become. Additional influence can come from distribution of these organizations over different locations with all the difficulties for communications and coordination.

1.6 Projects as Profit Centers

Some companies that perform customer projects exist only for the purpose of doing such projects as a business and making their living from it. Some exist for merely one project—may even have been founded for the specific project and are planned to be liquidated when the project has been finished. Others have a portfolio of customer projects that they are performing concurrently, and their business is a continuous process of market research, making contact with new prospects, business development with these prospects, bid/proposal writing, contracting, and post-contract services when the project has been finished. Several settings are common, among them:

---

[27] Even the degree of competitiveness of this education differs between cultures.

Table 1.3 Integrating and Fragmenting forces in PSNs

Integrating forces	Fragmenting forces
Cultural understanding	Cultural diversity
Legal homogeny	Legal diversity
Empathy	Disregard
Collocation	Virtualization
Business relations	Legal relations
"Mission success first" culture	"Own interests first" culture
Personal relations	Personal conflicts
Motivation	Rejection
Ambition to complete	Ambition to compete
Trust	Distrust
Trustworthiness	Dishonesty
Sharing liabilities	Avoiding liabilities
Positive feedback	Negative feedback

1.6.1 Small Consultancies and Development Offices

Corporations often give development and design work that is outside their area of expertise and strategic interest to small, independent providers. Sometimes, this work is the provider's only business, and while this is performed, there are no time and resources left to find other business, so consumed is the provider by the project.

The provider may consist of just one person or a small team of people, mostly experts in the field, and is often founded by a former employee of the customer organization, in which case the provider knows many internals of the customer. Consultancies and education providers supporting development projects or even performing complete projects for their customers are another example.

1.6.2 Large Consortia

A consortium, as the term is used in this context, is a temporary joint venture by two or more companies to combine resources in order to perform a project. Depending on the business situation, a consortium may also operate the project's deliverables for a limited time.

In construction, infrastructure, and other application areas of civil engineering, projects can become very large, and with them the firms established for the single purpose of managing and performing them. Take as an example the construction project of two new lock complexes for the $5.2 billion Panama Canal expansion in the years 2007 to 2016. Four companies founded a consortium as a temporary joint venture:

- Sacyr Vallehermoso S.A. (Spain)
- Salini Impregilo S.p.A. (Italy)
- Jan de Nul n.v. (Belgium)
- Constructora Urbana, S.A. (Panama)

The consortium was named Grupo Unidos por el Canal, and its leading venturer was Sacyr Vallehermoso. It was formed for the single objective of performing the lock construction.

If there are no operational tasks for the consortium to be done in the aftermath of the project, like handling claims, complaints, and warranties or post-project operations, such a consortium is either liquidated, which mostly includes selling all the equipment that has then become redundant and laying off the entire staff, or it finds a new project and remains busy for some more years.[28] I will discuss consortia again later as a form of teaming agreement.

1.6.3 Major Project Providers

The organizations described above are in most cases projectized organizations. For each of them, the single project (or a very small number of projects) performed consumes their resources fully, and the customer has chosen the contractor because of the expectation that the project will not have to compete with a functional organization or with a major number of other projects for resources. The customer expects that the contractor's focus is fully on the project, putting the customer in a VIP position, because the contractor's justification to exist is just this specific project. The project is not a profit center inside an organization; the entire organization's profit is generated by the project.

A different setting on the contractor side are corporations that perform a major number of customer projects concurrently. The number of projects performed at a given moment can become quite large, and many statements that I have made on customer projects and their typical strong matrix setting may then no longer apply in these companies.

Customers with their projects then tend to no longer be the focus of all attention by the contractor, and the customer's confidence and happiness may no longer be the highest objective of the project. When a provider of a customer project normally reminds us of a tailor's shop, where all activity is dedicated to the client and his/her distinctiveness in physical traits and taste, these businesses are rather similar to a fashion outlet, where the customer is just the person expected to take a ticket and wait to be served, or a lost soul standing in a waiting line at the cash desk to be allowed to pay.

---

[28] An interesting observation: Joint ventures that have been created and designed for long-term cooperation often do not survive their first two years of business, because they get too disrupted by different business interests and consequential distrust. Consortia, established as temporary joint ventures for specific projects, often outlive these projects and find new tasks. It seems that once the venturers have gone through hard storming times without the opportunity to split, because they had to meet contractual obligations with a customer, they had their *modus vivendi* normed to be successful together, and they then want to keep this modus alive. Some deeper research on this observation may be interesting.

An example is Hyacinthus Beetle,[29] a corporation that is active around the world, with a major number of national and regional service centers providing software and IT services, mostly in the form of development and implementation projects. Their largest service centers perform over 115 projects for different customers at any given time, mostly from the public sector, but also for privately held companies.

The corporation has a focus on a well-working customer interface, but even stronger is the desire to "operationalize" these projects in order to ease governance over the portfolio. Some measures that the organization took in the last couple of years include:

- Development and implementation of unified organizational process assets, including

  o a unified project management methodology including templates, forms, checklists, etc., with a focus on standardized cost reporting and forecasting

  o a company-wide project management handbook that was considered mandatory for all project managers in customer projects

- An electronic Enterprise Project Management (EPM) system, soon expanded with a team-ware solution to improve online collaboration.

- Definition of focal areas of interest, including fields of business and technologies, that were considered strategically attractive, and increased effort was made to win new businesses in these fields, to win new experience and gain reputation in them.

- Certification of all business development managers and proposal writers as Project Management Professionals (PMP® certificate holders).

The operationalization through standardization seemed necessary due to the low margins that the organization made on the projects, often less than 10 percent, and many projects ended with a clear loss.

Although it indeed helped to bring order to a previously chaotic company, it did not help to improve the margins and led to massive dissatisfaction of project managers and customers, who experienced the system rather as a straightjacket and a bureaucratic burden than as a means to improve efficiency and effectiveness.

Inside these larger provider companies, project managers are generally no longer in a strong position against a functional organization as the generators of corporate income. The line organization comes more in the form of a set of governance functions such as portfolio decision and review boards, project management offices (PMOs), quality departments, compliance officers, safety managers, and many more; and while the projects do not perceive much competition with operations for resources, they compete with one another.

We will discuss below how management attention can become the scarcest and most valuable resource, and project managers then compete for this resource much as children compete for the attention of their parents.

---

[29] All names in the example are changed.

1.6.4 Providers of Supportive Projects and Freebie Projects

Freebie projects or "razor-and-blade projects"[30] are free business endeavors for a customer —at least for the moment. The expectation is that the resulting business will pay back the investment of the provider and that the customer gets strongly bound to the contractor and cannot simply move to competing vendors of the same services or products any more.

An example are logistics providers that gain their income from services using trucks, ships, aircraft, and warehouse operations for customers. Modern logistics providers, sometimes also referred to as "Third-Party Logistics Providers" (3PLs) do much more—they manage the complete external logistics systems for their customers and, to ensure flexibility and scalability of their offerings, rely heavily on complex teaming networks with other suppliers. The complex systems they develop for that task must be tightly linked with the internal logistics systems of the client organization, and many complex IT projects are performed to ensure the collaboration of the vendor's shipping management with the customer's internal systems. Many of these projects are freebies for the customer, who gets a great product for nothing.

Well, almost nothing: The customer's business is made reliant on the logistics provider for years, and the logistics provider will use this dependency to get its investment paid back. It is similar to the traditional business model of razorblade manufacturers, who subsidize the razors to bring money home with the blades.

A similar business strategy is known from makers of instant cameras, inkjet printers, and capsule coffee makers, which are sold at a cheap price by companies that hope to make the genuine profit with the consumption materials. The business model comes with risks—for example, when other vendors hijack the business by copying the consumption materials, or when the earnings from consumption materials sales do not cover the initial investments.

Freebie projects are an example of customer projects that are not performed for their own sake and business purpose but are an add-on to an operational business to help its managers bring money home. Component manufacturers in automotive, aerospace, and similar industries support their customers with free development work, based on a business case that they will later sell the developed components to the customer.

Software companies have a similar business case, centered around the license that the customer will finally pay when the component has become a part of the final product. In my business—education—a contract trainer may develop a seminar for the customer free of charge, in order to perform the classes and live on the fee.

A second reason for a trainer to apply such a business model is copyright—the trainer is then free to perform the seminar for other clients, as the original customer has never bought the intellectual property.

Freebie customer projects come with high risks. One risk is that someone else may step into the business with a copycat product. The manufacturer of razor blades, who subsidized the holder to earn from the blades and the high margins that they brought, were at times

---

[30] Referring to the old business model of Gilette to subsidize the distribution of the holder and then make money with the razor blades. Manufacturers of instant cameras, inkjet printers, and other products applied this model later (Lehmann 2016b).

confronted with blades from other manufacturers that were compatible with the holder, but much cheaper. It may also happen that later, when it comes to the operational business that should refinance the freebie customer project, the buyer does not honor the investment of the vendor and does not become a customer.

A case that I have followed from some proximity was that of Cricket AG, a German manufacturing company that could look back on several hundred years of history, with many ups and downs, which took them finally into becoming an automotive supplier with a focus on brake components and some other automotive parts for passenger cars and trucks. They had a strong business focus developed on just one customer, who made almost half of their business.

Management found in the early 2010s that this situation entailed too much risk, and new customers should be found. It was the US company Locust Inc. that promised big business for them. Cricket developed a new type of hydraulic pump for the new customer and also developed the production lines necessary to make the pumps in the numbers projected and required by the customer. The business was intended to yield approximately €100 million ($130 million in the exchange rates of the day when the business was developed) over a five years' production cycle, which would pay back the original investment and allow for a moderate profit. Internally, the hope at Cricket AG was that the demand for the car and hence for the pump would exceed expectations. There were some warnings to Cricket that Locust was a very difficult customer to deal with, but the hope of new successful business was higher than the fear to fail.

In January 2017, half a year before start of production, Locust made a decision to terminate the contract with Cricket. There were several reasons discussed for this decision: The customer required a high degree of confidentiality on the business from the vendor, and Cricket AG, in desperately looking for more new customers, used the business with Locust as a promotional reference (without naming the company, but it was easy to make out who they were talking about). Cricket had recently had an IPO and needed success stories for the stock exchange to keep their share value high.

Then, Locust expressed frustration that Cricket was late in their development and would threaten the start of production of the new car for which the pump was developed, something the vendor denied. In private, one could hear from Cricket staff that a constant stream of change requests on the new product and also on the production lines dedicated for it made it impossible to meet deadlines.

Locust Inc. were known for some erraticism in their decision making and expected their vendors to swiftly follow their wishes without additional costs or delays. In other words, they expected a full service of implementing expensive and time-consuming change requests as part of the freebie project without questioning the deadlines. Locust also said that the pumps would not meet certain specifications, a statement rejected by the vendor.

Another influencing factor was the political change in the USA in January 2017, when a new president took office who followed a protectionist agenda and put automotive manufacturers under pressure to get cars and their components made in the USA.

Locust, the customer, re-opened the *make-or-buy* decision already made and decided this time to make instead of buying. How could they make such a decision half a year before the start of production? Normally, one would expect that it was too late to change the manufacturing strategy. Locust had most of the drawings, bills of materials, production flow plans, etc. that the vendor had developed for them. They were developed to a major degree as a joint undertaking, and for the customer, it seemed clear that they were under shared ownership. Therefore, they took the investment made by the vendor to develop the product and the systems to finally produce it and implemented them in their own house. For Cricket AG, the business model for the freebie project virtually fell apart. The company had to consider the project performed for Locust Inc. lost and had to write off all investments that they had made in the customer.

Investors at the stock exchange reacted immediately, and the company shares lost 6 percent value in less than a week. Customer projects are generally a high-risk business for all parties involved.

There is a second business model for operations supporting freebie projects. It can be found in projects to implement new software or equipment on customer premises and integrate it with legacy systems already in place there; then the actual income is often derived from the sales of the products, not from the implementation projects. The project team's job is to help operations by doing major parts or the entire implementation work for the customer, utilizing its insider knowledge of the product. The expectation on the project may not be a commercial one at all—the project may be fully paid for by the margins calculated on the product. It may also be that the project is paid for by the customer, but the expectation is not for the project to make a profit, just to cover its own costs.

Some supportive projects are required to make their own profit and contribute their share to the overall profit from the business with the customer. These projects are tightly linked to this business but are considered profit centers on their own, and their success is measured independently from that of the entire business with the customer.

Managers can be quite effective when it comes to leaving it unclear what kind of projects they are running for the customer. At the beginning of the project, the team may be told that their major objective is to support the product business, and later the question is raised as to why they are not making a profit on their own. In other projects, the team starts with the clear intention of making its own profit, to find out later that they have to subordinate to the strategically higher valued product business to a degree that makes it impossible for the team to run a profitable project.

1.6.5 Mixed Customer Projects

Some customer projects are performed in a simple one-to-one relationship between a customer and a contractor, and then it is relatively easy to describe the role of the contractor along the descriptions above. Often, we find complex and dynamic multi-tier PSNs with the consequence that the types described here come in much less clearly defined composite structures, and over time, the role of the contractor and its relationship to the customer is commonly undergoing change. It is also common that for different players in such a PSN, the project type looks different.

1.7 The Players in *Project Supply Networks*

During the course of a customer project, different actors come on the scene, some for the entire duration, others temporarily; the latter will leave the project again when their tasks are done. Such a course of action can be different from project to project, but an order of actions as is shown in Figure 1.11[31] is quite common. It gives a first understanding when certain players turn up in the process and leave it again.

The flow shown in Figure 1.11 is rather assuming project business with a new customer–contractor relationship. Business among incumbent parties can differ to some degree.

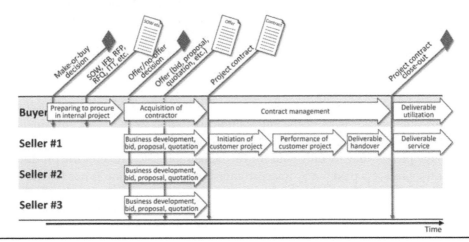

Figure 1.11   The complex flow of actions and documents in a customer project, assuming that Seller #1 wins in a competition against the other two sellers and becomes the contractor.

1.7.1 The Buyer

The buyer initiates the project. Often, the buyer is considered a "prospect" before the award of the contract and becomes the "customer" after that moment. As the customer (as the term is used here), the organization becomes the sponsoring organization of the project. The customer may contract out the entire project or parts of it that are large enough for contractors to define a new project for them.

In business-related projects, the buyer has some obligations to fulfill in a timely manner, including:

- Making information available to the seller(s) that allow them to meet their obligations timely.
- Paying invoices.
- Catering enabling services and delivering provisions to the seller, particularly after contract award.
- Being available for field changes (discussed later).
- If more than one contractor is used, coordinating them and protecting the project from conflicts between them.
- Listening to the expertise of the seller, where the buyer does not have that.

---

[31] Note: The diagram is descriptive, not prescriptive. It does not say "this is how the process should be" but describes what it often looks like.

Sometimes, organizations cooperate to buy together. The project would then have several customers. As a contractor, one must then take care that organizational and interpersonal snags on the side of the customers do not obstruct the project and damage their own business.

1.7.2 Sellers #1: The Contractor

When the contract has been awarded, a seller (or vendor) becomes a contractor. In a smaller procurement situation, a customer may just have one contractor allocated to do the outsourced work. The contractor may use subcontractors or not; from a customer perspective, this is often not transparent, and the customer may not be interested in that at all.

Obligations of the contractor include:

- Delivering what has been ordered and what the customer is prepared to pay for
- Making information and documentation available to the buyer in a timely fashion
- Making the buyer aware of issues early, so that measures can be discussed at a time when many affordable options are still available
- Offering expertise to the buyer in fields where this seems necessary

1.7.3 Sellers #2: The Prime Contractor

A prime contractor selects other companies as teaming partners and forwards major parts of the work, possibly the entire contract scope, to them. These companies then become subcontractors of the prime contractor, as shown in Figure 1.12. The prime contractor is in a hybrid role, as the organization acts as a contractor in the business relationship with the customer, but is at the same time a customer of the subcontractors. The prime contractor's job is mainly to pass on work from the customer to subcontractors and money in the opposite direction, similar to a dealer in product business.

The "sandwich" position of the prime contractor between customer and subcontractors can be financially attractive, because, located in the center, the organization has knowledge that

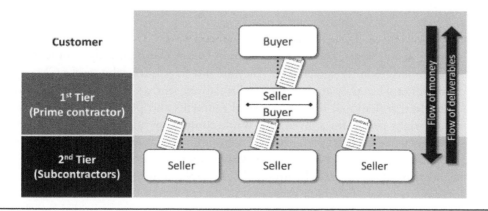

Figure 1.12 An example of a simple multi-tier contractor structure.

the other players do not have and that can be monetized. It may also be that the organization's margins get squeezed between the customer and the subcontractors, making the business uncomfortable, unprofitable, and risky. As the prime contractor is both seller and buyer, the obligations are a mix of both, as described above.

The prime contractor is commonly also the party that has to manage most contracts. Contract management can then become a challenging task. Here are some of the challenges:

- Answering the question "Does anyone know where the original of the contract with company ABC is?" can become difficult. It takes a lot of discipline by the prime contractor's employees to ensure that contracts do not get buried in stacks of other documents.

- Change requests need to be managed across several contracts, which all must be examined as to whether they need to be amended, and if they do, all parties involved have to agree.

- The prime contractor is the party responsible for the completeness, timeliness, and quality of the work done by the subcontractors. The prime contractor will also need to consider operational disruptions on the side of the client—a commonly overlooked hidden cost of a project that can lead to major conflicts with the functional organizations, both internally and on the customer side.

- When project managers must bring money home, they should understand the influence of contract types on the dynamics of success and failure. Differences in contract types with the various business partners can create a complexity, possibly an incompatibility, that is hard to predict and manage and can finally eat up all margins from the business.

1.7.4 Sellers #3: Subcontractors over Various Tiers

Some subcontractors will be at the bottom level of the PSN. They provide resources such as the people, space, and equipment to actually do the project work and deliver the results.

Other subcontractors may be in a similar situation to the prime contractor. They further outsource work to subcontractors at the next tier, and particularly in very large projects, the number of tiers can become quite large.

1.7.5 Freelancers

Self-employed freelancers are a special type of subcontractor. They act as small companies, sending invoices to customers, and they may also subcontract work, mostly to other freelancers with whom they have teaming agreements. They are not employees, who send a CV, a bio, or a résumé to a recruiter when they apply for a job. Freelancers send a CV focused not on a description of a complete professional life but on focal moments and periods that explain why the freelancer is the right person for a given job. While contracting brings flexibility into projects, this is particularly true for freelancers. They are commonly the fastest to obtain new knowledge, apply new technologies, and accept the customers' missions as their own.

Freelancers are not to be confused with mock self-employed people, who are actually normal workers forced into a role in which they are mostly left without entitlements such as social benefits. For the employers, this pays back in the form of tax savings, avoided social security contributions, ease of hiring and firing, and freedom from trade unions. To gain these advantages, employers then give people the choice to be either bogus self-employed or unemployed. This works well in asymmetrical markets, where a sufficient number of people chose being bogus self-employed, so that the employer could run the business with them.

Freelancers, in contrast, act in projects as true contractors who are mostly well-paid and fully self-managed professionals, who enjoy the boss-less life and the responsibility that they have taken for themselves. They decided for themselves to be self-employed in full awareness of the opportunities that this style of life and work brings, but also aware of the risks that they are assuming.

Customers are often unaware that their contractors use freelancers as subcontractors, and the freelancers are often obliged to not reveal their status to the final customer. In the eyes of the customer, they then seem to be employed staff. Alternatively, contractors may use the good reputation of the freelancer to enhance the value perception by the customer for their offering. The freelancer is then sold to the customer more like a top expert—a superstar or a guru—and the customer will then have to pay a surcharge for the work that will reflect this perception.

Very good freelancers are fast learners. They have their professional development under control, and as they know that their professionalism and competency are their foremost business assets, many invest time and money in themselves. Another asset is their preparedness to travel when this is necessary for an assignment.

A very specific problem with freelancers can come from the timing of their assignments to projects. Such an assignment can take some months, in which most freelancers are expected to be highly available for the project, often by 100 percent. About one to two months before the end of the assignment, they should look out for their next assignment, but this is just the time when the workload on the freelancer is commonly highest, having a deadline ahead and pressing for finishing and deliverable handover. Often, they go for a multi-week sabbatical between two assignments to relax and take care of their own professional development, but also because they do not have an immediately following next assignment. Some use agencies to ensure timely follow-up business, which comes with a cost but may also increase the number of billable days and possibly the daily rates they can achieve.

Sometimes, the opposite problem can also happen: A next assignment offers stable income for the next months but would need to be started before the current assignment can be finished. From time to time, I hear complaints about freelancers who did not fully finish their work in a project, and the lure of the next assignment is a common reason for that.

1.7.6 Other Players

Depending on the business environment and the project, there may be many more stakeholders more or less directly involved. Some contractors may not contribute to project work being done but provide infrastructure, special expertise, or help supervising the work. For projects that entail major workloads in construction, engineering, and similar areas, those who help disposing waste are also a kind of contractor, assuming that the project team is not simply fly-tipping its waste. Government agencies may play a major role in customer projects, particularly in larger ones. Even competitors can be important stakeholders who can influence the project positively—but also negatively.

It is generally advisable to have a stakeholder register to keep track of all the stakeholders, and it may be a good idea to allow the various organizations access to the document to allow them corrections where appropriate.

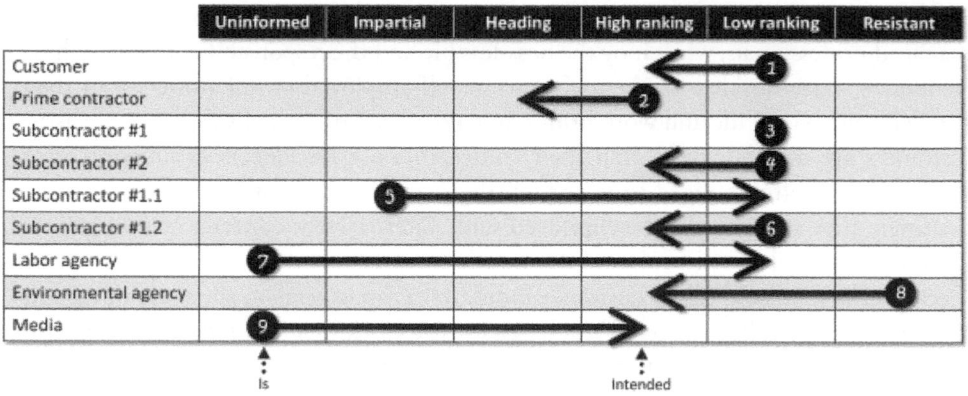

	Uninformed	Impartial	Heading	High ranking	Low ranking	Resistant
Customer				←	①	
Prime contractor			←	②		
Subcontractor #1					③	
Subcontractor #2				←	④	
Subcontractor #1.1		⑤ →				
Subcontractor #1.2				←	⑥	
Labor agency	⑦ →					
Environmental agency				←		⑧
Media	⑨ →					

Is · · · Intended

Figure 1.13 A stakeholder attitudes influence chart (SAIC) to be consulted when decisions are to be made, what influence they will have on stakeholder attitudes.

I also recommend using the Stakeholder Attitudes Influence Chart (SAIC, see Figure 1.13)[32] with a focus on contract parties of the project. I have customers who have printed the chart in a large format and put it on the wall in the team office. Whenever they make decisions on the project, they review the chart, asking if the decision option they are about to take supports the objectives regarding the development of stakeholder attitudes that they have outlined.

The following chapters will describe how the players act in Project Business Management, and what opportunities and threats they are facing from this coordination. Too often, actors in Project Business Management focus on competing more than on completing, jeopardizing the mission success of the project. Then, they often need help to get out of crises and develop joint success strategies.

---

[32] (Lehmann 2016b, pp. 216–217)

Chapter 2

The Difficult Way to the Contract

2.1 Bringing Order Into Chaos May Not Be Enough

Grasshopper Ltd.[1] considers itself a successful company in winning project business with customers. They have invested a lot in having a strong online presence with perfect search engine optimization (SEO), which has brought them a growing number of customer inquiries, from which they could generate a reliable stream of customer projects. The further growing number of concurrent customer projects seemed a perfect foundation for the company to do great business, and this is how it promoted itself in media and congresses: a fast-growing provider of a multitude of project services, packed with proficiency and experience, and always there for customers when these services are needed. In other words: a successful player in a highly competitive business environment.

An unsettling observation some years ago was that, while the company's turnover grew, the profit from the project portfolio did not. The number and overall financial value of these projects grew, but the margin made per project got less, and costs for organizational overhead to govern the growing project portfolio also went up, so that the overall profit remained static. The observation was made first seven years ago, and management did some analysis on what the numbers truly meant. They concluded that these numbers were not unsettling at all—they were just a sign that the company's young and wild years had gone, and with them the extremely challenging and stressful but also profitable projects, and that the business was steering into calmer waters. Grasshopper Ltd. considered itself an established and robust organization that had stabilized its business, from which it enjoyed a steady and reliable income. The company's management felt assured that the high turnover achieved so far was a good basis for

---

[1] The name has been changed.

future profit increases, and that the main task to achieve this would be to streamline processes and increase efficiency.

The whole project business got operationalized as far as possible. A strong project management office (PMO) was established, which unified the project management approach and communications. These were previously differing from project to project, and the implementation of strict rules was regarded as the most promising attempt to bring down costs for project governance and gain positive scaling effects. In addition, procurement and recruiting became centralized, and customers were forced to accept contracts with standard terms and conditions that covered most aspects of project management as it was done by Grasshopper.

The assignment of roles and responsibilities in the organization was clarified much more strongly, and the involvement of department managers in the projects, for which they had to provide the resources, was also strengthened. Bid and proposal management got streamlined by implementing a standardized process with gate reviews and a set of templates that eased winning new business and reduced the workload. Grasshopper's management was dedicated to bringing order into the perceived chaos, and they were very successful in that.

In the next year, the profit of Grasshopper Ltd. increased; indeed, these measures proved to be successful—Grasshopper's management loved to call them "Best Practices"—only to fall back again to the former poor level in a year later, and then to decrease even further. Something was going fundamentally wrong. To make things worse, the profitably of the projects varied strongly: There were highly profitable projects, but their margins were eaten up by projects with substantial losses, and these were often crisis projects in other aspects too, such as deadlines or poor deliverables.

Grasshopper's management tried to find a common pattern of these negative-margin projects— the loss projects—but wherever they looked, the data remained inconclusive. One idea was that they had good project managers and worse ones, but closer analysis showed that the same project manager who brought good money home in one project produced losses in the next, and vice versa.

Another approach was the implementation of software for enterprise project management (EPM) to increase the accuracy and reliability of data for management decisions, but, poorly understood and finally ridden with political restraints, the software delivered the wrong data. When managers base decisions on data, wrong data can lead to poor decisions.

This was the time when the company contacted me to use my training services. Discussions with seminar attendees gave me many insights into the processes—insights that helped me to identify the areas in which the project business went wrong.

The basic problem was the overall operationalization of the customer project business:

- The proposal templates focused more on the quick and simple proposal development than on offering customers what they asked for and needed.
- The hit rate was on average at 10 percent, which meant that one out of ten proposals led to a business contract—a number that is somewhat normal, but looking at the capture ratio, this was at under 5 percent, which meant that the majority of these won contracts were of low value.
- The high number of projects with negative margins showed that Grasshopper not only had no protection to avoid harmful business, but was instead successful in winning just that.
- The Grasshopper internal project management processes commonly collided with the processes defined on the customer side, forcing project managers to meet the demands of both process worlds when they should have focused on effective and cost-efficient delivery.

- Grasshopper simply did too many projects at a time. The company was great in winning business but did not have sufficient resources to perform them all. The problems arose particularly in projects that required special skills and knowledge that could not be easily found by hiring people from the street. Trained external resources were already booked by other companies, and internal staff would have to go through a time-consuming training process to build the expertise needed. This inability to provide the needed resources in a timely manner often disappointed just those customers who would have otherwise guaranteed profitable projects, and the soured relationships elbowed many projects from the profit to the loss side of the costing equations.
- The last-minute search for resources often forced project managers to accept as subcontractors vendors who had no record of successful projects in the specific fields and no customer-centered management approach, and in addition, some were financially in dire straits. This last group caused some of the greatest losses in Grasshopper's customer projects.
- The software solution added to the problem: Many calculations that would manually be done in a considered process, were now done by the software "under the hood" and were never reviewed. These calculations then too often did not reflect reality, but the results were considered sacrosanct and decisions were based on them.[2]
- The increased influence of the line organization had changed the organization's focus from extrospective effectiveness—seen from the customer's perspective—to an introspective process emphasis, converging around the simple question, "How can we do things more efficiently?" This led to unexpected cost increases when customers struck back with uncooperative behavior.
- The process emphasis came with a perception of project management as a closed-skill discipline, communicated in statements such as: "All our project managers need to do is follow the process; projects then will go perfectly well".[3] The ability of project managers to address the open-skill requirements of their projects with responsiveness and adapt ability were not valued at all.
- Project managers lost their joy in their profession; their approaches became instead grim and cheerless.
- Young people assigned to projects without sufficient preparation got consumed quickly, damaging their bios and their self-esteem.

I recommended that they change their basic approach and de-operationalize the company. I considered it beneficial to focus on bidding only for those projects that promised good business and ensured high capture ratios, then developing open-skill approaches to manage them in an environment dedicated to *complete,* not just to *compete.* I further recommended that they ensure an environment for the project managers that allowed for reliable mid- to long-term planning, enabling early booking of resources, and making sure that they had the backing of the organization.

My advice was not followed, and the actual company that is the role model for Grasshopper Ltd. in my case story is doing worse today than ever. They brought order into chaos, but

---

[2] This is similar to astrologers some decades ago, who strove to increase the credibility of their horoscopes by calling them "computer-generated". Interestingly, today, the human-made horoscope is considered more reliable by many of those who believe in horoscopes.

[3] The concept of open-skill vs. closed-skill disciplines is discussed in more detail in my book *Situational Project Management: The Dynamics of Success and Failure* (Lehmann 2016b).

improved neither their profitability nor the happiness of their customers, employees, and contractors. Project Business Management has many facets, but profitability and happy customers are probably the most important ones.

2.2 Introductory Questions

The following questions are written in the style of a certification test. They are intended to give you an understanding of the contents of the following text section and the questions that will be discussed in it. It may be interesting for you to answer these questions before you read the section, and then again once you have finished it.

---

1. What is the economic reason for a company to choose the *buy* option during a *make-or-buy* decision?

   a) Tapping the assets of another company
   b) Transferring the risk of project failure
   c) Reducing overall costs
   d) Keeping project details under control

---

2. How is the global market developing for customer projects?

   a) The market has mixed expectations.
   b) The market is robustly growing.
   c) The market is robustly shrinking.
   d) The market is generally static.

---

3. Online business-to-business (B2B) marketplaces bring together professional buyers and sellers in forum-like systems and help them develop the project business together. What is hard to develop in such systems?

   a) Comprehensive lists of sellers for highly specialized products
   b) A competitive situation with many bidders
   c) A contract that meets the requirements of the project
   d) Interpersonal relationship between buyers and sellers

---

4. If the terms are used accurately, how is a request for proposal (RFP) different from an invitation to bid (IfB)?

   a) The RFP describes the items or services to be procured in utmost detail; the IfB leaves them rather open.
   b) The RFP is generally not competitive, whereas the IfB is.
   c) The RFP describes the objective of the items or services to be procured; the IfB specifies them in detail.
   d) The RFP is used in industrial procurement, the IfB in public procurement.

5. A seller and a buyer have signed a memorandum of understanding (MOU) during negotiations as a baseline to document the momentary state of the agreement in order to make the further negotiations easier and more efficient. Two weeks later, the buyer ends the negotiations without a result. If the term is used accurately, what will this probably mean?

   a) An MOU is an obligation for the buyer to also sign the full contract when it is complete.
   b) An MOU is a legally binding document, but the seller will not be in a position to claim significant damages.
   c) An MOU is a legally binding document. The seller can claim significant damages.
   d) An MOU is a diplomatic document. The seller cannot claim any damages.

6. At certain points during project work, errors are being made. What is true for them?

   a) Over time, it gets cheaper to fix an error; it is therefore best to delay error fixing until it gets unavoidable.
   b) The cost and difficulty to fix an error grows with the local, temporal, and organizational distance from its origin.
   c) It is generally cheaper to let the error be fixed by another contractor than the one who made it.
   d) One can always convince a customer that the error is a valuable feature, and it will then not have to be fixed.

2.3 The Lure of the *Buy* Option

In November 2017, I had the opportunity to attend a global conference of project managers at a major manufacturer of high-technology goods in Europe. There must have been around 500 project managers present, almost all employees of the organization.

The initial keynote speaker started by asking attendees to open a website and type one word into a form field. He asked them to name the one thing that requires most of their attention and is most likely to keep them sleepless at night. The software then made a word cloud from these words. The frequency that a word was mentioned was reflected in its size—the more often, the larger the word; and its position—the dominant words were in the center of the cloud.

Figure 2.1 shows the results. The key word turned out to be *Resources*. One should note that the word is displayed more than once, particularly in French spelling as *ressources* and also in singular. If one puts them together, the dominance of resources as the number one hot topic in projects becomes even clearer.

The company purchases project services and products from outside, and the main driver for this decision is obviously just that: Resources. Project managers and their supervisors hope that they can make resources that are not available internally temporarily available from outside. The following paragraphs will dig deeper into this observation.

What are your project management hot topics?

Figure 2.1   The word *cloud* created by the software from the frequency of words named.

At the beginning of project business relationships, we still do not have a customer and a contractor. It is a good approach in project management to call them *buyer* and *seller,* because these are the intentions that the two have. In essence, both are prospects—a prospective customer and a prospective contractor. The stimulus to enter the business relationship lies on one of the two sides—commonly on the side of the organization that is going to become a buyer and later a customer. As discussed above, the focus of this book is on new business, unless it is stated that it is dealing with business between incumbent customers and contractors. Figure 2.2 describes where we begin with the analysis to get some first insights.

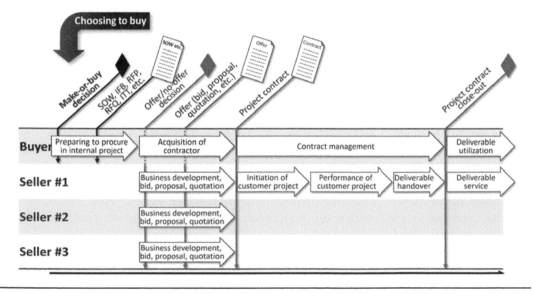

Figure 2.2   The standard process flow, as it is often seen in project business with non-incumbents, begins on the side of the organization that is to become the buyer in the subsequent process, when a *make-or-buy* decision has been responded to with *buy.*

On the buyer side, the business commonly begins with a need or a want: It may be a strategic decision by the organization whose management wants a project to improve processes, stabilize the business, gain financial benefits, or follow other considerations, monetary or not. These discretionary projects[4] are sometimes decided based on a business case—an elaborate document that, ideally, discusses the pros and cons of the project and possibly also those of not doing it. One variously observes projects without such a business case description, especially when the project has been initiated in an ad hoc decision. Another reason may be that the project is mandatory, initiated to meet a binding requirement like a law, a regulation, or another necessity that makes it inevitable to do the project. Another starting point for this project may be a contract with a customer.

While this project is being initiated and planned, a first decision presents itself, which is referred to as *make-or-buy*. An organization may decide to outsource a project entirely or in parts, or may consider itself able to perform the project with its own resources. The decision can be made based on many aspects, among them the following.

2.3.1 The Dilemma of Management Attention

The ability of management to pay attention to the projects that the organization is performing is the most valuable and often the scarcest resource in an organization. In my seminars, I generally recommend project managers to stay alert to the fact that the presence of management attention is no guarantee that other resources such as money, people, equipment, etc. will be available, but its absence should assure them of their dearth. There are many things that consume the attention of managers at any time in an organization, many of which are too important and too urgent to be left neglected. Then, they have families, hobbies, and other kinds of engagements. The amount of attention that managers can pay to their projects is limited, as all humans can pay attention to only a limited number of things.

Attention is time intensive, but it also consumes a lot of energy and mental resources. When I talk with project managers, almost all of them have had the experience of idle times in their projects caused by project sponsors and other key stakeholders who were unable to take the time to listen and make swift but informed decisions. The project managers will cc them in messages on issues such as delays, cost overruns, or extended times of operational disruptions that have become necessary, just to find later that these managers did not read the messages because others had higher urgency and because they were sent to them directly. Often, project managers notify their bosses directly of things that are about to go wrong and need management attention, but the message is in the sixth paragraph, sent to managers who do not read beyond the third one unless they are somehow forced to do that.

It is often said that the devil is in the details, and managers may therefore desire to know more about the background behind a decision that needs to be made by them—a decision for which they will later be held accountable by their superiors and other stakeholders, but for which they feel unable to reserve time to obtain this information. It may, for instance, be necessary to read a lengthy business case disquisition or to attend a meeting with experts. Another difficulty may be that they do not understand the special language and "cant" of these experts, who live in their own intellectual world and communicate in their own lingo.

---

[4] See the discussion on project types above.

Another factor that commonly limits management attention is the competition between the need to focus on internal structures and processes—where the requirement is mostly on corporate introspection, repeatability, and sustaining predictability and order—and the need to focus on external interfaces inside the various markets that the organization needs to deal with. Both areas of focus contribute in their specific ways to the stability and profitability of the organization, but while they are expected to complement each other, in business practice, they often compete for management attention.

Attention is definitively a resource, and as discussed above, in many organizations, it is the scarcest resource of all. There is an "economy of attention" in families, in organizations, and in society, and humans often value other people for the time and energy that others invest in them and, as such, for the attention that others "pay" to them. A corollary of this statement is that stakeholders often look negatively on managers who do not pay enough attention to them. Stakeholders here includes subordinates, but also superiors, shareholders, and many more. I am not trying to promote compassion for managers—they are mostly well paid for this kind of suffering—but rather to describe the nature of the attention dilemma in the environs of many managers.

Delegation of attention to subordinates promises help. These subordinates may have less distance to the people and things that need to be taken care of, and they may have more of the special understanding and language skills needed. There are unfortunately some caveats for that.

One is that certain tasks need to be dealt with by top management and cannot be delegated. There may be legal requirements, or it may be imposed by the largest customer or supplier of the company that the boss must put himself or herself in direct charge and control of a task. Delegating that task may be perceived as priority reduction and downgrading, or the risk of being finally being held accountable can be beyond a threshold that allows for delegation.

Accountability differs from *responsibility* in that the latter can be delegated, the former cannot. When a manager delegates responsibility to a subordinate, who in turn may delegate the task further, and so on, the accountability remains with the manager, as he or she was at the beginning of the delegation chain. The risks that come with this accountability, which the manager cannot delegate away, may be too high to do any delegation at all.

A second caveat, often overlooked, lies in the need for an increased investment in energy and time in developing the fitness for the task in employees who are expected to later take some burden off the manager's shoulders. Figure 2.3 describes this expectation.

If one waits too long to begin delegation, it may be too late to expect tangible relief. It takes a long time to bring the new staff member to performance, and the investment in time and energy may be no longer possible, as the workload on the manager and the organization is already crushing. Compare this with the increased speed of change concurrently happening at many dimensions of responsibility, as discussed below.

A third caveat is that it takes time to recruit people for the responsible position, and the task may be too urgent for that. An example: Some readers may remember an "Open letter" submitted 28 September 2012 by Tim Cook, CEO of Apple Inc., which was directed to users of a new and buggy mapping software launched only a week earlier. In this letter,[5] he wrote:

---

[5] (Savitz 2012)

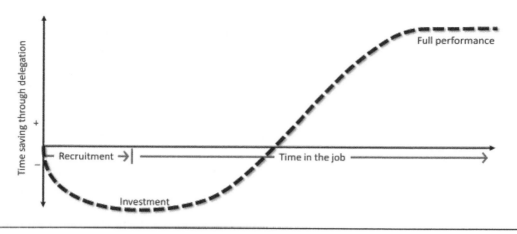

Figure 2.3   Investment and return expectations when work is delegated.

"To our customers,

"At Apple, we strive to make world-class products that deliver the best experience possible to our customers. With the launch of our new Maps last week, we fell short on this commitment. We are extremely sorry for the frustration this has caused our customers and we are doing everything we can to make Maps better".

[. . .]

"Everything we do at Apple is aimed at making our products the best in the world. We know that you expect that from us, and we will keep working non-stop until Maps lives up to the same incredibly high standard.

"Tim Cook

"Apple's CEO"

The letter was obviously intended to separate the highly valuable Apple brand from the software troubles. Apple's mapping solution came with a new sixth version of the iOS operating system used on iPhones, and it was not the only problem that users reported.[6] It would be interesting to run some scientific research into whether there is a measurable correlation of new product quality and involvement of corporate management in lawsuits, but when I observe corporations such as Apple and others, this seems to exist, and it is generally plausible. During the time when iOS 6 was developed, Apple was involved in patent wars with their competitors, including Samsung, Motorola Mobility, and others. By that time, there were over 50 lawsuits reported just with Samsung,[7] mostly on matters of patent infringement and similar intellectual property issues, and they must have consumed massive management attention that was then unavailable for their software development projects.

It seems that managers are generally aware of the attention dilemma. I've spoken with many of them in private about their perception of this impasse, and some considered it their strongest

---

[6] Excessive use of cellular data was another one, which could lead to high mobile costs (Rosenbaum 2012), and users also reported fast-draining batteries.

[7] (Mueller 2012)

personal shortcoming that they could not pay immediate attention to all things that necessitate it, and many feel guilt-ridden for not being able to give all people and things the attention they deserve. Giving the project work to an external organization is often linked with the expectation that the managers of that organization will provide the necessary attention to the project, and that these managers can be controlled by defining appropriate contractual conditions.

It also includes the expectation that the contractor's management has more time and energy available to pay such attention, particularly as this project provides income for the company, but also because it is performed inside the contractor's area of core competency. As I stated earlier, management attention is often the scarcest and most valuable resource in a project. Its continuous presence does not guarantee that the project will receive all the other resources it needs, such as people, money, infrastructure, and more, but its absence will generally lead to a shortage of these resources.

In a matrix organization, a project has to compete for management attention with operational activities. In a project portfolio and in a program—both kinds of multi-project environments that will be discussed later—the project competes with other projects. Outsourcing project work can indeed take the burden from management to pay attention to this work, but there is a risk that the *contractor's* management does not pay the attention expected, or pays more attention to aspects of the project that are not in the interest of the customer, such as cutting costs, where the contractor cannot see it immediately, but may later have a disadvantage from a cheap solution.

2.3.2 Agility and Speed of Change

Businesses are facing alterations at an accelerating pace in many dimensions: Technologies, political and social environments, paradigms of markets, and many other factors that influence their success and failure have become highly dynamic, and concepts that have been appropriate and successful in the past may fail today. An example: For over a full century, beginning with the first commercially available telephones and transmission systems in the late 1870s, it was a matter of course that a telephone had a cable attached, which limited the range of its use, which was normally confined inside private houses or at workplaces. When one family member used the phone, it was blocked for all others. When people were away from home or the office, they had to use pay phones in public phone booths or had to find other line-bound phones—for instance, in hotels.

Beginning with Motorola's DynaTac in 1984, mobile phones became popular, first mounted in cars, then getting smaller and lighter, so that they could be carried by users in suitcases and later, with further miniaturization, in pockets. With the growing popularity of these "cell phones", the paradigm of a phone that was linked to its location changed to one linked to the person carrying it, and interest in the old style phones decreased.

This process lasted until the year 2007, when Apple launched the iPhone. It came again with a shifted paradigm, in which a phone is no longer just an electronic device to allow two individuals to talk with each other, but one that connects its owner with the mobile internet. It allows reading newspapers, booking flights, using social networks—another new paradigm—to stay connected with family members and friends.

The next paradigm change is already on its way with wearables, mostly in the form of fitness trackers and smart watches, digital glasses, and other items that people can wear on their body without much hassle. In essence, these replacements for wristwatches and other items are small computers that can complement or even replace smart phones for a growing number of tasks.

This technological development has implications: The number of phone booths worldwide is in decline, and even in homes and offices, landlines are more and more replaced with smart and mobile items (see Figure 2.4). But not all of them. Some of the old technologies remain in existence, and new technologies grow alongside them. But sometimes, old technologies become extinct.

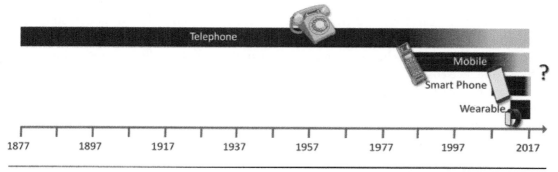

Figure 2.4 An example of the increasing speed of innovation and change of paradigms: telephones.

These new technologies and paradigms do not fully replace the old; instead, they give people more options to choose from. One should also note that the change is not limited to the handsets, but also includes the infrastructure that supports them and the handling of these items, which has come a long way from old mechanical dials to capacitive screens that allow swiping and to wearables that even respond to a person's pulse. With the changes of paradigms, new key players turned up, and the old stars of the markets lost a lot of their influence and relevance.

Large companies have difficulties developing the agility to cope with the speed of change. They are more like large tankships that require a long distance to accelerate to full speed, take turns, and finally to come to a halt again. Large companies hire small ones as contractors for their projects when they do not expect that their own personnel will be able to develop the necessary skills quick enough and instead rely on companies that have the developing process already done or are agile enough to perform it on demand.

2.3.3 Diversity of Skills

The growing diversity of technological options, which comes with the speed of change in technologies and paradigms, leads to an increase in the number of competencies that organizations must master today. While old technologies remain alive, in smaller demand but not fully extinct, and with them the skills necessary to use them for business, new technologies and new skills are needed in advance.

Henry Ford sponsored the project to develop the famous T-Model in the years from 1906 to 1908 in the Piquette Avenue Plant in Detroit, Michigan, USA. Colocated in a room only 5 × 4 meters (app. 16.4' × 13') small and secretly walled off from the assembly hall, Ford used a team of only four men:

- Charles Sorensen, a patternmaker and engineer
- Joe Galamb, a draftsman
- Gene Farkas assistant
- Louis Halmesberger, another assistant

A fifth role was Henry Ford himself, who spend a lot of time in this room in a rocking chair, ensuring that the car developed by the team would meet Ford's ideas of design simplicity, production-friendliness, and drivability.[8] For its time, the Ford T-Model was revolutionary in materials selection and technology and also in manufacturability, allowing for major scaling effects to reduce costs with growing production numbers, and also to lower price.[9]

Today, hundreds of developers have to work together, and they come from a variety of disciplines—including mechanical engineering, electronics, software, interior design—for the development of a vehicle and another team with a similarly broad spectrum for the design of the production. The small Sino-British car brand MG, for example, communicates that it has a "team of 300 engineers based in the European Engineering Technical Centre . . . responsible for developing the initial concepts for all new MG vehicles." The product development as such would then be carried out in China, but testing would happen in Britain again.[10] And MG's development facility is similarly small. BMW's Forschungs- und Innovationszentrum (FIZ – Development and Innovation Center) is the workplace of 20,000 BMW employees and additional 10,000 employees from contractors, and an expansion program named "FIZ Future 2050" is intended to increase the numbers by 50 percent to 80 percent.

The number and variety of disciplines involved in automotive development is tremendous, and the same applies to many other modern project environments. Organizations are often not able to staff all involved disciplines whose involvement would be necessary. They may also struggle with disciplines that they need only temporarily, when their own staff would expect to be hired for a fixed duration or indefinitely. These organizations therefore define core competencies that they want to control continuously and that they feel able to implement with their own people, and so, they will seek help from outside the organization for all other disciplines and buy other competencies from external partners, when they need them.

2.3.4 Unlocking Growth Potentials

Growth in the customer market requires expansion of internal structures. New products, productions, and services need to be developed, and they come with the need to provide additional space, hire people, and locate these people in an expanding organizational structure, define and implement processes, and ensure that all these systems do not get out of control when

---

[8] (Duncan 2008)

[9] Another name to be mentioned would be C. Harold Wills, who conceived the idea of using high-tensile steel and ensured its availability for mass production from steel mills. Furthermore, he developed the car's planetary two-speed transmission (Donnelly 2005).

[10] (MG Motor UK Ltd. n.d.)

they are ramped up too fast. There are factors that limit how fast this internal growth can be performed, including availability of capital, market conditions, quality and speed of recruiting, and, again, management attention. This internal growth comes with two kinds of risks:

- When the demand for the products or services is less than expected, the indirect and fixed costs of both running the business—administration, rentals, insurance, etc.—and of paying back credits and outlays for the initial investment may not be coverable by the margins generated by these products or services, and the organization's growth will lead to losses instead of profits.
- Limiting internal growth can in turn limit growth of sales, when the market demand for the products or services cannot be met, when their unavailability on the market leads to a lack of visibility on the customer side, and, worst of all, when the organization is outperformed by a competitor who does not have these limitations.

Sharing the burden of the development is a common solution for faster growth. Outsourcing development to contractors frees the customer from the need to internally keep pace with the external growth by using other organizations' resources, including, again, their competencies and management attention.

There is of course the caveat that this benefit can only be achieved if the customer's understanding of the goals of this development is shared by the contractors, and if their approaches, business interests, and cultures are compatible.

- Outsourcing project tasks can also support organizations in the opposite direction: Not all projects are tasked to develop something; projects may also serve purposes such as obsolescence management—disposal of outdated, redundant, or otherwise no longer wanted items. These items must have been assets in the past, but at one point have turned into liabilities. They may include building facilities, services, software, and other items.

    The same may even apply to people who are actively outplaced. Contractors will then be used by customers to divest themselves of these assets and to organize the process of making merited people redundant, people whom the customer organization does not want to simply fire, or where this is prohibited by contracts or law. The contractors would then take care of the obsolescence process, thus allowing the customer to focus its own resources, including management attention, on tasks that are relevant to sustain current operations or for the development of future business.
- Obsolescence management is rarely a thankful business for the sponsoring organization: One may decide to build a nuclear power station, run it for several decades, and make money on the electricity and possibly the excess heat generated. One can develop a business plan for the construction period, and then for 30 years' or similar operating time, and when all costs and income numbers have been estimated and calculated, make a decision based on the Return on Investment (ROI) or more sophisticated methods such as Net Present Value (NPV) or Internal Rate of Return (IRR) projections.
- Dismantling a nuclear power plant at the end of its lifecycle is an even larger project, and it takes decades to be finished. No one has yet knocked down a production reactor whose time has come and either turned it into a business center or a residential neighborhood or renatured the plant. While the investment in time and money will be significant, the business value of the project deliverables will probably not be very high—who wants to work or live in an area where a lot of contamination may still be present.

To make things worse, a lot of uncertainty surrounds the final disposal of tons of radi-ating waste that will pose a security risk and a source of costs for decades. It is no surprise that the operating companies of these power plants try to hand over the responsibility for these tasks to third parties—ideally, in their understanding, including the future risks that are so difficult to foresee. Demolition and disposal of obsolete nuclear power plants will be a safe business for a small number of experts in decades to come. It may also become a playground for organized criminals, who will dump the radiating rubbles in places where they will cause maximum damage.

2.3.5 Buying to Save Money or Time

An important factor for the outsourcing decision is, of course, the desire to save costs. In opera-tions, it has become a common calculation to balance the additional cost and delays of out-sourced productions and services against the savings from lower wages, lower environmental standards, and cheaper energy and raw materials. Similar calculations are done in project management, but, given the dynamics of success and failure in our discipline, with much less reliable results. In project management, each project is a new learning process, and part of this process may be the insight that the *buy* option brought more financial benefits than expected, or fewer, or possibly none at all.

External vendors may have cost benefits due to their location, being able to utilize lower costs for personnel, materials, or environmental protection. They may be smaller and therefore able to circumvent minimum salaries dictated by law or agreed upon with unions for larger companies, or they may be large enough to use their market power to reduce costs. A set of tasks may constitute a Mark 1 project[11] for the customer—new and unknown. For a certain contractor, the project may be a Mark *n* project, because the person or organization has done similar work in the past, found the knobs and switches that help save costs, but also knows the caveats that can lead to unexpected and massive cost increases.

This mostly comes with the expectation that the contractor will be faster. While the spon-soring organization would have to hire people first, purchase equipment, and possibly develop new know-how in house, the contractor is expected to have all that already. Decisions that would necessitate long discussions in an in-house project can be made much more quickly by the contractor, who brings much more knowledge about and experience with the specific task, at least in the expectation of the customer. The same is considered true when it comes to devel-oping staff and helping them learn new skills. The contractor is expected to bring educated people who can start working immediately.

2.3.6 Buying as a Political Decision

Crazy Ant, Inc. is a global consulting agency that works for large-scale corporations and gov-ernment agencies, performing organizational development projects that affect all aspects of the businesses of these customers. Their projects integrate organizational, technical, cultural, and business aspects of these changes, promising improved business success through streamlined

---

[11] The distinction of Mark 1 and Mark *n* projects is discussed in the previous chapter.

processes and increased efficiency. I was working temporarily for this organization on several projects in the late 1990s. One of the projects that I was involved with was the re-definition of sales channels of the Mantis Corp. airline in response to changing reservation behaviors by customers, which included travel agencies—some of them booking online over centralized systems (Amadeus, Sabre), others booking over the phone—but also included corporate travel departments and self-booking travelers.

It was the time when the internet was turning from a noncommercial information network, predominantly used by science communities, large corporations, and insiders, into an omnipresent business platform. The airline identified the risk that the booking behaviors of their customers was about to change and that the airline might not be sufficiently prepared to cope with this change, both marketwise and also technologically and organizationally. In addition, they hoped for cost advantages from an increased use of the internet for direct bookings, circumventing agencies and saving the fees they took.

The project seemed at first glance to have a small impact: The company already had a simple booking solution on the internet, and the basic infrastructure to link online bookings with the internal backend systems was also in place. The question was raised whether these systems would be able to cope with a demand that was expected to grow rapidly. The system was not built for easy scalability, and a major part of the processing still needed to be done manually and was often time-consuming. The change would have meant the need for many employees to learn new skills and accept a different approach to managing customer contacts. The project had the objective to make internet booking the leading system and to add manual agency booking as an alternative process, while so far the core system was manual and the online booking was the add-on.

Resistance inside the organization was immense. The internet was still new for most employees, and was particularly difficult for those with intensive contact to the travel agencies. They did not have to fear a loss of their jobs—at least not in the short term—but expected major changes to their job environments, including the contacts to the agency employees with whom they had developed close relationships over many years. The telephone and telefax as a major means for communications would be replaced with the PC, with which many people at that time were not yet familiar and confident. Some found that the manual processing time for bookings was reduced, so their jobs might still be threatened. One criticism actually came from employees who were concerned for the future of Mantis Corp.: They were afraid that customers would not change their preferred booking channels that quickly, and that neglecting the existing channels could damage the standing business before the new business was being developed.

For an outsider, it may be difficult to imagine how fiercely and angrily the dispute became over a time of a few months between the backers of the project and those opposing it, which caused rifts inside the organization that massively disrupted both the organization's operations and the project. Probably even more damaging to the project was a sublime refusal to go along with it by many employees. They slowed the corporation's processes down just to the point where they could not be held accountable for the loss in efficiency. Travel agencies already considered shifting their customers to competing airlines, where they felt processes moving swiftly, not frustrating as if they were stuck in a kind of jelly. The same happened in the travel departments of customer companies. In a project, you can have all the tools and techniques in the world, but slow and sticky resistance often finally wins out.

Mantis' management soon had to pull the emergency brake and halt the project to avoid further damage to the company. In discussions that followed, the bitterness of the employees was more and more focused on Crazy Ant, Inc., the consultants who did the project for Mantis. Depending on who one listened to, the consultants allegedly did the change too fast or too slow, too vigorous or with not enough dedication, listened not enough to people or spent too much time in interviews and meetings, and so on. Mantis Corp. terminated the contract with Crazy Ant, Inc. at favorable conditions and soon restarted the project more as a "submarine project" or "black project", which slowly came up with the same change but with less confrontational attitude to the employees and with much less politics.

By making the consultants the culprits, management saved face in front of their employees, and vice versa, and allowed the company to get back to normal operations. Between management and the consultants, there was an understanding that this was the best solution for the moment.

The *buy* decision may also be taken in order to deal with politics. Internal project team members often run into conflicts with other employees. Among the hidden costs of projects are operational disruptions, and, in addition to the economic consequences that these effects may have, they can lead to organizational and interpersonal disturbances.

The conflicting employees will still have to work together in the future, when the project is over. They will meet each other in the hallways, the company restaurant, and at other opportunities, and may have to go through more difficult situations in other projects, for which grudges and frustration from older projects may be a heavy burden. A project manager who was perceived to act against colleagues will be met with distrust in the future, and even if the person was highly successful in rebuilding rapport and mutual understanding, some of the mud will be certain to stick.

Management may decide in such situations that it may be better to buy the project team, or major parts of it, from outside the organization, as these can act with less considerateness. They can be more result oriented, because they will leave the organization again when they are finished and do not have to take long-term relationships, good or bad, into regard. These external resources are therefore expected to create faster progress and earlier results in environments in which high resistance is expected.

This approach comes with a corollary: When the project is on its way to fail, possibly because of this lack of considerateness and because resistance is growing too strong and compromising for managers, they can locate themselves at a distance and put all the blame on the contractor. It may well be that, for the contractor, the payments from the customer then become less remuneration for performance but a kind of monetary solatium paid to a scapegoat.

In such environments, the proclivity of managers toward the *buy* option to overpower or circumvent opposition inside the organization signals definitively a kind of weakness when it comes to implementing controversial decisions, and of muddled governance in conflict-ridden environments toward those who are finally expected to carry results home.

2.3.7 Tapping External Assets

One can subsume the luring elements described above, and many more, under the general heading "Tapping external assets". Sellers have hard assets such as facilities, equipment, and similar, or soft assets such as skills, licenses and patents, or management attention, that buyers desire to use as resources in their projects against payments. In some projects, these external assets have a rather marginal function. In others, they are combined with the internal assets of the buyer organization. There are also projects in which the external assets are the only ones used as resources for the project. Contractors can have different roles in projects, but the expectation is always that they are an external source of resources that the buyer organization needs for the project but is lacking.

Sometimes, assets turn into liabilities, and this is also true for contractors. I will discuss the risks that come with selecting the *buy* option later.

One may argue that all the risks that come with procurement of project work on the seller's side can be avoided by running projects with their own resources only, avoiding procurement whenever possible. There are organizations that follow this path on the asset-heavy projects, and some of them are very successful.

One may also argue that making a corporate living from offering oneself as a contractor to buyers is too high a risk to take, and one could find easier ways to bring money home than with projects. But then, this business is thriving worldwide, as I will show below, and many small to mega projects would not be possible without the tight cooperation of buyers and sellers, customers and contractors, simply because no one has all the assets that would be needed as project resources, including money, equipment, skills, licenses, expertise, and, most importantly, management attention.

Project Business Management has become an essential element of economics worldwide, and given its wide distribution, is it not astonishing that there is no literature, education, and research on the specific topic? I am happy to fill this gap at least partially and hope that others will follow, addressing the topics that I may be missing and correcting statements in which I may be wrong.

2.4 Customer Project Management: Where Does the Market Stand?

In the preparation for this book, I was interested in obtaining market data and information on customer projects from empirical research, but I could not find any. I decided therefore to explore the market by performing a micro-survey.

2.4.1 The Rationale of the Research

In the typology of projects and project situations presented briefly in the first chapter of this book, the distinction between internal projects and customer projects is one of the most obvious. Internal projects are commonly done to implement organizational strategies, to meet business goals, or to respond to mandatory demands from law or other binding conventions. If they are expected to create financial benefits such as cost savings or increased income, the majority or all of these benefits will come after the project lifecycle.

Customer projects are generally expected to produce this income while they are performed, and the timeliness and sufficiency of the income are the most crucial metric for project success.

Customer projects must support the organization's profitability and liquidity. Their benefit realization is commonly ended when the project is finished, possibly earlier, notwithstanding subsequent income from services that are then operational and no more projects.

Figure 2.5 shows how the lifecycles of projects and benefit generation commonly relate. An essential element of the benefit realization lifecycle is obviously the contractual payment scheme that the contractor has agreed to with the customer. The essential benefit from a customer project is the money that the project must bring home.

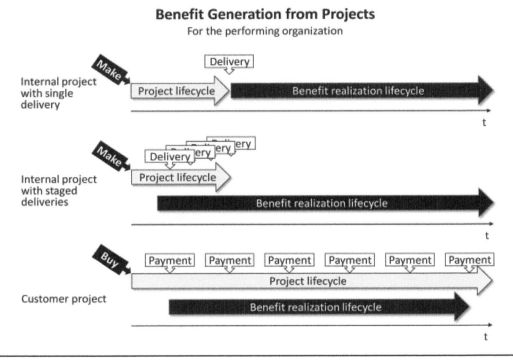

Figure 2.5   The different lifecycles around the project and the benefit realization for the performing organization.

When I tried to find information on customer project management in scientific and business literature, I found an astonishing shortage. It seems that this topic has not yet been the subject of research and professional contemplation. I wished to fill some of these gaps to support the future of customer project management with better data and a better understanding of their mechanics and dynamics.

A major factor that defines the future market for customer projects are *make-or-buy* decisions. Selecting the *make* option leads to an internal project. When a customer decides for the *buy* option, a "want" or a "need" is sent on a path that will finally turn it into a customer project on the side of a seller, who will later become the contractor. If the general trend goes for an increase in the *buy* option selected, there will be more customer projects in the future; a trend toward *make* would reduce their rate of occurrence. The research described here asked where the trend of the recent past and the future lies.

2.4.2 The Design of the Research

The research was designed around a micro-survey that allowed for a participant to finish in less than a minute. Brevity was indeed a major objective of its design, as one can observe a noteworthy survey weariness, which has evolved among project managers over some years. It can make it difficult to get a sufficient number of responses for a survey to consider its results meaningful. In the recent decades, project management has become a subject of academic discussion and research, and project managers are therefore often asked to respond to surveys—requests that they tend to ignore due to the time pressure under which they have to work in their projects. In addition to these requests, project managers are asked to respond to surveys by companies, associations, and other organizations, and many project managers therefore reject dealing with surveys at all.[12]

In my experience, micro-surveys are still accepted by an audience of project managers. They consist of a few questions only, and it must be communicated right from the start that they are easy and quick to answer. The survey therefore consisted of only four questions, plus a free-text field for comments and an opportunity to add an e-mail address for respondents who would be interested in the results.

The research presented here is based on such a micro-survey, open for responses over 17 days between 21 December 2016 and 8 January 2017. It received 590 responses during this time.

The first two questions asked for trends in *make-or-buy* decisions, using seven-step sliders with scales between −3 and +3, referring to recent experience and future expectations. Figure 2.6 shows how respondents could use these sliders to select their degree of agreement to one of the two contradicting statements.

1. On the slider below, which position represents best your *recent experience* for Make-or-Buy decisions?

2. On the slider below, which position represents best your *future expectations* for Make-or-Buy decisions?

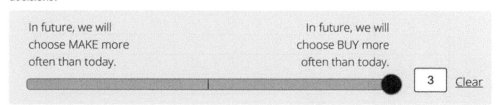

Figure 2.6   The sliders used to answer questions on recent experience and future expectations on *make-or-buy* decisions.

---

[12] US comedian Bill Maher has made a nice joke from that: "You know, I was actually pretty happy with your customer service, up to the point where you asked me to take a survey about your customer service".

Responses to two more questions allow for stratification of the results:

1. By vendor involvement in projects: "How much of the workload in the projects you are involved with is generally done by vendors?" The answer options were:

   a. Our own people do all the work.
   b. Vendors do less than 50%.
   c. Vendors do about 50%.
   d. Vendors do more than 50%.
   e. Vendors do all the work.
   f. I do not know.

2. By project size: "What is the most typical total number of people actively involved in your projects from start to finish?" The answer options were:

   a. 1–10
   b. 11–100
   c. 101–1,000
   d. 1,000–10,000
   e. More than 10,000
   f. I do not know

These two questions allowed only for one answer to help identify differences among projects with different degrees of outsourcing and of different sizes.

2.4.3 The Survey Respondents

The respondents were addressed in professional groups in social networks (mostly LinkedIn and Xing) and by directly inviting contacts by email. The 590 participants came from 85 countries, with the "Top Ten" by participating individuals shown in Table 2.1.

Table 2.1 Top Ten Countries with Participants in the Survey

Rank	Country	Responses
1	United States	113
2	Germany	92
3	India	63
4	Canada	24
5	United Kingdom	22
6	Saudi Arabia	19
7	United Arab Emirates	18
8	Australia	11
8	Italy	11
10	Pakistan	10

The approach was global and cross-industry. I would generally encourage repeating this survey with an additional focus on countries, industries, application areas, and other more specific areas of interest, but this would have been beyond the scope of this micro-survey. The questions in this survey related to the market of customer project management in its entirety: Can businesses expect it to grow, or is it rather static or even shrinking?

The survey also yielded a major number of interesting comments. Some of them will be quoted and discussed later.

2.4.4 The Overall Result

This research was undertaken to answer the question of whether the market for customer projects should be considered a growing, a static, or a shrinking market. As shown above, the respondents could make selections on two sliders what their recent experience in their project environments has been and what their expectations for the future are. Figure 2.7 shows how the responses were distributed.

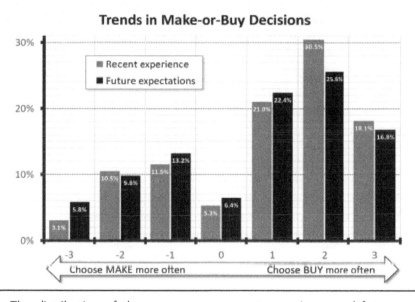

Figure 2.7 The distribution of the responses on recent experience and future expectations regarding *make-or-buy* decisions using the seven-step sliders shown in Figure 2.6.

The average values among the selected responses were:

- Recent experience: 0.95
- Future expectations: 0.74

Both the distribution of the responses chosen and the average values show a trend toward more frequent *buy* than *make* decisions during the recent past and for the future, with an expectation that the growth on the buy side will go on, but at a slower pace in the future. The reason for this result may lie in an often communicated desire by management to reduce dependency on external vendors, while business necessities force organizations to buy more and make less. It would be interesting to repeat such research from time to time to see whether the projected slowdown of the tendency to buy more signals a true future trend or is just wishful thinking.

2.4.5 The Influence of the Workload Assignment

I was further interested in discovering the degree by which the results were dependent on the workload that was already outsourced. Could it be that projects with fewer parts of the workload already outsourced are different in trends to those in which more work is given to contractors? Participants were therefore asked to state how significant vendors were in doing the workload in their projects. The participants could choose from six options. The answering options and the distribution of the responses among them is shown in Table 2.2.

Table 2.2 Answering Options for Workload Assignments in Projects and Distribution of Responses to Them

Answering option	Answers	%
Our own people do all the work.	55	9.3%
Vendors do less than 50%	204	34.6%
Vendors do about 50%	105	17.8%
Vendors do more than 50%	192	32.5%
Vendors do all the work	26	4.4%
I do not know	8	1.4%
Total:	**590**	**100.0%**

Figure 2.8 shows that project environments with more outsourcing were experienced to have a stronger growth in *buying,* and the expectation for growth in the future was also higher. The market for outsourcing is obviously the most dynamic when a lot of outsourcing is already done.

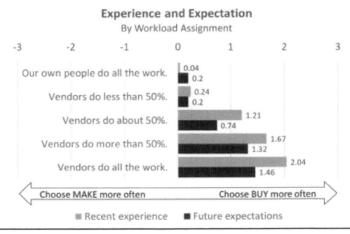

Figure 2.8 The past and future average trends for *make-or-buy* decisions grows with the workload assignment already outsourced.

2.4.6 The Influence of the Project Size

To what degree are trends on outsourcing dependent on the size of the projects? I further asked the participants to state how large their projects were, measured in number of people involved. The participants could choose from six options. The answering options and the distribution of the responses among them is shown in Table 2.3.

Table 2.3 Answering Options for Project Sizes and Distribution of Responses to Them

Answering option	Answers	%
1–10	187	31.7%
11–100	313	53.1%
101–1,000	72	12.2%
1,001–10,000	10	1.7%
>10,000	4	0.7%
I do not know	4	0.7%
Total:	**590**	**100.0%**

Figure 2.9 shows that project managers from environments with larger projects experienced a stronger growth in *buying,* and their expectation for the future was also that of a stronger trend. The market for outsourcing is obviously more dynamic for larger projects.

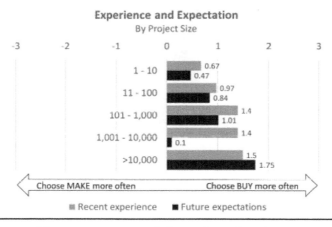

Figure 2.9   The past and future average trends for *make-or-buy* decisions grows with the size of the project.[13]

2.4.7 The Influence of the Project Location

Another question related to location: Are world regions different in their observations of past and expectancy of future trends?

---

[13] The outlier in the fourth stratum (1,001–10,000) for the future expectation may be explainable with the small number of only 10 cases.

Based on the IP addresses of the responses, I located the respondents in six world regions, as shown in Table 2.4. The table also shows how the responses were distributed over these regions.

Table 2.4 World Regions Defined and Distribution of the Responses Across These Regions

Region	Answers	%
Africa	27	4.6%
Asia	110	18.6%
Australia	10	1.7%
Central & South America	21	3.6%
Europe	218	36.9%
Middle East	62	10.5%
North America	141	23.9%
Total:	**590**	**100.0%**

Figure 2.10 shows the distribution of the responses over the six regions. The average results are positive in all regions, which shows that the expectation for growth in outsourcing business is unbroken in all of them, only expectation on the speed of future growth is different.

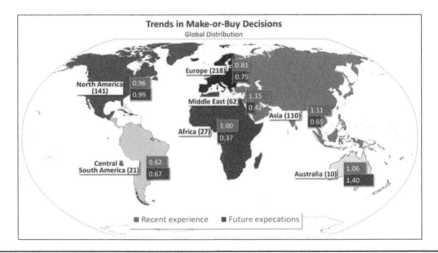

Figure 2.10   The average trends for the recent experience and future expectations on *make-or-buy* decisions are different over the world regions, but the trend is toward further growth in all of them.

2.4.8 Conclusions from the Research

The business of bringing money home by performing projects for paying customers is obviously a very robust one, which shows no signs of decrease. The global trend for this type of business is a growing demand, driven by an increasing number of *make-or-buy* decisions in which the *buy* option is preferred over the *make* option. Differences could be found in the speed of this growth for the past and the future, but the growth as such is not broken. It may also be interesting to repeat the research in the future, to better understand why the growth expectations for the *buy* option (averaging at 0.74) were smaller than the experiences in the past (average: 0.95). This may indicate an upcoming saturation in the market place. It may also signal management strategies to insource, and it would later be interesting to see if managers were able to implement these strategies or whether changing market conditions dictated an increase in the intensity of outsourcing beyond what they had intended.

Another observation from the author's seminar business is interesting in this context: Project managers are commonly not sufficiently prepared to take over the responsibility for customer projects, which should be considered temporary profit centers in complex and highly dynamic environments. Project managers in customer projects have to consider the interests of two or more parties.

These interests are sometimes in alignment, but at other times are contradictory. Particularly when the survival of the organization depends on the financial success of its customer project, and when the satisfied customer is necessary to provide a reference for winning future business, the requirements on the business acumen of the project managers are beyond the education and training contents delivered to them in classical project management seminars. There is a future need for specific Project Business Management education with a focus on customer projects, helping their project managers to ensure that they bring money home with their projects and bring a smile to the face of the happy customer.

Project managers in customer projects are not a minority group inside the entire project management discipline, as I described in the previous chapter. In my survey, 51 percent of the responding project managers selected the option that they do a project for a paying customer, and that the project provides income to the project manager's employer. For these project managers, it is not sufficient to be technically competent—they must understand the highly complex dynamics of the customer–contractor interface.

Because many customer projects are part of *project supply networks* (PSNs), with a multitude of organizations involved, and in which self-employed freelancers contribute additionally as one-person contractors, the number of interfaces is growing further. Let the PSN extend itself over different countries, cultures, time zones, business styles, and legal systems, one can easily imagine what requirements are put on project managers.

In a further globalizing world economy, in which distances across the world are shrinking and the wind of change has turned into a class 7 storm, educators in this field are also not sufficiently prepared to instruct project managers to cope with these challenges, and while I am writing these lines, I do not exclude myself from that statement. Whether an educator is a trainer or coach in business or a teacher in academia, this statement is probably true for all of us.

The description of insufficient preparation of project managers inside the contractor team is also applicable to their colleagues who sit at the opposite side of the negotiation table: They develop and manage these complex PSNs, often across country borders, as described above, spanning cultures, legal systems, time zones, business interests, and many more environmental factors that lead to fragmentation and siloing of the project team. Understanding and managing such complexity—which is generally not included in the preparation of professionals for project management—is a key challenge for these project managers, and I have to repeat that the educators doing such preparation are still not in the marketplace.

2.5 Risks of the *Buy* Option

The survey above had a free-text field for comments by attendees. Out of a total of 590 responses, 185 used the field to leave a comment as an additional response to my questions—that is, 31.4 percent. The focus of these responses was on the reasons for organizations to choose *make* or *buy*. Some responses discussed the risks that come with the *buy* option, and the following paragraphs elaborate these risks. From a seller's perspective, these and more risks must be considered, but also from the buyer side, it is important to understand these risks; they are a driver for the behavior of the prospective customer, and addressing these risks in a way that brings the customer peace of mind may be essential to finally winning the business.

2.5.1 Time Losses

The lure of choosing the *buy* option is set off by many risks. The most obvious is the time lost for procurement activities, particularly when they prove futile and when one finds out late that the *make-or-buy* decision needs to be put on the desk again and that one needs to choose the *make* option this time or find an entirely different solution.

- An example was Tesla Motors, the maker of electric cars, in the years 2005 to 2008 in Menlo Park and later in San Carlos, CA, USA.[14] Tesla, by that time still rather a small but fast-growing garage development company than a well-organized manufacturer, had so far built two prototypes of a fully electric car based on the British two-seater sportscar Lotus Elise and was going to develop a production car from that called the Tesla Roadster.

  They identified the need to include a two-speed transmission to better sell the car to wealthy early adopters—a high gear capable of speeds of over 200 km/h (130 mph) for normal cruising and a low gear for maximum acceleration (0 to 100 km/h in 4 seconds). The latter was considered a major sales argument for the intended consumer group; it would enhance the driving experience of the car and make it competitive in the Sprint discipline with expensive traditional sports cars. A British manufacturer of transmissions was asked to develop and later manufacture the gearbox, which so far was considered an unproblematic item.

- The launch date of the Tesla Roadster was originally envisioned for early 2006. This was then shifted—and communicated—for middle of 2007, but it soon became obvious that unexpected problems with the transmission needed to be resolved, and it was at that point clear that this would take longer than expected.

  The electric motor ran at much higher revolutions than a traditional combustion motor, typically over 14,000 rpm, and the spread between the two speeds was much larger than between two neighboring speeds in a conventional multi-speed gearbox, so that each switching action created an increased mechanical stress on the transmission. The average running distance of a transmission before it failed was at 3,600 kilometers (2,000 miles), far too short, especially for a car in the $100,000 price range.

  Tesla repeatedly changed the manufacturer, but without improvement. It turned out that a root cause of the problem was that the manufacturers seemed not to have put their

---

[14] (Vance 2016)

best efforts into the development for Tesla because they did not consider them a great future investment. Tesla's strategy was to start out with a high-price, low-volume product and then over time increase the volume and reduce the price. For the time being, Tesla's output numbers were too small for the suppliers to take them seriously, and they did not believe in the production numbers predicted for the future.

- Tesla finally changed the basic design of the motor to allow for a simple one-speed transmission. In combination with a reduced weight of the car, they could still get basically the same performance data without the need of changing speeds. This change took them about ten months to implement, and cars delivered earlier to customers received it in the form of a free upgrade.
- The delay from the inability of vendors to deliver the transmission with the necessary robustness jeopardized Tesla's basic funding. Investors developed doubt whether Tesla would be able to deliver finally, and this doubt was reinforced by the negative press that came with the delays. It was a critical time for Tesla. Finally, they came out with a better solution, replacing a mechanical system with an electronic one, which allows easy updates, ideally over the air, while an update or a repair of a mechanical part generally needs a visit to the repair shop. The solution is more aligned with the basic paradigm of Tesla, but it took the detour over the non-delivering vendors to finally arrive there.

2.5.2 Unclear Scope

Before one goes shopping, one should have a clear idea of what one actually wants and needs. Otherwise, *caveat emptor*[15]—one will pay more money when one has bought the wrong thing and has to buy again, or needs to rework the purchase to make from it what was originally desired. Buying over the internet, depending on applicable business terms and legal requirements, one will possibly be able to send the product back when it is found unsuitable. When customers procure services and products as projects, the same applies.

Transport for NSW (TfNSW) in Australia was founded in 2011 through the merger of a major number of previously separate public transport agencies of the state of New South Wales.[16] It had around 25,000 employees and had also inherited around 130 software systems, which they wished to replace with just one new enterprise resource planning and asset management (ERP/EAM) system from SAP. They hoped to save AUS$100 million (US$73 million) annually from the consolidation of the existing systems. The investment for that seemed appropriate: AUS$151 million.

The project should have started in 2012 and been finished by the end of 2015. After this time, the new system should be set up, the data migrated into it, and operations of the old systems should then have been discontinued. In June 2014, the project underwent an audit, which came out with a warning that the business transformation aspect on top of the technical aspects seemed insufficiently addressed and that frequent changes of scope and team threatened timely delivery.[17]

---

[15] "Cautiousness is the buyer's job" (not the seller's).
[16] (Wallbank 2017)
[17] (Cowan 2014)

Migrating a single IT system can be a tedious task. Legacy systems rarely have the consistency and quality of data that modern systems require, and data structures and processes implemented in them differ from those in the new system, so a lot of manual work will be needed. Migrating and consolidating 130 old systems, each with its own internal structures and processes and without interrupting agency operations, sounds like a nightmarish task. To make this task possible with contractors takes a high degree of discipline in requirements management on the customer side, so that the contractors can develop and implement a clear plan and do not get disrupted by changes due to unclear requirements, specifications, and processes to be implemented. In August 2016, it was reported that various contracts with project contractors grew in cost by over 100 percent and that major delays were to be expected[18]—a common result when a project has to many scope changes, which in turn is a common sign of insufficient requirements management.

2.5.3 Fragmentation

I already mentioned Conway's law in the first chapter and referred to projects that ran into crises due to its disregard. Conway's law says that "Organizations which design systems [. . .] are constrained to produce designs which are copies of the communication structures of these organizations".[19] Written by a computer scientist, this sounds complicated, but it can be put into two simple statements:

1. Team structures must be compatible with the structures of the system they produce. If you ignore the intended system architecture when you structure a team, the team may not be able to build the system. In project management practice, people mostly follow this principle.
2. If sub-teams communicate well while they develop and integrate system components, the system has a chance to work well too. If the sub-teams do not communicate well, the system they make will be flawed and destined for failure. In project management practice, this principle is commonly ignored.

The *buy* option gives many examples of Conway's law in action. You buy a complex software program and expand it with an add-on from a third-party vendor. When the add-in crashes the program, the vendor of the add-on will tell you that the blame is with the program maker ("They changed something in the software without telling us"). The program maker will tell you that it is of course the buggy add-on, and that its developers have ignored timely communications by the program maker. As the customer, you fall between two stools and are left with the disruption of your work, the difficulties of finding a solution to the problem, and with all costs involved.

Fragmentation also occurs when the parties involved—customer(s) and vendor(s)—follow different business interests. The customer wants the ready-to-use solution at the lowest costs possible, while the vendor wishes to deliver the most simple and easy-to-make product at the maximum price. At the moment of contract signature, they had an understanding and

---

[18] (Coyne 2016)
[19] (Conway 1968)

a common ground, often called a *meeting of the minds*, but as time passes, this commonality may get lost.

Fragmentation can also happen between contractors in a project. A certain task needs to be done, and both can do it. The task may be financially or otherwise attractive, and both would like to make it. Or it is tedious and badly paid, and both avoid it. Contractors may prefer not to take a risk and instead transfer it to someone else. When all of them do that, the risk will remain unaddressed—often the most certain way to make the worst occur.

The greatest risk for organizations and people comes with trust. If one trusts the wrong people, one will be deceived. But it also holds true that distrusting good people will deprive a project of many opportunities and can finally bring it into crisis. Trust as the foundation of actions in project management comes with the risk of failure; distrust makes this failure certain.

Between the two monsters, project managers have to navigate a way that protects the projects from getting damaged. In order to maintain a sound and mutual trust relationship, it is helpful to focus on one's own trustworthiness first, as Stephen R. Covey said: "If you want to be trusted, be trustworthy".[20] A second step would then be to look for clues for the trustworthiness of the other person or organization. In project management, where we often have to deal with new contacts, this might be difficult.

Selecting the *buy* option increases the relevance of the trust/distrust dilemma. Both contract parties, customer and contractor, select to work with someone with separate business interests, possibly with someone with whom they have no experience. Reference organizations can be helpful in building trust:

- Reference customers support the contractor's claim: "We have worked successfully for companies A, B, and C and would now like to work for you too".
- Reference contractors: "Past contractors were the companies X, Y, and Z, and they have profited from working for us".

Personal relations, rapport, and demonstrated empathy can also help build trust, but during the lifecycle of the joint work in the project, events will happen that challenge this trust. In a PSN with several parties involved, one event may damage trust in one business relationship, and this may be the first domino piece to tumble and fall, taking others with it.

When the *buy* option is taken, it should be clear that solidification of the project is a continuous endeavor that consumes time and energy. Fragmentation into silos can occur quickly at any moment when this endeavor has been disregarded. Then, the project has parties involved that are able to *compete,* but unable to *complete.*

2.5.4 Unattainable Cost Benefits

Many years ago, I was working as a workplace coach for Centipede AG, a German software company, and had to go to another building to get to the workplace of my next coachee, when it rained heavily. Someone recommended me to use the basement to go to the other building, as the houses were interconnected underground, and as my guest badge would allow me to take this way.

---

[20] (Covey 2004, p. 51)

Hurrying along the basement corridors, I noticed many open doors leading into makeshift work rooms, in which I could see people apparently from Southern Asia sitting at PCs and coding software. They were obviously very surprised to see a person passing by and greeted me in a friendly way. Later, I asked my direct contact officer in the company about my observation and was then told that these developers worked for a number of contractors who were officially located in their home country. At one point in the past, the company found itself unable to manage the developers over a distance and colocated them in their basements. They failed to keep the team working across continents, time zones, cultures, legal systems, etc. and made a decision to secretly colocate.

Projects are continuous learning processes, and part of these learning processes may be that outsourcing can be hard to manage and may not yield the expected results. Among the software offerings of Centipede AG were products to support offshore work and virtual project teams, and it was very embarrassing for the company and its project managers that they could not overcome the team fragmentation from team virtualization and had to get back to what they considered "old-style" colocation. Their problems were not a lack of technology but of team cohesion and interpersonal performance.

Corporations often have expectations on outsourcing that reality may not meet. The example of Centipede above is one, where cost benefits were expected. Buying instead of making can open new growth options by tapping external assets and using them as resources for an organization's own projects.

Managers who have these expectations must be aware of the strong likelihood that they may be disappointed.

2.5.5 Narrowing Strategic Options

The *buy* option can constrict the space for strategic decisions quite dramatically.

Buying is generally used to free management attention, the scarcest and most important resource in an organization, for tasks that are considered core tasks, are defined as strategic, or are commercially attractive. Sometimes, these core tasks are neither strategic nor commercial but are more or less clearly mandated by law, and outsourcing of other tasks can again allow more focus on meeting them. So organizations define core competencies which will be managed in projects using the *make* option. Others will be seen as fringe or supportive competencies, and project work linked with them will then be bought from outside.

It may happen that the strategic core competencies defined are later found to be the wrong ones. While contracting helps the customer tap the resources, competencies, and agility of the external contractor, it may be difficult to fully exploit them as desired. Harvestman, Inc. was a provider of equipment for home security devices such as alarm systems and burglar-proof door and window locking mechanisms.

The company's CEO regularly visited congresses and seminars that dealt with strategic management, and the lesson communicated in these events was mostly the same: Focus on meeting strategic objectives. There, David Ogilvy, the guru of advertisers, was quoted saying "The essence of strategy is sacrifice",[21] and attendees of these seminars were taught to

---

[21] Quoted in Hoffman (2012). The quote is also variously attributed to the French emperor Napoleon and to Prussian general Moltke.

either entirely give up non-strategic and off-focus activities or to hand them over to third parties. Harvestman, Inc. had a research and development department for future products, which included a small team of three software developers working on a web-based "Internet of Things" (IoT) alarm system that would use sensors and cameras located in private homes, which would send alarms over an internet connection to a monitoring center that, in turn, would be able to forward these alarms to the nearest police or fire station.

At one point in time, the CEO of Harvestman, Inc. adopted a view that the company's core competency was in the development and the manufacturing of the hardware, and considered the internet connectivity a dispensable add-on to that, which consumed resources and management attention beyond its business value. He preferred to say, "Our job is making profit. We have to focus on the essentials, not the bells and whistles". He also found that the small number of customers who were prepared to pay for the additional online service would not justify setting up the alarm centers that would have to be manned around the clock. He sacrificed the development team, laid off its developers, and gave the web part of the business to a contractor.

The development team members left with "Golden handshakes", payments that enabled them to start up their own company, which became financially self-supporting very quickly. The focus of this company was interconnectivity solutions for home alarms. They bought the systems from their former employer, as these were the systems that they understood best, they grew fast, and after 24 months, they made more profit than their former employer, who was by then only one contractor to them among others, and the price pressure on Harvestman Inc. was high. Outsourcing the most future-relevant parts of the business deprived Harvestman Inc. of growth potentials and forced them finally to accept the role of a low-margin vendor of commodities.

The interests of a customer and a contractor will have many commonalities, such as creating great deliverables, avoiding undue stress, and having fun together during the project while both are mastering major challenges. There are also natural differences. A first one is of a financial nature: The amount of money that is available to both parties is limited, so the money that one of them claims cannot be used by the other one. Other conflicts of interest deal with assignment of risks, liabilities, workloads, causes of delays, and many more.

A further example of the often divergent interests of contractors and their customers is the need to exchange knowledge. This coincides with the desire of the contractor to make the customer dependent on the services and supplies provided to ensure a steady source of income for a long period, while the customer often wishes to keep the contracting relationship temporary and remain fundamentally independent. The contractor will then just communicate the amount of knowledge that is inevitably necessary and contractually mandated for the project. The customer may have hoped that contracting out tasks that are considered peripheral would allow a stronger focus on core competencies while ensuring that these tasks are professionally done. Instead, the customer may find that the outsourcing of fringe competencies can impact the core competencies to an unexpected degree and damage future business instead of improving it.

2.5.6 Remoteness of Error Fixing

A general principle of any production, both project and operational, is that the cost, workload, and difficulties of finding and fixing an error grows with the distance to its origin. This is probably even more true in projects, where work goes through different hands and locations during the course of the project. The person whose hands made the error is mostly in the best position to fix it, particularly when not much time has passed between making the error and finding and fixing it. This individual knows what has been done and has the skills and the infrastructure available to fix it comparatively easily. The further the erroneous item is transported along the subsequent production process, given to other people with different skills and competencies, in other locations, using different methods, infrastructure, and tools, the harder it gets to fix. It gets most expensive when the error is identified on the customer's site, which may necessitate sending service there or recalling the product. Then, the error can result in loss of reputation and, with it, market share.

This principle gets amplified when a number of contractors are involved. In addition to the local and temporal distance and the change in people, at some point another organization has the item, with its own processes, culture, and business interests. This organization may fear that its work will be disrupted while error fixing is being done and may then decide not to allow it. And so the temporal distance will grow further.

Another common source of conflicts in PSNs is the question of who needs to take responsibility for fixing errors, particularly when this cannot be billed. Error fixing is a form of rework: something has been considered finished and must now be taken again and work must be spent on it. Contractors therefore prefer either to have someone else do the work or to do it against additional payment. The further the distance of error finding and fixing is, the harder it gets to identify who should take the responsibility for it. One contractor will point to the other one as the origin of the error, and no one will take the responsibility to simply fix it and let the project move ahead.

This does of course not occur in all projects, and in a *project supply network* culture based on a "Mission Success First" attitude among all parties, such conflicts are quickly resolved. I will describe this "Mission Success First" later in detail in this book.

2.5.7 Speed Blindness

Another hazard that comes with the *buy* option is speed blindness. While speed blindness can turn up in any project, it is especially threatening in projects that span several organizations.

Tiger beetles are a group of fast insects.[22] Very fast. They are among the fastest runners in the world, at least when one puts their speed in relation to their size. One of them, the Australian *Cicindela hudsoni,* runs 2.5 meters/second (over 8 feet/second), which is 125 times its body length of 20 millimeters. This would translate to an average human—1.75 meters (5.74 feet) tall —running at a speed of almost 800 km/h (500 miles per hour (mph), not much under the cruising speed of a jet airliner. The beetle lives a predatory life, and its speed, together with its good vision, helps it to run down any animal that is small enough to become its prey, but also to run away from most predators that would in turn consider the beetle a nice meal.

In the past, scientists were confused about the running style of the beetle: It approaches the prey at high speed, then waits for some seconds, and starts running again. It can do this several times before it finally strikes its prey. Then they found out the reason: While the beetle is running fast, it is virtually blind. During this time, the beetle keeps its antennae forward

---

[22] This time, the example is really about insects.

and uses them to feel obstacles in its way before it runs into them, just like a person moving in the night does, who stretches the hands in front of the body to feel impediments before stumbling over them.[23]

Vision necessitates a complex chain of events: Photons reach a photosensitive surface in the eye, generating nervous signals that need to be transported to the brain and be processed to gain information from them and build a mental representation of the observed objects. When the tiger beetle is running, it loses sight of its prey and of anything else around it. The prey may have moved meanwhile to another place, and the beetle is no longer running in the right direction. The beetle also needs to re-assess the distance to the prey. So it stops, looks, and when it has refreshed the representation of the situation in its small brain, sets off again and repeats the run and stop activity until it finally catches the unsuspecting victim.

Humans are astonishingly effective in managing high speeds. One reason may be that our eyes have much more optical performance than those of beetles. Our brain adds to that, as it is much better in processing the data coming in from the eyes. But at certain speeds, we experience the same phenomenon: It begins with tunnel vision, partial speed blindness that drivers of fast motorbikes, cars, trains, or aircraft perceive at very high speeds.[24] The vision field narrows to focus the brain's resources on the small spot in front that is important for immediate survival. In a car at 800 km per hour, for most people, the spot would become so small that we would no longer be able to actively drive at all—we would be left speed blind and simply black out.

It is surely no coincidence that the current holder of the land speed record (at 1,228 km/h, 763 mph, or Mach 1.02) is a former RAF wing commander, Andy Green from the United Kingdom. He has been trained intensively to travel at high speed on Phantom and Typhoon fighter aircraft without getting speed blinded. When an author of the magazine *Wired* made a flight with him in an acrobatic plane, he found his "vision starts to narrow and turn grey at the edges", as he later reported, while the pilot remained "completely calm".[25]

Another kind of speed blindness happens when information is available, but people are blind and deaf to it due to an overpowering intention to be quick. This occurred when the *Titanic* was on its maiden journey from Southampton to New York City with 2,208 crew and passengers on board in the night of 14 May 1912, just before it crashed against an iceberg at 11:40 PM. Two hours and 40 minutes later, the vessel broke in two and sank, killing 1,514 people in the cold, arctic water. Before the collision, the ship had run full speed at 21.5 knots (40 km/h, 25 mph) to meet its scheduled arrival time in New York. It had repeatedly received warnings of pack ice and major icebergs on its way[26] but kept on at full speed, and as even the crow's nest crew had no binoculars that might have helped in the black night to spot the iceberg earlier and negotiate it safely, it was running blindly into disaster. It is reported that the wireless operator of the *Titanic* responded to a warning sent to him by his colleague on another ship, the *Californian*: "Shut up. Shut up. I am busy. I am working Cape Race".[27]

---

[23] (Zureck and Gilbert 2014)

[24] (Pozzi 2014)

[25] (Franklin 2014)

[26] (Box 2004)

[27] The message meant that he was busy sending a backlog of private passenger's messages to a wireless station at Cape Race in Canada, and that he felt disrupted by the warnings sent by his colleague to him (Titanic Inquiry Project 1912).

Speed blindness is an effect seen in many projects, both internal and customer projects. Teams run fast, creating deliverables against pressing deadlines, just to find out later that they developed the wrong deliverables. The reason may be misunderstandings in what the requesting people or organizations actually wanted or needed, or what degree of operational disruptions from the project they consider acceptable. Speed blindness can also happen when these stakeholders' wants or needs are changing, but the project teams, busy with work, are not notified or were notified but did not take notice.

Speed blindness commonly takes the form of technical, organizational, or interpersonal carelessness. Then, things go wrong in the project, which the teams do not perceive and that threaten their success. Highly focused and working industriously on the project, the project manager and the team can overlook that a team member or a contractor is on the way into a crisis and needs help, or that the project needs to make adjustments for the lapse. Speed blindness in projects can be present when a team should be acting cautiously, taking care of obstacles and threats in its way, but instead ignores all hazards and steams on at full speed, possibly into irreversible disaster.

Speed blindness increased by a complex PSN was an experience that Horsefly, Inc.[28] recently made. The company builds heavy premium motorcycles (over 500 ccm) at a costly production location and sells them at a premium price. To expand into the market of beginners and of countries with lower purchasing power, five years ago they made a decision to launch a mid-size model.

While they made product development mostly by themselves, for the development of the production facilities and later the production, they teamed with Ladybug Company, a manufacturer in a low-cost country, who was highly successful too, but only in mopeds and small bikes (up to 200 ccm).

From Horsefly's point of view, the cooperation was dedicated to making production affordable and expanding into a smaller segment; Ladybug, in turn, expected to expand into a larger segment and a premium market which promised higher margins but also had higher expectations regarding reliability, durability, and application of technology. The companies had planned to launch the motorbike in the summer of 2014, and Horsefly's management emphasized in an early announcement its importance for the company's product strategy and international outreach. The schedule for the development of the motorbike and its production was challenging but seemed feasible, if project tasks were done right the first time.

An important aspect of speeding up development was the inclusion of third parties into the process to distribute the load on more shoulders. Ladybug, the contractor for the development and the operation of the production, further sub-contracted the six-speed transmission to Earworm, who were also located in a low-cost country. Then, Earworm built some first rough prototypes, which gave an initial understanding of what the later product would look like, but when they were shown to Horsefly, Horsefly was not fully happy but accepted that these were first prototypes.

Most worrying were the difficulties of changing the gearbox into neutral, but as these were the first prototypes, this seemed acceptable—something to improve in further development. More prototypes of the gearbox were delivered and soon were built into prototypes of the full

---

[28] Insect names here are again aliases for existing companies.

motorcycle, but the problem with the gearbox persisted. Well, Horsefly thought, the prototypes were made with development tools, and as soon as the tools for mass production would be used, the problems should be resolved. They were not.

Summer 2014 approached, and while brochures, pricelists, and marketing materials for distribution were already handed out, and production of the motorbikes was already started, these motorbikes could not be finished. There were still no gearboxes delivered by Earworm that could be used. The semi-finished bikes had to be stored in a warehouse—a nightmare for a company used to applying just-in-time production and management—to wait for the gearboxes to be delivered to allow for final assembly.

The problem with the gearboxes gradually improved, and in December 2014, some selected bikes could be given to journalists for testing. Still then, some complained about the problems with switching to neutral—one called it "a bit of a notchy tranny action".[29] The time lost before the motorbike finally hit the market was almost a full year, and the main reason was that Horsefly's engineers, performing their part of the development work in high speed, were speed-blinded and did not pay enough attention to the subcontractor, who desperately needed help their help.

Speed blindness in projects is a general problem. One of the most difficult tasks for a project manager is to know what is going on in the project—in any project. Project work given to vendors tends to become a kind of "black box". As the customer, you do not really know, and sometimes you do not want to know, the details, making sure that the contractor remains responsible for the results. While for the customer side, this releases managers from a heavy supervisory burden and frees attention for other tasks, it comes with the risk that no one knows when things go wrong.

2.5.8 Bringing Strange People into the Project

From time to time, the *buy* option will bring your project team members into business relationships with people who can compete but not complete: big egos, true bullies, psychopaths, and sociopaths.

Like it or not, in project business we often have to deal with people of questionable education, character, behavior, and trustworthiness. With violent people—people who take joy in humiliating, scaring, and in scapegoating others for their own faults. These are the kind of people that get you into a cooperation with high hopes, but when they ring the division bell for your project, you will find that all the early successes get lost after a while and your project turns noisily onto a path of failure.

The official term is "Antisocial Personality Disorder" (ASPD).[30] It is used for people who commonly pay no attention to laws and social rules. In normal language, they are commonly called big egos and bullies. They have contempt for the rights of others and do not regret when they hurt people with their behavior. They make their own rules for life and believe that others have to follow them. They can be astonishingly successful in business, and also in politics,

---

[29] "Tranny" for transmission.
[30] A great explanation of the commonalities and differences between sociopaths and psychopaths can be found in a *Psychology Today* article (Bonn 2014).

sports, and culture. Often, their success is not built on cooperation and win–win solutions but on driving others into bankruptcy to easily take away their assets. Project managers commonly meet these people as sponsors of a project, or just among the people whom they have to trust and rely on to a certain degree in order to make the project proceed.

Astonishingly, many of them have traits that others consider leadership skills. They will therefore have their followers, often deeply fanatic followers, on whose support they can reliably build. They love to build walls where others would build a bridge, and while the followers will discredit bridges, they will acclaim the walls.

If egomaniacs sponsor your project, you should expect deep troubles. They will come with change requests at a frequency that cannot be managed with reasonable impact analysis and sound implementation processes; even the term "change *request*" is inappropriate—these are clear orders that must never be questioned.

Things can become even worse. Berlin-Brandenburg Airport is a troubled construction project in Germany, over which the project teams lost complete control. The project had three sponsors—the mayor of Berlin, the prime minister of the state of Brandenburg, and the German Federal Minister for Traffic—and their frequent change requests included many without a good business case, but were claimed by the politicians competing for authority and political weight.

The opening date was originally planned for 30 October 2011. In summer 2010, the date was delayed into 2012, then later to late 2013, and so on. While I am writing these lines in January 2017, the airport is still not open, and recent news says that it is highly unlikely to be opened in this year.[31] Unofficial sources say 2020 to 2023 are more realistic.

In any case, three sponsors is two too many for a project; a project can be successful with more than one project manager, but more than one sponsor is a setting that is most likely to kill the project. They make things worse—all three were career politicians and, as such, dominant males. Change requests were frequently used not to improve the airport but to display dominance, and warnings that these changes might be detrimental and that the impacts needed to be assessed before a final decision could be made were ignored.

The main area of change requests was non-aviation infrastructure, particularly the shopping areas. A modern airport is much less a traffic-centered piece of infrastructure, but a shopping mall with attached aviation facilities.[32] The greatest impact were fire protection systems. An example: Locations that originally should be empty as waiting zones for people (empty zones do not burn, and humans also have low flammability) were turned into clothing shops, and textiles burn easily. One big ego in a project can be a problem, several egos in competition spell massive disaster.

The *buy* option brings together people who often have not met each other before. There is not much time for them to go through the process of team building and find a common mode to work together. Among these people will be some who are much less interested in completing than in competing. They can become a major risk to the project. Inside their own organizations, it may be possible to get rid of them, but when one is bound to them by a contract, it may be difficult to separate them from the project.

---

[31] (DPA/The Local 2016)
[32] (von Gerkan 2013)

2.5.9 Opening the Project for Corruption

In my classes, students are sometimes surprised when I discuss this topic. There are "clean" cultures in matters of corruption and others that are massively polluted with corruption. *Cultures* here refers not only to different regions and countries, but also to industry cultures and even specific corporate cultures. People who have never been in contact with corruption cannot imagine its disastrous effects, and those who have often do not want to talk about it.

I still remember a procurement manager of an automotive company many years ago, who generally sat in confidential negotiations wearing seven expensive rings on both hands. All non-thumb fingers had a ring on, apart from the middle finger of his right hand, as a unspoken signal to me of what he expected from me. When the man was fired one day without notice and his practices researched by his employer, I was glad that I had never surrendered to the temptation of making my work for the company easier by responding to his signal.

Like it or not, corruption is a problem. In summer 2005, I had an opportunity to talk with Peter Eigen, a former director of the World Bank and, in 1993, founder of Transparency International, the global association against corruption,[33] during a joint journey through Bavaria along a German autobahn. I asked him why he had founded the association, and he explained to me that he observed, in his role at the World Bank, how money that was made available to save lives and help people out of poverty instead oozed away in sinister channels. Lots of money. "Corruption kills children", was the main reason for him, and as there were no mechanisms in place to stop this oozing, he decided to develop them.

Many project managers do not care about corruption. They consider it as a greasing-the-wheel technique, just like lubricating a bearing. In their understanding, project managers must ensure that the wheels are turning, and it does not matter how they do that. There are several reasons to rethink such a position, among them:

- **Loss of freedom.** While corruption seems to be an effective addition to a project manager's toolbox for some people, it actually dispossesses them of personal and professional freedom. To quote South-African lawyer Guido Penzhorn in a report on a massive corruption case in Lesotho: "Clearly once you involve yourself in the murky world of bribery it is not open to you to simply opt out whenever you like".[34]
- **Dysfunctional processes.** When an organization creates a milieu that accepts corruption for its benefits—for example, in order to win business—it should not be surprised if this is taken as a precedence and is then also applied against its own business interests, often on a massive scale. A common example is when it sub-contracts work to vendors. If an organization does not want to pay excessive prices to vendors, who then kick money back to procurement executives, the organization has to actively fight corruption in any form.
- **Hampered communications.** Penzhorn further wrote on the Lesotho case: "There is a wall of silence which is very difficult to penetrate. The reason for this seems to be that everyone who is in a position to talk cannot do so because someone else in turn has something on him. Once corruption creeps into, for instance, a department of state it

---

[33] (TI 2016)
[34] (Penzhorn 2014)

is very difficult to know how far the rot has spread. There is no obvious victim as, for instance, in an assault case. The victim is society". A key aspect of great projects is open and trustful communication. Corruption makes this impossible. In turn, I recommend that, if communications are a problem in a project, to research deeper; it may be a signal for corruption occurring, threatening the integrity of the entire project.

- **Punishments and public humiliation.** Venice, Italy, is a tourist marvel, but also a city slowly sinking into the soft ground of the lagoon in which it was built on wooden poles in mediaeval times. This is not the only problem of the city. In July 2014, the *Newsweek* magazine published an article entitled: "Venice is Sinking Under a Tidal Wave of Corruption".[35] The background of the story was a project called Progetto MOSE (for *MOdulo Sperimentale Elettromeccanico*, Experimental Electromechanical Module), a system of 78 barrier elements in the form of hollow steal containers that would normally be filled with water and rest on the sea ground at the entrances of the lagoon, but in case of predicted high waters would be filled with air and float on one end to the sea's surface while the other one is hinged at an undersea concrete foundation. The project is intended to prevent the city grounds from being flooded during "aqua alta", floodings that occur several times a year.

  A consortium named Consorzio Venezia Nuova (CVN) was founded for the work on the project, which was led by Italian construction company Impregilo. CVN has been working on the mobile gates since the year 2004, and they are currently scheduled to become fully operational in 2020. The project suffered from severe cost increases, starting at an original budget of around €2 billion,[36] but expected to finally come out at costs of €5.5 billion. It was also hampered by technical problems, such as barrier elements rusting in salty sea water and one of the barriers being damaged by a storm. It had to be replaced.

  The most massive technical problem is the sinking of the heavy concrete foundations by 4 centimeters each year into the silt underneath, compared to a yearly sinking of the city by 2 to 3 millimeters.[37] The gates were intended to protect Venice from water tides up to a height of 3 meters, but sinking at this speed will naturally reduce this protection in just 25 years to only 2 meters. The *Newsweek* headline mentioned before referred to another problem: A major anti-corruption raid of Italian finance police in June 2014 resulted in arrests for 35 persons, among them the mayor of Venice and other politicians and high-ranking public service agents. The mayor stepped down from his position in the ensuing days.

2.5.10 Coercing Behaviors by Contractors

When one observes that a couple fights out its conflicts in front of their children, common friends, and other third parties instead of seeking solutions in private, one knows that the issues between the two have become very serious. The same is true for organizations, when they communicate their conflicts with each other over press releases, as happened in the following

---

[35] (Manera 2014)
[36] (PMI 2004)
[37] (Gatti 2016)

case story in early 2014. As it all has been published by the two organizations in the form of a temporary avalanche of press releases, I am not telling secrets inappropriately here.

I mentioned already the Panama Canal expansion project, one of the largest infrastructure projects in the world. It had started in the year 2007 and was originally scheduled to be finished in August 2014. By the end of 2013, the project had made a progress of 70 percent, when the prime contractor, Grupo Unidos por el Canal (GUPC), reduced working intensity to 25 percent of what was agreed upon. In January 2014, they stopped working completely. The canal was by that time scheduled for opening in 2015; this then had to be delayed to June 2016. The main cause behind the delay was that in late 2013 GUPC had run into massive technical problems with considerable monetary consequences. The consortium had been hired as a contractor to build the new ship locks, which were the most important work item of the expansion project.

The first canal project in Panama done was conducted from 1881 to 1894 under the lead of French diplomat Ferdinand de Lesseps. It failed due to several reasons, among them geology: The soft ground under the channel, which additionally become soaked during rainy seasons, made it impossible to dig out the channel route as planned. During the rainy season, which in Panama is eight months a year, landslides repeatedly came down and smothered the construction site with mud, and while the angles of the slopes needed to be reduced repeatedly, the amount of earth that needed to be moved grew beyond what the project team was able to do—technically, but also financially. They had to give up the project, which was later taken over and concluded by US American engineers in the years 1904 to 1914.

The technical and financial problems associated with building on soft ground came back in the expansion project, where GUPC said that "unforeseen geological conditions" (among other causes) led to cost increases of "$1.6 billion".[38] By the end of 2013, GUPC wanted to step into negotiations with the Panama Canal Authority (Autoridad del Canal de Panama, ACP) over the coverage of these additional costs. ACP rejected these negotiations, pointing to the contract, which did not have clauses for ACP to cover such increases. In press releases, ACP was then pressured, for example on 20 January 2014: "Failure to reach an agreement on co-financing of the unexpected costs will result in a serious delay and it will mean that the works will not be finished in 2015 causing damages to all parties involved".[39]

An aspect to increase the pressure were the 16 lock gates that were made by an Italian sub-contractor. A first delivery had already been made in 2013, but the bulk of the deliveries was scheduled to be starting in late 2014. Without locks sufficiently completed, the gates could not be put in place directly from the transportation vessel, and would have had to be temporarily put in storage—at dimensions of 58 m × 28 m × 10 m (190' × 92' × 33'), no easy task.[40]

Beginning on 20 February 2014, the work was ramped up again, while attempts were run concurrently to resolve the situation in negotiations.

The contract parties in a project become strongly dependent on each other. If GUPC would have terminated all working finally, or if they would have gone into insolvency, the state of Panama would have had the largest industrial wreckage in the world.

---

[38] (GUPC 2014a)
[39] (GUPC 2014b)
[40] (Anon. 2016)

2.5.11 Opening New Doors for Malware

I mentioned already my past assignment to work at Centipede AG, a major IT and software corporation, for which I was hired to perform a series of structured and on-the-job coaching sessions for about 80 project managers. The measure was part of a major reorganization and qualification project. I was a subcontractor of a training and service company who was in turn contracted by Centipede. When I started working with Centipede, it soon turned out that I needed to bring my own hardware. The PCs they could make available for me as their temporary human resource were a bit oldish, left over from when employees were given new PCs and not able to cope with the technical demands that the project came with. Before I could use my own PC, I had to show that antivirus software was installed and up to date and that the general settings of my PC were so that it would be difficult for malware to infect my PC and, once connected with the customer's network, jump over to other computers. This check was done once. I worked for the company for over two years, and the risk that I would bring in malicious programs was only addressed once.

One could argue that by the time I worked for Centipede, the threat from malware was much smaller than it is today. Professional observers such as AV-Test observed a low and constant threat from malware by numbers of viruses and other malicious programs until 2005, and almost an explosion in the years after that.[41] Malicious software has ripened over the years from a nuisance developed by bored and mischievous youthful hackers into a genuine threat with massive economic impact for individuals and organizations, and behind this malware stands a professionalized industry, including a number of governments, with a lot of criminal energy. It is used for industrial espionage, theft of data and blackmailing. In times of "Industry 4.0" and the "Internet of Things", the attack surface of organizations is growing. Any smartphone, printer, observation camera, roboter, and any other item that is mostly or permanently linked to the internet can become an entry gate for malicious software into the corporate network or be used by such software to shovel critical data out of the organization. Malware may simply add corporate computers to so-called *botnets,* where they act for criminals. While they do not cause damage to the hosting system, they consume valuable resources such as processing power and bandwidth.

The threat was still at the lower level when I worked for Centipede, but they should have nevertheless put more energy and dedication into protective measures—too much value lay in their own data and the data of their customers, which they also had. The corporation has meanwhile changed their attitude toward malicious software and has one of the most professional cyber-protection teams in the world. Most companies would not be able to have such people at hand. The use of contractors can add to this risk, when they are given access to the customer's corporate computer networks.

2.5.12 The Reputation at Stake

Still some years ago, a customer, who bought products or services from sweatshop manufacturers and other dubious sources could still dispute responsibility for illegal, immoral, or dishonest

---

[41] (AV-Test 2017)

actions of a contractor, asserting that the accountability for activities of the contractor are solely with the contractor. Times have clearly changed. The following examples are from companies that understood the challenge of maintaining integrity in their PSNs and addressed it actively:

- Apple Inc., seller of the iPod, the iPhone, the iPad, the Macintosh series of PCs, and other premium level electronic consumer articles, buys products from original manufacturers in China and other countries. In August 2006, the British newspaper *The Daily Mail* visited the production facilities of Taiwanese company Foxcon in Longhua, China, and reported harsh working conditions with 15-hour labor shifts, low payment, group dormitories for workers, and military-style drills.[42] In the following years, more reports turned up, adding allegations of exposition of workers to hazardous chemicals and high risks of injuries in the production environments across Apple's network of suppliers.[43] Apple responded with two documents titled *Supplier Code of Conduct*[44] and *Apple Supplier Responsibility Standards*[45] and audits against these rules.
- Since 1994, Sweden-based furniture store chain IKEA[46] was repeatedly confronted with allegations of child labor and other forms of exploitive work conditions in its global PSN. The company upholds a cozy image before customers and maintains internally a standard of environmental and social responsibility that is communicated to customers as an additional sales argument, in addition to low prices.[47] IKEA is among the small number of corporations that have a chief sustainability officer, giving social and environmental matters a voice in the executive team. For a company that bases its marketing and corporate culture on such high standards, these allegations were deeply detrimental. In response, IKEA not only published a Code of Conduct but teamed with NGOs and committed to influencing and educating suppliers actively to change their practices. They also put in place an intensive auditing process to identify violations of their rules and identify the need for corrective actions.[48]

The examples are not taken from project management but illustrate how highly successful corporations address the reputation risks from their PSNs and influence contractors to comply with the high internal standards of these corporations.

To my knowledge as the first country in the world to do so, France adopted an update to its *Code de Commerce* in February 2017 to hold corporations accountable for "serious violations of human rights and fundamental freedoms, the health and safety of persons and the environment" not only by actions of themselves and their subordinate companies, but also of contractors and suppliers.[49] The law requires organizations to develop vigilance systems for the protection of humans and the environment and allows for damage claims as well as for fines. These new rules of law do not explicitly include projects, but as these are also not excluded, one

---

[42] (*Mail Online* 2006)
[43] (Sacom 2016)
[44] (Apple 2017a)
[45] (Apple 2017b)
[46] Changes in the financial structure has turned IKEA into a de-facto Netherlands-based group.
[47] For example, in IKEA (2013).
[48] (Christopherson and Lillie 2005)
[49] (L'Assemblée nationale 2017)

should take care when working in a project with contractors under French law. Other countries are likely to follow the French role model.

Projects, as impermanent endeavors, from time to time run into similar troubles. The temporary nature of the projects as such, and even more of the business liaisons between customers and their vendors, makes it difficult to identify the problems in a timely manner and act proactively. In a short time, contractual relations must be developed and resources made available, and while ensuring integrity and adherence to standards takes a lot of time and attention to details, these are rarely available when deadlines are pressing.

To give an example: In September 2013, British newspaper *The Guardian* published an article entitled "Revealed: Qatar's World Cup 'slaves'", in which it reported of Nepalese workers in debt bondage relations and under a de-facto enslaving system called "Kafala"[50] working on the construction sites that build the infrastructure for the 2022 Football World Cup in Qatar.[51] The articles was further subtitled:

> *"Exclusive: Abuse and exploitation of migrant workers preparing emirate for 2022—World Cup construction 'will leave 4,000 migrant workers dead'—Analysis: Qatar 2022 puts FIFA's[52] reputation on the line".*

FIFA, ridden by a massive corruption scandal in its top ranks, rejected in November 2013 to actively respond to the allegations, pointing to amendments of laws announced by the Qatari government.[53] By that time, the government had assigned a consulting firm to research the allegations, which were confirmed in the final report by the firm, dated April 2014.[54] It then took the government of the Emirate until December 2016 to enact the announced new laws. Kafala and other mechanics of forced labor have been formally forbidden, but the punishments for corporations still applying them is limited to QR50,000 (~US$13,750)—virtually peanuts compared with the fortunes made by the contractors based on forced labor.

The US American sports television channel ESPN broadcast a 17-minute documentary on the construction work for the World Championship in May 2014. The focus of the documentation lay on workers from Nepal, but also from India and the Philippines. The documentary focused on young men who left their homes in healthy condition, looking forward to their new jobs that would bring income for their families far away, and were returned some months or years later in coffins, with papers saying that they had died from heart attacks or committed suicide. The documentary did not put the blame on the contractors but on the Emirate of Qatar and on the FIFA, their contractors. Both are considered corrupt in major parts of the public opinion, and their construction contractors strongly contributed to this opinion.

The examples show how, left unmanaged, the ethical shortcomings of contractors can leave marks on the reputation of a customer.

---

[50] Kafala is a form of forced labor based on the confiscation of the passport of a migrant worker by an employer in order to trap the worker in the job contract for a long time, making it impossible for the person to resist massive exploitation.

[51] (Pattisson 2013)

[52] FIFA: *The Fédération Internationale de Football Association,* the international governing body of association football ("Soccer") based in Zurich, Switzerland.

[53] (Al Jazeera 2013)

[54] (DLA Piper 2014)

2.5.13 The Contractor as a Data Leak

The following case story happened to a training customer of mine some years ago; I therefore have to anonymize names again. The corporation, which I will name here Millipede S.A., used a contractor, Termite Ltd., for different development tasks in the field of automotive motor management software. They had some new algorithms developed that they considered worth patenting, but a deeper analysis showed that the protection from a patent would be insufficient, and the risk of laying open the functions of the software to allow for easy copying would be high. The algorithms were able to dramatically reduce the fuel consumption of an engine in low-throttle situations, and as car motors spend over 90 percent of their running time in situations with such low output demand, even small savings can add up to numbers that are attractive for car owners; the update would be much less costly than changing the hardware of the engine.

A joint team of Millipede and Termite staff did the development work together, and while a cost reduction program at the customer, Millipede, made some of the developers leave with a "golden handshake", actually without considering the potential effects on the project, Termite was happy to recruit these people, use them in the same project, and bill the customer for their work at quite an expensive rate.

The project became more expensive for Millipede, but not many people on their side seemed to notice that. Termite, in turn, made good money on development as a contractor, lending Millipede its former staff at high rates, and employees seemed much happier in their new position, where they were not given the impression of being no longer welcome. As the project team could go on with its work based on existing relationships, everyone seemed to be happy with the solution.

This changed when Charles, an employee of Termite, the contractor in the business, sent out an email to a major number of its own employees and of another customer, for whom the company was working on a similar project, in order to coordinate the software development around some key components and their interfaces. He had attached the files that included original source code for these software components, and which would be easy to read with just the programming environment that was used to make them originally.

Erroneously, Charles added a developer of Millipede to the distribution list and sent off the message. A minute later, he noticed his mistake and sent a second message to the Millipede developer, asking him to delete the message without reading it. The developer was a trustworthy person and had no intention of reading a message that was not his business.

However, at the moment that he went to delete the message, he noticed something familiar: The files had the same endings as their own source files, a signal that this competitor had used the same development environment.

Even more familiar were some names of the files. It seemed highly unlikely to him that two independent development teams would use precisely the same file names, so he opened one of them, and his curiosity was rewarded: It was software that he himself had written—for his company's project, of course, not for the benefit of competition. He informed his management, who then contacted the contractor and the competitor. The whole story ended at court, where the three parties finally reached a settlement that helped liquidate some of the damages on the side of Millipede S.A.

It could never be finally clarified whether the theft of the intellectual property of Millipede was arranged by a Termite developer who wanted to make his life easier, or by Termite's management to reduce the workload for development and meet a tough deadline, or was even mandated by their customer. Millipede S.A. found Termite Ltd. as a contractor no longer trustworthy and ceased business with them project by project, and they also made sure that the market knew, which made it hard for the contractor to win new business. Today, the company is expected to not survive the next five years.

Once bitten twice shy—Millipede's *make-or-buy* decisions today are more often responded to with "*make*" than in the past, but corporations tend to forget lessons like this one over time, and they will then probably return to their old practices.

2.5.14 Other Risks

There are more risks for the customer that come with outsourcing, and many exist in the opposite direction as well:

- **Bankruptcy.** A contractor can go bankrupt and deliverables built so far become part of the managed insolvency assets, which makes them inaccessible for the customer. The problems with contractor insolvency can happen even long before this formally occurs, when the contractor's management—desperately trying to prevent the company from bankruptcy—makes a sequence of seemingly strange decisions, and while the contractor has explanations for this behavior, the needs of the customer no longer turn up in these explanations. A company on its last breaths can be highly dangerous as a business partner.

- **Thieving magpies.** The contractor may have employees or subcontractors who do not respect the customer's privacy or property. When these employees or subcontractors steal items, data, or other tangible or intangible assets from the customer, it may be impossible to identify the culprit; the customer does not have the same freedom to run investigations on the contractor side as the organization would have internally.

- **The complexity trap.** While a team works on a project to build some kind of system, in the normal course of events, complexity on technical, social, functional, and interpersonal levels goes up. At one point in time during the course of the project, the team may no longer be able to manage this complexity. By that time, the number of interfaces and interdependencies has become too large to overlook, and a minor change at one point has the power to wreak havoc at many others.

  The complexity is easy to identify: A failure search run has been performed and a number of errors identified. The errors get fixed and another error search is performed, which yields the same number of errors, or an even higher number. When error search and fixing no longer reduce the number of errors in a system, one sits in the complexity trap. Examples of the complexity trap are the legal systems and the tax systems in many countries—both are highly complex, with many inner contradictions and inconsistencies that permit abuse by those who understand and exploit them. Whenever politicians try to fix these issues, they create new ones.

  When work is outsourced, some intricacies may be eased, but at the same time, a new dimension of complexity is added—the legal dimension. In the project, the different parties have to act as one team, but when conflicts embitter relations beyond a degree that parties can handle by themselves, they may seek remedy in legal action, and then they turn into competitors.

In international contracting, the different particularities of legal systems add further complexity: A certain behavior that is perfectly appropriate in one country may be wrong in another one. Under US American "Common law", one must take care what one documents in the project; the documentation may one day be required by the judge to be handed over to him or her. Judges in central European "Civil law" do not have such power. Here, project teams should document any detail that is related to the contract and its enactment; the lawsuit will be won by the party with the stronger written evidence.

Contracting adds a new layer of complexity to the already existing ones. With this complexity come new risks. Contractors and their employees' smartphones, PCs, and other items are additional entry gates for such criminal software, often without the contractor knowing. Customers often take measures to make the systems safer, but remaining risks will still endure.

The tight contractual and organizational bond between the organizations involved makes it possible at any time that deficiencies or problems of the contractor turn into a problem for the customer—and vice versa. Project Business Management can be highly profitable, but is also high-risk business for all parties involved.

2.6 Finding and Approaching Sellers

In the previous chapter, I described a standard process that turns buyers into customers and sellers into contractors. The first step was the *make-or-buy* decision on the seller side. When the *buy* option is chosen, the process enters the next stage, still on the seller side, to identify potential contractors, make contact with sellers, and ask them for responses. Figure 2.11 describes how this stage is embedded in the common procurement process for new sellers.

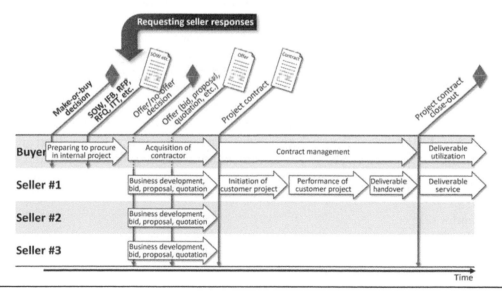

Figure 2.11   Identifying sellers, approaching them, and requesting responses from them generally follows the *make-or-buy* decision.

The following paragraphs focus on the development of new seller–buyer relationships. A lot of project work is actually done by incumbent contractors, which simplifies and accelerates the processes to start up the business. Because it builds on an existing relationship and mutual familiarity, the contract parties know each other, and the mutual expectations are much clearer. There are several reasons that a buyer may wish—or need—a new seller. The kind of work may be new, and the buyer has no experts among the incumbent contractors.

Another common reason is shortages in capacity of existing contractors, who are not able to obtain new resources fast enough. Dissatisfaction with a contractor can be the reason, or the desire to try out another seller, who may be cheaper, faster, or simply a better communicator. It may also be that a business unit that needs a certain seller is not aware that another unit has already contracted with them. Organizations are not monoliths and internal communications between units are rarely perfect.

2.6.1 Personal Recommendation

Many sellers are found by personal recommendation. A member of the project team may have worked with them, or a contractor already in business with the customer has had positive experience with this company. Also, corporations search the internet to find vendors in a specific field.

2.6.2 B2B Marketplaces

Online business-to-business (B2B) marketplaces have become a great trend in wholesale business and other operational business networks. They can become quite a gamechanger when they move their business model into the domain of project procurement. They connect sellers and buyers that, under other circumstances, would be rather unlikely to come together by comparing catalogues of listed sellers' offerings with lists of demands and requirements prepared by the buyers.

These marketplaces can be horizontal, benefitting general suppliers and service providers who are free to offer a wide variety of products and services and are not limited to countries or regions. Vertical B2B marketplaces focus on a specific industry or a particular type of solution. There are, in addition, national, regional, or local marketplaces. Some buyer or seller organizations have proprietary 1-n markets for their specific business, or have teamed to run such markets in collaboration. The horizontal B2B marketplaces are often quite large, such as Ariba in the USA and Ali Baba in China being among the largest. The vertical B2B markets are more limited to niches, but they can be very successful within them, with their focus on and understanding of certain industries and services.

Typical functions of online B2B marketplaces include:

- Vendor side:
  - Profile page
  - Multi-stage role models: Who can submit and access what?
  - Integration with ERP systems
  - Catalogue management
  - Management of reference lists

- Submission of RfIs, RfPs, IfBs, RfQs, ItTs and SOWs,[55] etc. from buyers
- Development of seller responses to these enquiries
- Question and answer management
- Contract negotiation and award
- Contract management
- E-invoicing
- Handling and securing of payments
- Customer rating (e.g., on general communications, handling of provisions and enabling services, timeliness of payments, disruptions of the vendor's business)

- Buyer side:
  - Profile page
  - Multi-stage role models: Who can submit and access what?
  - Integration with ERP systems
  - PSN management
  - Development of RfIs, RfPs, IfBs, RfQs, ItTs and SOWs, etc., and submission to sellers
  - Submission of seller responses to the buyer
  - Reverse e-auctioning (The vendors bid their prices openly in the system, trying to undercut each other until they feel forced to drop out, one after the other; the last and therefore cheapest bidder in the auction will then make the business.)
  - Question and answer management
  - Contract negotiation and award
  - Contract management
  - Order and shipment status
  - E-payments
  - Handling and securing of payments
  - Vendor rating (e.g., on general communications, on handling of change requests, timeliness of deliveries, disruptions of the customer's business)

Vertical B2B marketplaces that focus on a specific market or a specific technology may add particular functions that are of importance in the specific field. Marketplaces dedicated to project management may, for instance, include access for sellers to bios or profiles of the professionals that the seller intends to offer as project managers, which can contain information that is considered relevant for business development by either side. Regional or local B2B marketplaces can in turn include information on the geographic area of service and delivery of a particular seller.

To some degree, professional social networks can become B2B marketplaces, mostly for self-employed freelancers. Examples are Viadeo from France, German Xing, and of course US-American LinkedIn. They do not have the specific business functions for reverse auctioning or payment handling, but rather they connect the self-employed one-person contractors with the departments in organizations that may wish to hire them. Some have specific add-on services for dedicated groups of freelancers, such as Coaches.Xing.com for business coaches

---

[55] These acronyms describe different types of requests for information or offers. They will be explained later in detail.

and trainers, that give students of these persons an opportunity to evaluate the instructor. A similar approach can be found in the British CourseConductor.com portal.

B2B marketplaces have something in common with traditional farmer's markets, flea markets, and trade fairs: They need buyers to lure sellers and sellers to lure buyers. If one group is missing, the other group may either turn up but then leave the market again very soon in disenchantment or not turn up at all. Another similarity: Either one or both parties must be prepared to pay a fee to make the business model work. Often, providers have a free entry offering with limited functionality, with the goal that soon interest will be sufficient for marketers and buyers to be prepared to pay a fee. Additional fees are then charged to make offerings or requests stand out by placing them on top of results lists, adding images and additional texts, or similar means.

B2B marketplaces can speed up procurement processes and open up the process to more players. A caveat is that many of these new players are unknown, not only inside the buyer's or seller's organization but also inside the industry. They come from other countries or industries or are just new on the marketplace, and one of the shortcomings of electronic procurement is that it is not an effective tool to reconcile cultural differences and build teams across different business entities.

Some cases have also been communicated that B2B marketplaces have been abused by fraudulent behavior—for example, when great business is promised to freelancers, but the interested person is told to first submit a payment to the person promising it, and when the payment is made, the one who received it is gone. B2B marketplace providers work hard to keep such swindlers out, and the cases are not common.

2.6.3 Public Tendering

In government settings, one of the most important objectives is complete fairness to all competing vendors. This has a moral aspect, of course, but depending on the applicable law, probably also a legal one: A rejected vendor may take legal action against the procuring agency at court, and if the court accepts the case, the consequences for the project may be dramatic. The project will, depending on the national law, not be allowed to move on and create a *fait accompli* that would create a precedence for the subsequent decision by the judge. The project will have to be put on hold until the court has made and communicated its decision on the case.

- Courts generally do not act quickly unless an emergency situation is about to occur from which no recovery is possible, and projects are rarely considered to be among such situations. The waiting time will probably take some months, possibly years. While this time will be mostly idle time for the project, costs can still occur—for example, to pay infrastructure that needs to be sustained, or for damage payments to contractors and other business partners who have resources blocked for the project that will not be needed, at least not at the time for which they were booked. And the possibility of losing the lawsuit and being penalized is still another sword of Damocles hanging over the project's head.
- For decades, the established process for public tendering was classical advertisement in special magazines, governed by laws that defined in which magazine the advertisement needed to be placed, how often, at what size, and containing what information. This

form of advertisement still exists in some jurisdictions, because the old laws are still in force, but most governments have turned to e-procurements, and their business-to-government (B2G) marketplaces strongly resemble the B2B marketplaces in industry.

A difference is that they are often (not always) free for vendors, but vendors may have to make a small investment in a digital signature that is needed as an identifier to access and use the online marketplace or, in a similar measure, to protect the procurement process from fraud.

2.6.4 Bidders' Conferences

Bidders' conferences are commonly used for major projects such as infrastructure or production development, where hundreds of sellers from different industries are invited to gather in a conference center or a major hall in a hotel. Sometimes they are also held in smaller settings, and the term is also inaccurately used for open B2B reverse auctioning sessions, at which sellers are requested to bid for business in an open meeting. Modern bidders' conferences may also take place as online meetings to save costs and add flexibility.

Bidders' conferences are used for several purposes:

- Introduction to the intended project in one event for a greater number of potential bidders simultaneously to save cost and, even more important, to save time. The project is explained, some PowerPoint slides and possibly a little video is shown,[56] a question and answer (Q&A) session is held, and the bidders are handed a packet with all relevant information for them to offer their services, including a catalog of services and deliveries that need to be procured, details to contract types that the project intends to use, contact data of the relevant managers on the buyer side, terms of business, and possibly a code of conduct.
- Ensuring a protest-proof process for public projects. The bidders' conference setting allows for well-documented fairness, with the intention to ensure that a bidder cannot later protest a rejection based on allegations of discrimination and unfair treatment. A bidders' conference allows later verification that all bidders (1) have been provided with the same level of knowledge on the project and the bidding process, and (2) have been present during the same Q&A session; it is then documented that they got the same treatment.
- Making it obvious for the vendors that they will be bidding under competition to ensure that they will calculate competitive prices and offer high-quality service. Their competitors are also present during the bidders' conference, and all know that they can easily lose the competition for the business.
- Making participation in the project alluring for bidders on more than plain financial interests. The bidders' conference is also a sales event, used to give vendors a buzz of enthusiasm for the project, which may increase the chances for the buyer to obtain great offerings. A critical moment will be the bid/no-bid decision that the vendor will make, and the buyer will promote the project to avoid selection of the no-bid option.

---

[56] These videos can be later extended with real-life footage from the actual project and then be used as a documentary to promote the business of the companies involved.

The buyer also desires offers developed with care, consideration, and diligence, and not just based on reusable templates. Bringing money home with projects is the basic driver for vendors to be in the business of customer projects, but being part of a great undertaking, together with other great vendors and creating an outstanding result, is also an inspiring performance driver for many people—during and after the project, mentioning the great reference in promotional materials can make winning new business much easier for a vendor.

The bidders' conference will try to create a realm of such inspiration for vendors. And because this is much easier in a physical setting than in a virtual one, it is unlikely that online bidders' conferences will fully replace the meetings in the congress hall.

2.7 Requesting Seller Responses

2.7.1 Traps in Terminology

In my first book, *Situational Project Management: The Dynamics of Success and Failure,*[57] I discussed ambiguities in terminology and how they can lead to misunderstandings and confusion, even for experts. Such terminology traps are present in many cases in the project management context, but they are probably worst in the realm of bid and proposal. Critical misunderstandings are commonly found among the top reasons for project failure; they are also a major cause for distrust and team fragmentation, which can also become a reason for project failure—even when the misunderstanding as such was not critical for the project, the discord it caused may be.

These ambiguities are especially common in the two complementary fields of bid and proposal management on one side of the contract development table, and project procurement management on the other. Terms are often mixed up, even when the language used for such expressions is actually very clear, and all that people would need to do is simply to look at the meanings of the words that they are using. Public standards should clarify these terms but are often not truly helpful.

The project procurement management section in the *PMBOK® Guide,* 6th ed.,[58] for example, provides some clarification on the terms "bid", "proposal", and other similar terms to then state that the terminology used may differ depending on the industry and location. Although this second statement is generally correct, one would expect from such a standard that it would create clarity where so far confusion reigns.

I will try to develop some clarification in the following definitions, but I recommend to not expect from colleagues and business partners that they understand and share these definitions. They may have their own definitions, use different terms for the same things, or have no definition at all. The last is the most common one—people in the bid and proposal management discipline seem not to care much about clarity and unambiguity of language.

---

[57] (Lehmann 2016b, 22–27)
[58] (PMI 2017, 477)

Request for Information (RfI)

Typical timing:	After the *make-or-buy* decision, at the beginning of the procurement process.
Submitted to __ sellers:	Many
Typical purpose:	A shortlist of sellers to be included in the further procurement process and possibly information to give the buyer a better understanding of what this process should look like.
Response desired:	Information on the seller's preparedness and ability to offer services and deliveries that the buyer wishes to procure, plus some basic data on the company.
Freedom for the seller:	High
Degree of formality:	Low to moderately formal

Prequalification Questionnaire (PQQ)

Typical timing:	With the RfI as an attachment or following it as a next step.
Submitted to __ sellers:	Many
Purpose:	A shortlist of sellers to be included in the further procurement process, assuming that the buyer already has an understanding of what this process should look like.
Response desired:	A filled-in form with basic data on the seller.
Freedom for the seller:	Low
Degree of formality:	Very formal

Request for Quotation (RfQ)

Competition type:	No formal competition, but alternative quotations may be asked for.
Typical timing:	After the RfI/PQQ or at the beginning of the procurement process.
Submitted to __ sellers:	One to three
Purpose:	Finding a suitable seller for a smaller procurement item that does not justify a formal competition.
Response desired:	A quotation: A confirmation that a seller is able to deliver products or services that satisfy the described wants and needs, a description of what they will look like, and a price for that.
Freedom for the seller:	Low to high
Degree of formality:	Moderately formal

Invitation for Bid (IfB)

Competition type:	Price competition
Typical timing:	After the RfI/PQQ or at the beginning of the procurement process.

Submitted to __ sellers:	Three to five
Purpose:	Opening a (mostly) price-driven competition for the procurement items that are well understood by the buyer and that the buyer can easily describe in a statement of work (SOW) or a similar context.
Response desired:	A bid: A confirmation that a seller is able to deliver the procurement items and a price for that.
Degree of formality:	Very formal

Invitation to Tender (ItT)

Competition type:	Price competition
Typical timing:	After the RfI/PQQ or at the beginning of a (commonly) public procurement process.
Submitted to __ sellers:	Three to five
Purpose:	Opening a (mostly) price-driven competition for the procurement items that are less well understood by the buyer and that the buyer cannot easily describe in an SOW or a similar context, but for which the buyer is desiring to identify the cheapest solution.
Response desired:	A tender: A confirmation that a seller is able to deliver products or services that satisfy the described wants and needs, a description of what they will look like, and a price for that.
Degree of formality:	Highly formal

Request for Proposal (RfP)

Competition type:	Solution and price competition
Typical timing:	After the RfI/PQQ or at the beginning of a procurement process.
Submitted to __ sellers:	Three to five
Purpose:	Opening a price- and solution-driven competition for procurement items that are not fully understood by the buyer and that the buyer cannot easily describe in an SOW or a similar context. The buyer describes in an SOW or a single document the needs or wants that the seller would be asked to meet, and the decision will be made based on the attractiveness of the solution and its price.
Response desired:	A proposal: A confirmation that a seller is able to deliver products or services that satisfy the described wants and needs, a description of what they will look like, and a price for that.
Degree of formality:	Very formal, depending on the procurement items and the people involved.

Invitation to Pitch (ItP)

Competition type:	Solution competition or non-competitive
Typical timing:	After the RfI/PQQ or at the beginning of a procurement process.
Submitted to __ sellers:	One to three
Purpose:	Opening a solution-driven competition or a non-competitive decision process for procurement items against a generally fixed budget that are not fully understood by the buyer and that the buyer cannot easily describe in an SOW or a similar context. The buyer describes in an SOW or a single document the needs or wants that the seller would be asked to meet, and the decision will be made based on the attractiveness of the solution and its adherence to the budget.
Response desired:	A pitch: A confirmation that a seller is able to deliver products or services that satisfy the described wants and needs, a description of what they will look like, and a confirmation that the budget will be met.
Degree of formality:	Informal, depending on the procurement items and the people involved.

2.7.2 Statement of Work (SOW)

Any of the documents above can be accompanied with a statement of work (SOW), a detailed description either of the procurement items or of the needs and wants that these procurement items will have to satisfy. There are different types of SOWs:

- **Internal SOW.** Used during initiation of an internal project. It may be developed by a functional department to describe its identified requirements including needs, wants, and expectations to a project team inside the same organization.
- **Procurement SOW.** Sent with a request for a seller response such as an IfB or RfP to give the seller an understanding of the prospective customer's wants, needs, and expectations that the response is intended to address.
- **Contract SOW.** In most cases a procurement SOW that has become part of the contract. This means for the contractor, that if any requirements stated in the SOW are not met, the contractor is immediately in a breach of contract situation. The desire to make an SOW an element of the contract is mostly desired by the customer.
- **Terms of service (TOS).** In consulting and similar service businesses, the term TOS is sometimes used instead of SOW. The TOS specifies the service mandate that the buyer intends to give to a seller and allows the seller to offer the desired services.

SOWs used for a price competition, mostly using an IfB or a similar request for sellers' responses, are, at least in theory, different to those used for a solution competition with an RfP. The first type will focus on a complete, detailed, and unambiguous description of the products or services desired. The intention is to tell the seller precisely what they are expected to offer, so that the decision can be made on price as the only difference between them. The SOW for a solution competition, in contrast, accepts that the buyer is not able to describe the product or service in sufficient detail and focuses on the buyer's needs, wants, expectations, problems, visions, etc., allowing the seller a high degree of freedom in offering a solution for this problem.

The literature assumes that requests for seller responses commonly have an SOW attached. In practice, many procurements come without this document, and it will then be the seller's responsibility to explore the needs, wants, and expectations of the prospective customer. Many SOWs come in poor quality: They have been written by people who do not have the necessary process knowledge on the customer side and have only a limited understanding of the environment in which the customer and the contractor will have to act after award. Employees on the buyer side with such knowledge are too busy with other tasks to write an SOW. Availability is often no sign of competence, and the unavailability of capable people that led to selecting the *buy* option in the first place may also impact the quality of the procurement and, finally, of the entire customer project. I will come back to this problem, which jeopardizes so many customer projects, later.

2.7.3 Thresholds for Procurement

In public procurement, but also in larger companies, there is often a ladder of requirements on the procurement processes to be involved, depending on the total value of the procurement items and possibly also on the risks involved in the procurement process. One of my customers uses the numbers and rules shown in Table 2.5.

Table 2.5 Example of a Set of Rules with Staged Requirements on the Procurement Processes to be Applied, Depending on the Value of the Procurement Item

Thresholds			
Procurement Item Value	Procurement Method to be Used	SOW Needed	Description
Under $10,000	No requirements	No	Informal and non-competitive selection of a suitable seller
$10,000 to under $100,000	RfQ	No	Formalized selection of a suitable seller
$100,000 to under $500,000	RfQ sent to three sellers	No	Formalized selection of a suitable seller with informal competition
$500,000 and more	RfP or IfB	Yes	Formalized competitive selection of a suitable seller

A problem with such thresholds is that many procurements tend to grow over time—whether through higher bidding prices than expected, change requests by stakeholders, or successful benefit engineering by the contractor—and then may fall into a higher category. Laws in some jurisdictions and internal rules in corporate procurement may then make it necessary to step back into the procurement process and start it again. Parties involved should be aware of this risk.

2.8 The Offer/No-Offer Decision

The next section is more interesting to sellers than to buyers, but it may be good for the latter to better understand the former for great procurements, so go on reading.

A word to vendor companies: Don't expect a buyer, once turned into a customer, to throw money at you out of delight and appreciation. Project business is hazardous business, and the future of your organization as a vendor has many uncertainties. One of the few certainties that one can name is that the customer will be happy for every dollar (or euro, sterling, yen, etc.) that can be saved at your expense. Be diligent with the decision for whom you are prepared to work, as much as buyers should be diligent in selecting vendors. The better your name and your position in the marketplace, the easier this gets.

Another essential question has to do with advance payments, services, and deliveries made. Each party naturally tries to put the burden of such outlays on the other party. These advance outlays come with risk, and they also impact an organization's liquidity. Liquidity and profitability are tightly connected but not the same. Profitability in a customer project refers to the money that the project brings home. Liquidity refers to the question of whether this money will be at home in time when rents must be paid. Repeated losses within an organization will reduce its liquidity, and liquidity problems will force an organization to make emergency decisions that diminish profitability. A subject in any *offer/no-offer* decision, on top of profitability, should be liquidity: "Can we afford the outlays that we need to make for services and deliveries that we will have to make in advance?" One of these advance services will be offer development, and for new customer business, the likeliness is on average higher that it will not return business and with that income at all, as I will discuss later.

There is a lot of literature available on how to write persuasive and winning bids and proposals. The discipline of bid and proposal management has developed its own set of practices and skills, and it even has its own professional association, the globally active Association for Proposal Management Professionals (APMP).[59] They have some relevant standards—namely, the *Shipley Proposal Guide*[60] and the *APMP Body of Knowledge*.[61] There is also a multi-level certification system for its experts.[62] The discipline is well prepared to satisfy a market with growing relevance.

One point addressed in side notes only in this literature is the question of whether it would be good business to win a particular buyer as a customer. The approach in the bid and proposal management literature is that one would decide not to offer when the chance of winning the business is considered low. There is a second caveat: Customer projects are high-risk business, and the decision of whether one wants to make an offer or not in response to a request for seller response is also an act of risk management. It may often be much better to leave the bad business for others, keeping competitors busy with difficult customers, and focus on easier and more profitable prospects elsewhere. An essential start on the seller's side is therefore the *offer/no-offer* decision.

---

[59] See www.apmp.org

[60] (Newman 2011)

[61] (APMP 2014)

[62] I have in the past prepared candidates for the certification.

Figure 2.12 shows how the *offer/no-offer* decision is positioned in the common workflow described here. The location of the events is now moving from the buyer to the sellers, who have received the request to submit an offer and have to make a decision on whether they are going to comply with this request or not.

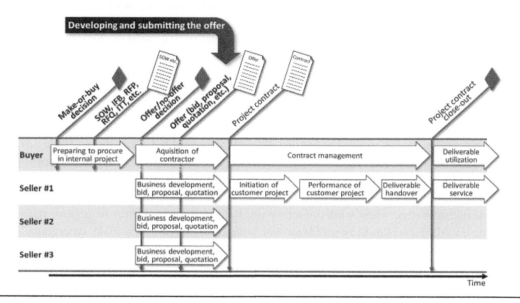

Figure 2.12   The venue is now changing to the sellers from whom the buyers would like to receive an offer. The decision by the seller to make such an offer or not has many facets.

2.8.1 Risks of an Offer Development Process Without Input from Project Managers

There is a strong difference between small and large vendors: The smaller the organization, the more likely it is that at this moment a project manager will be on-boarded to the new business opportunity and bring project management understanding early into the process. It is also likely that this will later be the person who manages the project for the customer. In large organizations, it is rather common to first win the contract with the customer and then get a project manager assigned to perform the contract.

The second approach is cheaper and faster. The offer team does not have to wait for input or even approvals from project managers. They do their job of submitting the offer right before the deadline set by the customer, win the business, and when this job has been finished—successfully or not—they will be free for the next prospect. These benefits come with high risks: The dedicated proposal team has only limited project management competence to assess the feasibility of the project and consider the restrictions on the resources of the seller and on the means and infrastructure that the company can make available for the project. Such matters should be assessed by a person who is experienced in delivery and would ideally later have the responsibility to manage the project, if the contract is awarded. Part of the offer may be a commitment to deadlines set by the buyer, each of which are possibly linked with financial penalties. Practitioners who are used to navigating inside the resource restrictions of the organizations are commonly in a better position to assess the achievability of these due dates.

Figure 2.13 is a repetition from the previous chapter as a reminder of the different exposures of project managers and the teams to time pressure under deadlines. Many of these deadlines have not been put into place by the seller, but by the buyer's organization. The promise to submit an offer and, even more, the offer itself are commitments by the seller to accept these deadlines as binding for the project, and missing them constitutes a breach of contract situation. The deadlines may be enforced with contractual penalties (common in civil law countries) or with either liquidated damages or not-payed incentives (typical for common law countries).[63]

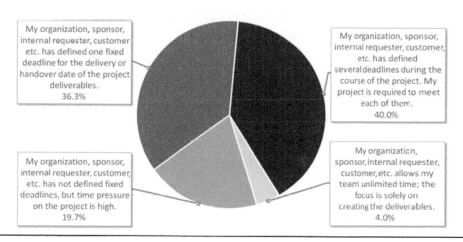

Regarding time pressure, which statement is true for your current project?

My organization, sponsor, internal requester, customer etc. has defined one fixed deadline for the delivery or handover date of the project deliverables.
36.3%

My organization, sponsor, internal requester, customer, etc. has defined several deadlines during the course of the project. My project is required to meet each of them.
40.0%

My organization, sponsor, internal requester, customer, etc. has not defined fixed deadlines, but time pressure on the project is high.
19.7%

My organization, sponsor, internal requester, customer, etc. allows my team unlimited time; the focus is solely on creating the deliverables.
4.0%

Figure 2.13   A survey among project managers on deadlines revealed that a three-quarter's majority of projects has time pressure, and that most of these projects have not only one.

2.8.2 Templates

Many organizations do not send requests from buyers through a formal *offer/no-offer* decision. The fear of missing a business opportunity is great, and every enquiry, as remotely promising it may be, is responded to with an offer. The tool for responding to a greater number of buyers' requests—and doing so in the short time typically available before the submission date—is a set of templates that can be easily reused. This approach comes with major problems:

- The greatest enemy of a good offer is the template. A template leads to offering what has been offered before and what the seller is happy to offer. This may differ widely from what the buyer wants or needs. When I am asking project managers, who do a lot of procurement, their most common complaint is that vendors offer goods and services that were not asked for, and do not offer what is required.

---

[63] The difference between common law and civil law and the consequences for Project Business Management will be discussed in the next chapter.

Templates are among the reasons that many contracting organizations have poor hit rates, and these hit rates force companies to simplify processes by using more templates, which in turn reduces the hit rates again and so on. This is definitely not considered a good practice, at least when templates lead to reduced preparedness to diligently identify customer requirements and to match them honestly with the seller's own capabilities, as this can cost a lot of time and blocks the best people.

- Things that can be done easily and quickly are often done at the last moment. The urgent is the most vicious enemy of the important, and there are more pressing things to do than working on the offer. When a lot of time has been wasted that could have been spent asking questions, discussing details, and elaborating a winning offer based on the customer's stated needs, templates come in handy to develop a quick and cheap offer for the prospect.

- Because the companies that build their bid and proposal management around templates do not select the attractive businesses from the unattractive, they will have too many of the second group at the end—projects that consume their resources beyond what is acceptable for a profitable business and that is also satisfactory for all people involved, including employees and the customer. When many project vendors in a globally growing market are "just about managing" (JAMs) today, the fundamentally unselective approach toward bid and proposal development and submission is among the main reasons.

2.8.3 TRAC: Influencing Factors for the Bid/No-Bid Decision

A variety of factors influences the *offer/no-offer* decision (see Figure 2.14). Often, different people or business units focus on different factors, so the decision can become quite difficult. To make matters worse, the decision requires forecasts in an uncertain business environment, under immense time pressure: If the chosen option is "Yes", one will have to meet a submission

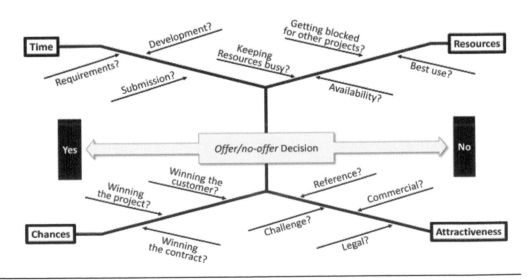

Figure 2.14   TRAC subsumes many common influencing factors for the offer/no-offer decision.

deadline, and, because the procurement activity on the buyer side can be part of a larger project, the deadline is likely to be short, possibly less than two weeks. I recommend using the TRAC[64] model to ensure a 360° approach to the decision. TRAC stands for Time, Resources, Attractiveness, and Chances. Each of these dimensions has criteria involved that one would want to assess in order to make a decision. These criteria can point to the "Yes" or the "No" side and can be weighted or not. I will show examples for that later.

Time Dimension

- **Requirements.** Is sufficient time available to study the SOW (if one exists) and speak with the buyer's staff to get answers to open questions?
- **Development.** Is sufficient time available to develop a convincing capture strategy on which to base an offer?
- **Submission.** Is sufficient time available to submit the finished offer to the buyer before the deadline, if one has been given? How fixed is this deadline, actually?

What would be addressed in these dimensions?

Resources Dimension

- **Keeping resources busy.** Will it be necessary to avoid idle phases for human resources or for facilities and equipment?
- **Availability.** Will the resources be at hand to meet the requirements from the business? If not, would they be easily acquired from the marketplace, or would a persistent resource shortage jeopardize the project?
- **Getting blocked for other projects.** If the project will block resources for other projects, what will the seller then be losing?
- **Best use.** Will the project actually make best use of the resources, or could other projects better use and further develop their skills?

Attractiveness Dimension

- **Legal.** What threats and opportunities will derive from the contract type (and the applicable legislation in international projects)? In case of crisis, what options will be left to the contractor to withdraw from the business, and at what costs?
- **Commercial.** What impact will the project have on the seller's profitability? How can layouts impact or promote the seller's liquidity?
- **Challenge.** When parts of the project become stretch assignments, they will overpower the team's skills and abilities for the moment, but the team can use them to develop new skills useful for future projects. They may also lead to disaster if the team cannot pass the learning process.
- **Reference.** A customer with a good name may be a valuable reference, particularly when the seller enters a new market, in which a reputation of proficiency still needs to be built.

---

[64] Not to be confused with Trac, an open-source software for web-based software project management.

Chance Dimension

- **Winning the contract.** Who are the competitors and what is the seller's chance to win against them and get the contract awarded?
- **Winning the project.** How likely is it that the seller will win the business but, unaware of certain details, later will lose during implementation when these details arise and prove damaging.
- **Winning (or losing) the customer.** How likely is it that the project helps win a great customer and becomes an incumbent? How likely is it that a failed project will sour an existing relationship? What consequences can a won or lost customer have for share values and creditworthiness?

Further criteria can be added that are of particular relevance in the specific organization or market. Some projects may be detrimental to people's health, and this question may also influence the *offer/no-offer* decision. In other environments, such as IT or organizational development, this question will not have such significance.

The TRAC model allows a systematic approach to evaluate the pros and cons of the project for the organization before an offer is made, which consumes resources for its development and often comes with first binding obligations towards the customer, such as a letter of intent. I will show two easy ways to implement it: a weighting system and a force-field analysis.

2.8.4 A TRAC-Based Weighting System

The process has four steps:

1. Assign a weight to each of the criteria above (or ask management to do that). I recommend using a scale of 1 to 10.
2. Rate the customer request and the subsequent project on a scale with positive and negative numbers. I recommend a scale from −10 to +10.
3. Multiply weight × rating to receive a weighted score for each criterion.
4. Sum up the weighted scores to develop the total score of the project.

Table 2.6 gives an example of a weighted score developed for a customer project prior to the *offer/no-offer* decision in a small workshop or by jointly filling out a form.

A score of 175 makes the project look, in general, quite positive, particularly given its commercial attractiveness. Emphasis should be placed on the urgency of the offer process, on the high likeliness of losing it anyway, on legal risks, and on the possibility that resources may not be assigned in their best way. The score may also be a helpful tool when the organization has received two requests from customers for responses and wants to make a decision about which one to respond to, if any, and on which one to put more emphasis on when resources for offer development become scarce.

Experience shows that such a tool, used for the first time, leaves many people somewhat helpless on the numbers to add, but it gets easier over time. It is also advisable to record the workshop or other kind of session during which the table was developed. Important aspects of the projects may appear for the first time during the meeting that should be remembered and taken into consideration during the offer and the later project phase.

**Table 2.6 TRAC Used in a Weighting System to Develop a Total Score
That Helps Make the *Offer/No-Offer* Decision**

Criterion	Weight (1 to +10)	Rating (–10 to +10)	Weighted score (Weight × Rating)
Time			
Requirements	3	–4	–12
Development	5	–4	–20
Submission	1	–4	–4
Resources			
Keep busy	4	3	12
Availability	7	6	42
Blocked	1	–3	–3
Best use	8	–5	–40
Attractiveness			
Legal	10	–2	–20
Commercial	10	9	90
Challenge	8	6	48
Reference	4	0	0
Chance			
Winning the contract	4	–4	–16
Winning the project	9	7	63
Winning the customer	7	5	35
		Total Score	175

2.8.5 A TRAC-Based Force-Field Analysis

Force-field analysis[65] is a great tool to see in a situation of change how driving and restraining forces balance out, and the moment when the offer has been submitted to the enquiring buyer changes a lot in the organization: Resources will be dedicated partially or in full to the development of the offer and to winning the contract, and in case that this will be successful, the organization will be busy doing the project for the customer, possibly for a significant period of time. If the contractor is a small company, doing a major project can be their primary source of income for some time, possibly the only source.

Figure 2.15 shows an offer force-field analysis (OFFA[66]). The focus here is much less on accurate data but on visualization. Adding a separate weighting field would probably overcharge the method, so the weighting should already be included in the numbers estimated.

On top of visualization, there are two numbers of interest:

---

[65] (Lewin and Weiss 1997)
[66] See also StaFFA, the Stakeholder Force-Field Analysis (Lehmann 2016b).

- The average value of the numbers, interpreted as an equilibrium of forces. A positive value signals a dominance of driving forces; a negative value says that restraining forces dominate—a hint that it may be better not to develop and submit an offer.
- The span between the strongest driving and restraining force, which signals stress in the system.

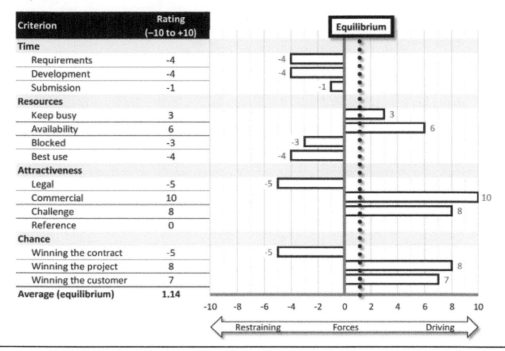

Criterion	Rating (−10 to +10)
Time	
Requirements	-4
Development	-4
Submission	-1
Resources	
Keep busy	3
Availability	6
Blocked	-3
Best use	-4
Attractiveness	
Legal	-5
Commercial	10
Challenge	8
Reference	0
Chance	
Winning the contract	-5
Winning the project	8
Winning the customer	7
Average (equilibrium)	**1.14**

Figure 2.15   TRAC used for a force-field analysis.

The example in Figure 2.15 has an equilibrium of 1.14, a positive number, which is a recommendation to develop and submit an offer. The span of 15, however, on a total scale with 20 steps signals tension in the force field. The hope for commercial success is great, but be prepared to not win the contract and be also prepared to avoid or fight costly legal battles.

Methods such as the two described here often bring new answers as a benefit. A much greater benefit may be that people ask the right questions first. They direct attention to the influencing factors of a difficult decision, which may have strong implications on the future of the organization but are too often neglected. The results are poor hit rates and disastrous customer projects.

2.9 Winning the Contract

2.9.1 AIDA—Singing for the Business

AIDA stands for an old model in sales that is perfectly applicable to the development of the contract business with the buyer, who will then become the customer. AIDA stands for phases that a prospective customer passes through on the way to the final purchase. It is simplified and idealized, but nevertheless quite realistic:

- **Attention.** The buyer has become aware of the vendor. When the *buy* decision is being made and the buyer is seeking contact with appropriate sellers, this vendor will be approached and not ignored if there is some overlap between the products and services that the buyer wishes to buy and those that the vendor can sell.

  In reality, the presence of the vendor may very much influence the *make-or-buy* decision, and often, vendors actively participate in developing the procurement documents. This is particularly true for incumbents, but great account managers can also be very helpful to procurement people, reducing their workload and, at the same time, making sure that the procurement documents ask for products and services that this particular vendor can successfully offer.

- **Interest.** This phase is driven by the aspiration of the buyer to learn more about the vendor's organization and its offerings, but also about those of its competitors, if these have been also invited to quote. This phase is mostly driven by the questions of the buyer and the responses of the vendor. The responses to the questions are important for the buyer, but the buyer will also be influenced by the timeliness and responsiveness of the seller.

  A second important behavior during this phase is proactivity by the seller. It means communicating knowledge that the buyer didn't ask for, but which should be relevant for the decision process. Communicating such information is a signal of competency and of interest in the customer. During the information phase, it will also be important to offer the information in a way that is understandable and informative for the different groups of formal and informal decision makers on the buyer side. More on that point later.

- **Desire.** The desire may be simply to buy as cheap as possible. The more experienced the purchasing staff on the buyer side is, the more likely it is that other factors may get higher priority. The foremost is whether the vendor understands the wishes, needs, and constraints of the buyer.

  Another is the overall attractiveness for the buyer in quantitative and qualitative aspects. The process tries to lead the buyer through a three-step process, in which the buyer begins by liking the offer, then develops a preference for the offer over other options, and is finally convinced that this is the offer to accept.

  Well-communicated *lifecycle costs* or *total net benefit of ownership* (TBO) forecasts may make an expensive price look attractive. I will discuss this later in more detail.

  Another important element of desire building is the communication of a customer-centered and proactive attitude that makes the vendor a great partner to work with.

- **Action.** In the development of a contract, this is the phase in which the customer is expected to sign the contract. This is also the phase during which objections on details are raised again, and everyone is waiting for the approval of management to finally start. In reality, most contractors have already begun their work during that time, because resources were just freed and the deadline(s) of the contract would otherwise become unachievable.

  The risk that the signature will finally not be made was weighed against the risk of delay, and the second was considered greater. In other projects, it is impossible to start without the signature of buyer management, so that every day that gets lost makes it more difficult for the vendor to meet the deadlines.[67]

The AIDA sequence does not fit all project business development situations, because life is more complex than what can be described in four letters, but it is often a great help to structure the process and track progress. It helps further manage the process with the skills of a project manager, understanding that the way to the contract is a project in itself, and given the often low hit rates, one for which the risk of failure is higher than the risk of success.

2.9.2 Hit Rates and Capture Ratio

There is another interesting aspect of low hit rates: how much time, money, and energy can a non-incumbent vendor invest in identifying and understanding the customer's requirements and those of other key stakeholders relevant for project success and profitability? If a statement of work (SOW) exists, a first step for the vendor is commonly to analyze it, list the keywords found in the document, and develop an assessment to validate whether the vendor organization is positioned to do the project; then finding out how the project can be performed as a profitable business with a satisfied customer at the end is a major investment in this business.

In the case that no SOW exists or an SOW has been provided but does not go into the level of detail necessary for a good offer, spending time with the customer to research this information takes also time—typically unpaid time. Access both to locations to obtain first-hand knowledge and to people on the buyer side who have this detailed knowledge may also be limited. There is still no contract between the seller and the buyer that would make such access mandatory. It is also likely that the people on the buyer side who would have the knowledge needed will be busy with other tasks and may not be available for the seller at this early time. In the absence of hard facts, the first understanding of the demands of the buyer will, to a large degree, rely on assumptions and estimations, sometimes even on pure guessing. During this time, uncertainty is highest for the vendor, and gaining certainty is difficult.

One can often use internet research to refine the understanding of the prospect's business situation, corporate culture, imminent challenges, and strategic setup. Whether this rather broad information will be helpful for the specific business case that is underlying the enquiry is yet questionable, and research is time intensive. The resources for bid and proposal development are generally limited, and because many offers are developed "last minute" before the submission date, there may not be sufficient time left to do the necessary research.

In highly siloed organizations, one may also find that other departments have the knowledge needed, because they have a business relationship with the buyer or know the organization as a cooperation partner, competitor, or otherwise, but there is not enough communication to pass this on to the offer team. I observed this effect even for account managers, who have maintained contact with the prospect for years, and who passed information to a proposal team writing the offer to the buyer only parsimoniously. The persons considered this knowledge their personal assets, not organizational assets, and were not prepared to give it away freely.

---

[67] I generally recommend for every contract that has deadlines included to make the obligation of meeting them conditional on timely signature of the contract.

They were measured by the quantity of requests for offers coming in; the success of these offers was someone else's success metric.

To make things even more difficult, the majority of offers developed do not lead to contracts at all, at least not in business development attempting to win new customers. When a vendor places a bid, a proposal, or another kind of offer, the company faces four kinds of competitors:

1. Other vendors, who want to do the same business.
2. The option for the buyer to rethink the *buy* decision and turn it into *make,* now that the vendors have released know-how to them. Each vendor may have been parsimonious with the know-how communicated, but the buyer, who has just ceased to be a buyer, can put the mosaic pieces from the various offers together to create a fairly complete picture.
3. The option to do nothing. As an example, the customer may have hoped to get offers in a range of $1 million but received offers starting at $2 million and even much more. The company may not have the budget to go ahead with the project, or it may not have enough value for them.
4. The buyer's decision is already made. The decision makers on the buyer side have already decided with whom they want to make the project and need competitive offers as "fillers" only to satisfy the requirements of strict internal procedures. In such a case, the seller's chance to make the business is near to zero, unless the offer is surprisingly convincing or coincides with the souring of the incumbent seller's relationship with the customer. Given the often significant investment in a bid or proposal by a vendor who has near to zero chance to win the business, one should consider such behavior outside the boundaries of fair and appropriate business ethics.

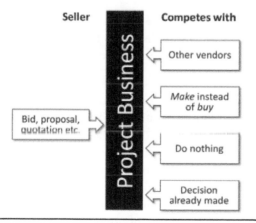

Figure 2.16   The competition for an offer is more than just the other offerer.

Figure 2.16 illustrates the various "competitors" that impede high hit rates in business with new customers. It also shows why incumbents have much easier business development chances than new vendors—these competitive forces are much smaller for them or do not impede their business development at all. At least, the incumbent knows the customer much better and should be able to better predict how strong these forces are.

Another detrimental factor for the seller may be re-organization initiatives on the customer side. The process of inviting seller responses on the buyer side was initiated in a specific demand situation by individuals with tasks assigned and with the responsibility for particular corporate goals. Then, while the procurement process is conducted, the organization is changing its structure, the business goals are no longer valid, and the persons who performed the process have been moved to new positions or have been fired. When you hope to conclude the business with the customer, this business no longer exists. I personally had a case some years ago when, by the time an offer was ready, the entire buyer organization no longer existed; the holding to which it belonged had closed it down for lack of profitability.

When I ask vendors for their hit rates in new customer business, particularly in competitive offer situations, the numbers are mostly somewhere between 5 percent and 20 percent, with a most common value at 10 percent. This number stands in sharp contrast to the commonly named 90 percent hit rate for business won by incumbent vendors. Hit rates of 10 percent means that nine out of ten offers do not lead to a contract. Things are getting worse when online B2B marketplace systems, which I described earlier, are used, which allow managing a larger number of sellers concurrently than what is possible in traditional procurement. Having a greater number of vendors involved in the procurement process is particularly desired for reverse bidding, the aim of which is to buy something at the cheapest price possible, and as vendor management is simplified, the auctions can be done with far more than the three vendors that commonly come from a traditional pre-selection system. For the vendors, this means that while their overhead for developing offers gets reduced, their average hit rates will also go down.

This leads to the problem that the overwhelming majority of offers will not lead to success. Sellers have two ways to respond to this dilemma:

- A *quantity*-based approach. Simplify the selling process to allow more offers with the limited resources given.
- A *quality*-based approach. Apply an elaborated *offer/no offer* process as described above, and then focus on those prospects that seem business-wise attractive, trying to increase the hit rate by developing a small number of convincing bids, proposals, and other kinds of offers.

From a pure selling perspective, both approaches may be successful. From a project management perspective, in which we not only want to win the contract but also to win the project for the organizations involved—seller and buyer—and finally come home with a profit and a happy reference customer, the second is clearly superior.

How can one increase hit rates? I again used a survey to ask practitioners for their observations and experiences regarding not only their own companies, but also the customers for whom they work. The survey was open from 9 February to 2 March 2017 and received 551 answers.

The respondents came from three groups of organizations:

- Sellers only: 203   39.6%
- Both buyers and sellers: 217   36.7%
- Buyers only: 131   23.8%

There was a question dedicated to respondents from selling organizations, and another one for respondents from buying organizations. Respondents from organizations that do both answered both questions.

The questions were:

Question	To be answered by
Please rank the decision criteria of your prospective customers, which you would expect them typically to apply when they select a vendor from the options they have.	Sellers of products and services for projects
Please rank the decision criteria of your own organization, which it commonly applies when it selects a vendor from the options it has.	Buyers of products and services for projects

The criteria that respondents were asked to rank were the following:

- Price
- Operating costs of the results
- Operational disruption on customer side
- Openness for change requests
- Reference customers/projects
- Profile of the intended project manager
- Seller's reputation
- Success record
- Appeal of the offer
- Licenses and certificates
- Proximity
- Bribes
- Understanding of customer's needs, wants, and expectations
- Others (please describe below)

The criteria were presented in randomized order to prevent this order from influencing the results. The respondents could rank a criterion as first place, another one as second, and so on. They could also select that a criterion is not applied at all. For the "others" selection, I offered a free-text field. Many additional criteria were named in the field, including past experience, whether the vendor is listed, and more. There was not much repetition in these criteria, confirming that the 13 criteria named above were the most common ones.

Figure 2.17 shows the percentage that each of these criteria was given in the survey as top rankings for both the expectations by the sellers and for the criteria actually in use by buyers.

Among the two most common top criteria applied by customers was price, but it came second after "Understanding of customer's need, wants, and expectations". Balancing these two criteria is obviously a major task during the development of the offer. Developing a good understanding can be expensive for the development of the offer, but the customer also does not want to pay more than what is unavoidable.

A common response by sellers is "low-balling": Offering a price that is expected not to cover the costs, hoping that during the time of the project a situation may occur for which additional money may be billed to the customer, turning the project to into a profitable one. Such opportunities include change requests, claims, and an approach that I call *benefit engineering*. They will be discussed later.

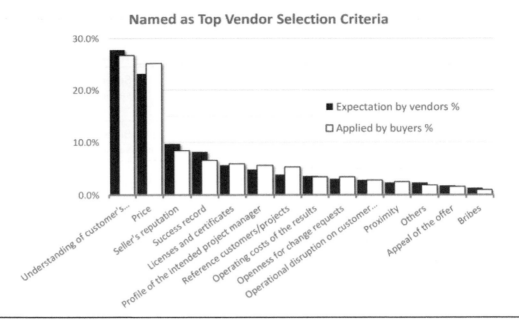

Figure 2.17   The top rankings of selling organizations' expectations of buyers' selection criteria are quite well aligned with the top-ranked criteria actually applied by buyers.

Hit rates, quantity of offers, quality of offers, costs of developing offers, and the revenues from the projects won through these offers build a complex network, which adds more risk to the already high-risk nature of Project Business Management. When too many providers of project services today are JAM (just about managing) companies, this is the root cause from which it originates. Most project service providers have many experts in the technical aspects of their business, plus some people for sales and marketing, but the two groups do not talk enough with each other, and their activities are not sufficiently integrated.

Sometimes, experts in bid and proposal management recommend using a number called *capture ratio* instead of the *hit rate* as a better metric of success. Capture ratio asks how much of the monetary value of offers submitted finally led to contracts. The number can become quite misleading. When a customer makes a budget available of $10 million for a project to be outsourced, and the vendor offers to do it for $25 million and therefore does not win the business, the capture ratio would consider the second number as a business value not captured, when the actual business value is described by the first number. Capture value based on customer budgets would be a great success metric, but in most cases, customers do not communicate their budgets, because they expect that this would prevent vendors from given their best price in a competitive setting.

2.9.3 Being an Incumbent

As discussed above, life is much easier for the incumbent: The customer is known and knows the contractor, personal relations have been built over time, a team that consists of customer and contractor already went through the difficult phases of team building and should be performing well, and the understanding on applicable contract terms and conditions should be developed on both sides. The hit rate is commonly nearer to 90 percent rather than 10 percent, which makes developing offers much more cost effective. In addition, the much higher hit rate brings an economic benefit, as is shown in Figure 2.18.

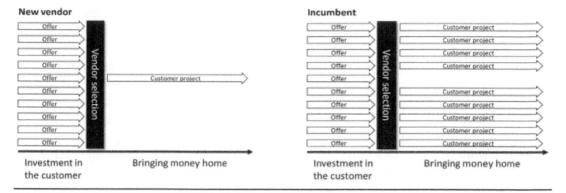

Figure 2.18   The cost calculation of the incumbent is commonly easier, because the investment in the development of the offer is paid back by the combined earnings from more customer projects.

The comparison shows a new vendor with a 10 percent hit rate and an incumbent with 90 percent, numbers that are commonly achieved. The solitary project of the new vendor in the example must not only pay back the cost of the bid or proposal development that preceded just the project that was successful, but also the nine futile offers; otherwise, the company would one day go into insolvency—even if the project is profitable—from too high costs invested in winning the business. The investment for the incumbent is more predictable, as is its outcome, and with a 90 percent hit rate as in the example, only one failed offer must be paid from the income of nine projects. So, the world of Project Business Management seems very gloomy, and the incumbent seems always the winner. But is this always true?

A Case Story

Snail, Inc. is a major engineering company in the field of energy, with its business mostly in oil and gas. Some years ago, they saw the need to diversify into sustainable energy sources and decided to outsource their complete IT needs on a two-year framework agreement to have heads and hands free for the new challenges. For their first two-year contract as an IT customer, they selected Slug LLC over other information technology service providers in a highly competitive invitation-for-bid process, which included two rounds of pre-selection and a reverse auction. The framework agreement covered virtually all day-by-day project services and, in addition, the quick implementation of changes and amendments to Snail's computer networks and server landscape, which Slug operated for its client in the form of mostly minor projects performed under the existing framework agreement.

The expectation by Snail was that Slug would provide a team of over 100 specialists to both ensure reliable 24/7 operations of the IT systems and respond to project requests by immediately producing a quotation, which, if accepted, would be instantaneously implemented. The contract promised, on the one hand, a secured stream of income for Slug LLC, and, on the other hand, a contractor perfectly prepared and integrated with the customer to run operations and projects on a maximum service level for Snail, Inc. It looked like a perfect win–win solution.

After mastering a small number of initial problems, the contract went well for the most part for the first nine months, but at about half-time of the contract, Snail, Inc. found that its contractor was becoming slower and less responsive to its requests. At the beginning of the contract duration, Slug did its best to exceed the contractually agreed service level and was very swift in resolving problems and answering requests from the customer. But after 12 months, Slug often took all the time allowed by the contract to resolve operational problems and send quotations requested by the customer.

The start of projects ordered by Snail also took much longer, and the performance during the course of the projects dropped significantly. The behavior of Slug was not infringing the framework agreement, but the spirit and motivation of the early months was lost; Slug staff worked rather as clock-punchers—by the book, testing and sometimes even stretching the contractual limitations.

After a while, the reasons for the reduced service quality became apparent. In order to save costs, Slug had made staff members who were working on projects for other customers redundant, and the employees who were originally fully dedicated to support the contract with Snail had to work on these other projects in addition to their work for the Snail contract. Slug management assumed that these staff members had idle times that Slug could make use of—an assumption that was only partially true: Although many staff members were at times on standby for the requests and necessities of Snail, using these standby times for other work meant that, when Snail needed them, they had to finish the other work and were not immediately available.

A second problem with these staff members was a rising burnout effect. They developed a growing feeling of *effort–reward imbalance* (ERI), the perception that they were putting considerable effort into both their own company and the customer's, but that this effort was not sufficiently rewarded. They expected not only monetary rewards, but appraisal of their proficiency at work and its value for the organization, as well as assistances such as training and other forms of professional development. In addition, they got increasingly exhausted from the many hours of overtime required to finish the work for Snail in a timely manner, plus the additional work for other customers.

The combination of ERI and exhaustion is known to be a common cause of burnout syndrome. People got sick much more often than normal, one left the company, and the availability problem that Slug already had was amplified by absenteeism and people quitting, and also by a loss of energy when they were at work.

To further increase the problem, the pressure at Slug led to communication problems. Their staff did not have the time that it takes to communicate and socialize, and as technical problems were increasingly responded to with finger-pointing instead of searching for solutions, the ability and willingness of people to cooperate for the common goal also decreased. It took less than two months to turn the contractor organization from a high-performing and reliable service provider to a disintegrated and unsteady "wild bunch"—the term was actually used by the customer in reference to the 1969 Peckinpah movie, which was more driven by violence and chaos than by a somewhat logical plot and a purpose to entertain or inform.

As the responsiveness of the contractor for project quotations under the framework agreement declined in both timeliness and quality of the offers, the customer gave more and more of the projects to other providers, who used them to show superior service quality and build a relationship with Snail, Inc.

During these projects, the providers developed a deep knowledge of the customer, its culture and processes, and also of the formal and informal decision makers. At the end of the two-year period of the framework agreement, it was very clear to Snail that the agreement with Slug would not be renewed, giving another vendor the opportunity to replace the incumbent—at a higher price, one should add.

As the trainer of Snail, Inc. in the engineering field, I was not directly involved in the business with Slug, but I did notice the overall dissatisfaction with their services. When Snail managers asked their colleagues at Slug for a clarifying meeting toward the end of the contract, I was asked to attend, as both an observer and as a witness, in case Slug made a statement that could be used against them by Snail to claim back payments or, in a worst-case scenario, to sue them at court.

Slug's managers did not talk openly about their problems—one should never expect this in the presence of the customer—but made statements that allowed for conclusions, given the messed up situation by that time. The highly competitive procurement process by Snail, which included two pre-selection steps and a reverse auction, left Slug with a contract that was not sufficiently profitable. They would have had to recruit new staff in a very short time from a market that was rather empty.

They were then expected to have a major number of people on standby during the contract duration to be able to respond to problems and requests immediately, but they could not afford that from the income stream from the customer. In order to save costs, not only did they not hire sufficient new staff, they also fired some old employees whom they considered unqualified for the business; they also laid off two middle managers, whose job would have been to ensure coordination, communications, and work appraisals to the employees involved. Slug would have been able to subsidize a small project for the customer here and there, but this business would have required them to accept permanent losses accumulating over two years in both projects and operations, and the contractor's top management was neither willing nor able to do that.

I also had the impression that the desire to win every competitive bid lay deeply within Slug's business genes. It seemed plausible that this was not their only business as an incumbent in which financial problems turned to organizational and interpersonal problems. They entered every bidding competition for new business with the hope of finding the money to help them sort out existing problems, but instead they slipped deeper and deeper into trouble. With the distance of some years, my impression got even stronger that their managers enjoyed the thrill of competitive bidding and of implementing successful capture plans, but then got bored and distressed when they found themselves tasked to ensure reliability and dependability of services. They were much better in competing than in completing.

2.9.4 Conclusion

The case story shows how an incumbent contractor can suffer from the requirements of sustaining consistency in purpose and service, continuity of interpersonal relationships, and coordinating business interests for a successful cross-company mission. I should also note that the new engineering tasks that Snail wanted to focus on—developing marketable solutions for sustainable energy—were massively impeded by the inability of the IT systems to provide them the agility and adaptivity needed. The business of customer projects is a high-risk venture for all parties involved.

My recommendation for Slug would be to collect and communicate knowledge on customers early, develop a sound *offer/no-offer* decision process, and understand the business with customer projects and operational services as a form of portfolio management, where one tries to avoid overwhelming the organization with too much work and weakening its foundation with unprofitable business. I would further recommend hiring managers who can sustain long-term working relationships instead of helicoptering from one competition to the next and from one project crisis to the other.

My recommendation for Snail would be to reduce the competitive pressure on contractors to give them the financial resources that they need to provide great service. The money they saved in the bidding process was much less than the costs of poor projects and services. I am aware that it is often difficult to change such overly competitive behaviors, when the cost savings are considered the success of one department, but the additional costs must be borne by others.

2.10 Offers: Bids, Proposals, Quotations, etc.

On the following pages, we assume that the *offer/no-offer* decision was made by choosing the "offer" option, so that the procurement process can move ahead. I am sometimes asked if one should tell a buyer when the *no-offer* option has been chosen. My answer is generally "yes", even if the buyer is rather unimportant for the success of one's own business, just out of basic respect; besides, one never knows if one will meet the people involved again in a different context. A short but friendly message should always follow a *no-offer* decision.

It can become quite frustrating for a buyer when all vendors approached decide not to develop and submit an offer, or start developing an offer but cannot finish and submit it in the time left. This can delay a project massively and make it impossible for the buyer to meet a deadline. Project Business Management is high-risk business, as I wrote before, and it may be a good idea to have a contingency plan for such a situation.

2.10.1 Developing and Submitting the Offer

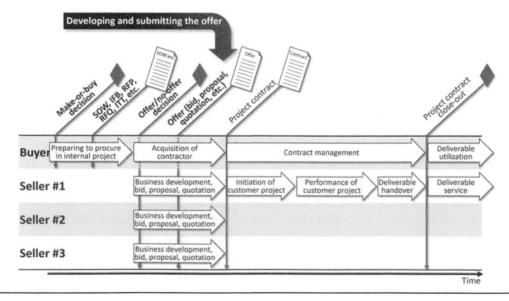

Figure 2.19   Sending the offer to the buyer is a signal of preparedness and capability by the seller to do the intended project work, and also a commitment to satisfy the customer's needs, wants, and expectations.

2.10.2 Types of Seller Responses

When I explained above the different forms of requests for seller responses, I warned that one should not assume that the people involved understand them accurately and use the correct terminology. This is somewhat surprising, because the plain wording as such is clear: In normal language, no one would confuse a bid and a proposal, but in offer management, this happens very often. Many proposal managers who never write a simple bid, but whose job it is to develop complex proposals that also promise meeting complex customer needs, nevertheless have the title "Bid Manager" on their business card. The terms as used here are in compliance with the definitions of requests for seller responses above.

Information (In This Context)

Responds to:	Request for information (RfI)
Typical timing:	Before the *bid/no-bid* decision, at the beginning of the business development process.
Typical purpose:	Helping the customer to develop a shortlist of sellers and ensuring that one is on the list.[68]
Response desired:	Note that the seller is included in the further procurement process.
Freedom for the seller:	High freedom to present their own organization and its offerings.
Time to be invested:	Mostly short
Degree of formality:	Low to moderately formal

Filled-in Prequalification Questionnaire

Responds to:	Prequalification questionnaire (PQQ)
Typical timing:	With the response to the RfI, as an attachment or following it as a next step.
Purpose:	A shortlist of sellers to be included in the further procurement process, assuming that the buyer already has an understanding of what this process should look like.
Response desired:	Note that the seller is included in the further procurement process.
Freedom for the seller:	Low
Degree of formality:	High
Time to be invested:	Significant
Degree of formality:	Very formal

Quotation

Responds to:	Request for quotation (RfQ)
Competition type:	No formal competition, but alternative quotations may be asked for.
Typical timing:	After the *offer/no-offer* decision.
Response desired:	Contract

---

[68] Another purpose may be to get excluded from the shortlist early to avoid wasting time if the business is not regarded as desirable or promising even at this early stage.

Freedom for the seller: High
Time to be invested: Significant
Degree of formality: Moderately formal

Bid

Responds to: Invitation for bid (IfB)
Competition type: Price competition
Typical timing: After the *offer/no-offer* decision.
Response desired: Contract
Freedom for the seller: Low, often limited to the freedom to name price.[69]
Time to be invested: Can become a major project.
Degree of formality: Very formal

Tender

Responds to: Invitation to tender (ItT)
Competition type: Price competition
Typical timing: After the *offer/no-offer* decision.
Response desired: Contract
Freedom for the seller: Low, often limited to the freedom to name price.[70]
Time to be invested: Can become a major project.
Degree of formality: Highly formal

Proposal

Responds to: Request for proposal (RfP)
Competition type: Solution and price competition.
Typical timing: After the *offer/no-offer* decision.
Response desired: Invitation to a presentation, contract.
Freedom for the seller: High, different sellers are expected to offer different solutions.
Time to be invested: Can become a major project.
Degree of formality: Very formal, depending on the details of procurement process and the people involved.

Pitch

Responds to: Invitation to pitch (ItP)
Competition type: Solution competition

---

[69] Even this freedom may be restricted: The customer names a price, and vendor bids are mostly stating whether they are prepared to do the business for that price and what the customer will get for it.

[70] Even this freedom may be restricted: The customer names a price, and vendor bids are mostly stating whether they are prepared to do the business for that price and what the customer will get for it.

Typical timing:	After the *offer/no-offer* decision.
Response desired:	Invitation to a presentation, contract.
Freedom for the seller:	High, but limited by the budget that the buyer has available.
Time to be invested:	Can become a major project.
Degree of formality:	Informal, depending on the details of procurement process and the people involved.

2.11 Binding and Non-Binding Offers

2.11.1 Binding Offers

What does it take to make a contract? Many people will say that a document is necessary, but to the extent that I know international legal systems, none of them requires that. One can say that there is a distinction between the commercial and the legal understanding of a contract. The commercial understanding of a contract is commonly either a document with two or more signatures or a system of two or more documents that together make the contract. In a legal understanding, most—presumably all—jurisdictions allow for verbal contracts, and often a contract may not be concluded by words but may be construed based on actions by a party. A supermarket presents goods on a shelf, and a customer takes them to the cashier to pay for them. The legal offer in this example is made by the customer, and the cashier by registering the purchase and taking the money concludes the contract, while both may talk about the weather or the children or may not talk with each other at all.[71]

Contracts are made by offer and acceptance and are mutually binding. If one party offers a performance in a legally binding form, and the other party accepts the offer, also in a legally binding form, a contract has been made, and unless there is a clause that allows one or both parties to easily terminate the contract for convenience, or one party allows the other one to terminate the contract as a gesture of goodwill or against a damage payment, it will be difficult to get out of the obligations promised in it.

Both the offer process and the project stretch over long periods, and changing business situations during these times can make it difficult to meet contractual obligations. Not being able to meet such obligations is often considered a "luxury" problem. It is certainly no such thing.

The case of Tiger Moth Ltd. is such an example. Tiger Moth sent a bid to Butterfly AG in January 2015 for the renewal and expansion of their communications and IT infrastructure. Butterfly makes heavy use of high-definition video-conferencing on a global scale, and the bandwidth used by this data-intensive technology had repeatedly brought their existing infrastructure to a breakdown. Butterfly sent out an invitation for bid (IfB) to a number of service

---

[71] Please note that this information, as are all statements in this book relating to contracts in project management, is given without warranty and is not intended as legal advice; different national legal systems may have different rules deviating from the broad descriptions given in this book. The topic here is to teach project management in a contractual environment, and before applying any statements made here as recommendations, it is advisable to double check with a lawyer.

providers to help them replace their hardware and implement software with higher performance on a global scale.

At the moment when Tiger Moth received the IfB, and also later, when the decision was made to send a bid and when this was actually done, Tiger Moth's ability to perform the project seemed obvious. They were doing a project for another customer that was coming to its end, so that resources would be free again for new tasks. They also had a small recruitment project to find three new specialists in digital communications, and as they got responses from 16 well-educated applicants, it seemed easy to hire new staff when they won the contract.

In order to limit risks, Butterfly put a deadline for approval of their offer into the bid, which would be invalidated if it was not accepted eight weeks after submission. Some days before the end of the eight weeks, Butterfly sent a message to Tiger Moth that they accepted the offer and welcomed the supplier as their new contractor.

Butterfly had no idea by how much Tiger Moth's business situation had changed in just under two months, and even Tiger Moth's management spent some time analyzing where they actually stood resource-wise inside their own organization. They found themselves indeed unable to do the business. The other customer project that should have been finished meanwhile had had some lengthy delays, caused (1) by unexpected technical difficulties that were only identified by the team in the last weeks and needed to be fixed, but also (2) by the rejection of the customer to accept the results of the project because they only partially complied with the agreed-upon specifications.

Of the 16 applicants on the job adverts, only five were finally considered acceptable; the others had promised experience and knowledge that they actually did not have. Tiger Moth found that they had beautified their bios in a similar way, following an article in a popular online magazine that described how this could be done so that an employer would not notice it—unless, of course, the employer read the same article. Actually, it was a member of the HR department who spotted it.

Of the five remaining candidates, three had found other jobs meanwhile and were no longer available, so that Tiger Moth could only hire two new employees. Instead of expanding the work force, Tiger Moth found it actually shrinking, when three employees left the organization, obviously to set up their own company, and another one had to leave as a result of conflicts with colleagues that seemed irresolvable.

To make things worse, at the end of January, Tiger Moth received an acceptance of another offer previously made to another company. They now had a loss-making project with missed deadlines that they were unable to close out and a new project that they would find difficult to perform, and they definitely had no free capacities when the signed copies of the contract with Butterfly was finally submitted to them, placing obligations on them that they were not capable to meet.

Tiger Moth's management was a victim of the corporation's great performance in winning the contracts and much less great performance in also winning the customer project. They had to visit the customer to ask for release from the contract, and only when they offered a payment to Butterfly to make up for loss of time and money for the futile IfB did Butterfly agree to the termination. By that time Tiger Moth had a loss-making project, a major payment to be made to a non-customer, and another new project that they had to perform, for which they also had no resources free.

In most jurisdictions, contracts are made by two identical declarations of will called *offer* and *acceptance,* a process often referred to as "meeting of the minds"; but in project business, the minds may in reality not meet at all. It is more like two people at the same location on different days; they do not meet, but nevertheless the contract is created. There can be weeks between the two declarations of will, and a lot can happen on both sides during that time. How can the seller gain protection from having a valid contract with obligations that they can no longer fulfill? They can mark the contract as an *invitation to treat,* a purely commercial, non-binding offer.

2.11.2 The Invitation to Treat

To understand the character of an invitation to treat—a non-binding offer—it is helpful to remember the different definitions of the term *contract* by commercial people and by those coming from a legal discipline (see Figure 2.20).

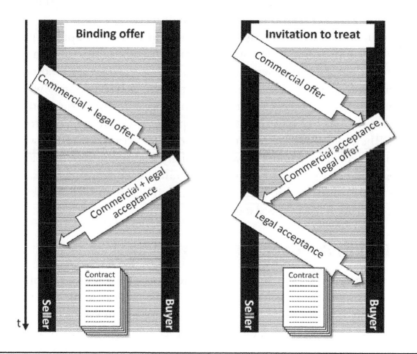

Figure 2.20 An offer can be binding, which makes it an offer in both a commercial and a legal sense. The acceptance of the offer by the other party then makes the contract. An invitation to treat is an offer in a purely commercial sense. When the buyer accepts it, the contract is still not made in a legal sense—it is the acceptance of the seller that concludes the contract.

The invitation to treat is not an offer in a legal sense. It is an invitation to the other party to *make* such an offer. The items displayed in a supermarket shelf or in a shop window are invitations to customers to offer the seller a business transaction in which the customer buys them and pays for them. If a bid, proposal, etc. should be understood as a non-binding invitation to treat, it should be made clear in the text that the contract only becomes valid after the *seller* has formally accepted the *buyer*'s legally binding offer to buy.

This allows the seller at the moment of contract conclusion to verify that the resources to do the project are still available and to possibly step back from the business if this is not possible. The seller will then have to write off the costly and essentially successful investment in offer development, but this prevents the company from adding more cost on top of that for running a loss-making zombie project for the customer.

Invitations to treat must be used with care:

- In some civil law countries such as Germany, sending an invitation to treat (*Freibleibendes Angebot* or *unverbindliches Angebot*) can be considered a legally valid offer, when the buyer accepts the offer and the seller does not immediately and unambiguously reject the conclusion of a contract. This can even be true if the buyer has changed significant parts of the original non-binding offer, such as reducing the price. In these countries, it is necessary to react immediately when an unwanted contract is being concluded by the other party, but the good news for the seller is that it is still possible to say "No" and avoid stepping into obligations that one cannot meet.

- Before an invitation to treat is used in response to a request in public procurement, one should verify with a legal expert whether this can lead to the exclusion from the further procurement process. Some legislations consider accepting non-binding offers a breach of the principle of fairness in public procurement, because offerers with *binding* offers would have a disadvantage. One should remember that this fairness principle is not just an ethical principle but must be upheld by public buyers to avoid protests from rejected vendors. These protests may take the form of legal action, and if a lawsuit for discrimination is accepted by a court, it will definitely lead to massive time losses and budget overruns while the project's progress gets stalled over many months, waiting for the decision of the court, before it can start again. It will then take further weeks, possibly months, to bring it up to speed again.

2.11.3 Bid Bonds, Performance Bonds

Whereas the invitation to treat *reduces* the risk for the seller in case the offer wins but the performance is no longer possible by that time, bid bonds *increase* the risk for the seller. Bid bonds are commonly used in major construction projects and in infrastructure and are mostly unknown in most other industries. A bid bond is either a cash deposit or an insurance sum that the buyer can use as a guarantee in case the winning offerer does not perform what was offered. It is commonly used to pay the buyer the price difference to the second lowest offer, which will then be taken instead, if it is still open by that time. Depending on the jurisdiction, there may be caps on bid bonds—10 percent or similar—but these do not exist everywhere. The rejected offerers can generally expect either to get their deposit back, or, in case an insurance has been used, that the insurance contract is closed down without a payment by the insurance.

The last point may not always be true. I remember a case in which cash deposits were required to be made by the sellers of 10 percent of the offer value (fixed price); the customer would otherwise not have considered their offer, but no return payment was promised to the seller rejected in the process. In such a model, 10 offerers' deposits, or even less than that, would be sufficient to finance the entire project, because the cheapest seller would be selected.

In another case, the buyer was no buyer at all, but a group of fraudsters who built a Potemkin village of an intended infrastructure project, then cashed in on the deposited bid bonds and ran away. In such environments, Project Business Management turns into pure gambling, and vendors should ask themselves if this is the kind of business they want to participate in. It is, in any case, recommended to discuss the matter with a legal expert when a customer requires a bid bond, just in order to understand the magnitude of the risks and to request clear and legally enforceable commitments to protect the deposit. The high time pressure during bidding and the expectation of high profits from working in a great project should not make managers blind to the risks they are entering into.

A *performance bond* is similar but is typically used later in the project as a guarantee to protect the customer from poor contractor performance or from no performance at all. It commonly is a much higher percentage of the project value, but because the project often grows during its course, the percentage number may get smaller over time. In some business environments, bid bonds and performance bonds are confused or even considered the same.

2.11.4 Questions Forbidden

The information that sellers receive from buyers before the contract has been signed is commonly incomplete and leaves many questions unaddressed. Statements of work are rarely written by experts in the prospect's organization, and those who write them are rarely given sufficient time for research and writing. Environmental conditions may also change during this period and are then rarely captured in writing and communicated to sellers. One should also not forget that with a statement of work, the buyer hands out information on current or upcoming problems, future plans, long-term strategic goals, and many more facts that they prefer to be kept invisible to their competitors and other stakeholders.

Another problem is that the authors of statements of work are rarely trained in the skills of developing a complex document that is both precise and concise. The development of the statement of work may be contracted to a specialist, who in turn may be a great writer but does not know enough about the buyer's organization, its desires and needs, and technical details, especially when these are about to change. For these reasons, one should not expect a statement of work to be well written, unambiguous, and truly up to date.

Sellers will commonly need to make contact with the buyer to ask questions and to get uncertainties clarified. This question and answer handling may be an important element of the pre-contract process, but some buyers disallow sellers to contact their staff—again predominantly in public project procurement.

As before, the reason is documented fairness with the purpose to make the project protest-proof. A rejected seller may be able to verify that the winning seller had more knowledge and take this to court as a case of discrimination. If the court decides that the case will be accepted, the project will have to be put on hold. The project will lose a lot of time and will have to bear further recurring costs such as wages, office rents, etc. while it is idle, and contractors who have blocked resources for the project may still bill the project for their availability or may claim damages for lost profits. Project managers in public projects must always be acutely aware of this risk, and not allowing vendors to talk with their staff is a common solution for them.

Given that many statements of work are ambiguous and incomplete, this turns into a risk for vendors: The prospect requires sellers to commit to prices, deadlines, and other obligations without sufficient clarity on what these obligations are. In training and coaching for both proposal management and project management, I have repeatedly been asked by students how to behave in such a situation, and when I offered my services in governance projects for professional development, I was in the same situation myself. Here are my recommendations:

- Propose to the customer that they have an anonymized platform, ideally in the form of a simple password-protected website, on which sellers can place their questions and the buyer can answer in writing, so that all sellers can watch the Q&A communications and take part in them. The customer should make sure that all vendors involved have access to this website, and this should be documented to avoid later disputes.[72]
- If this is not possible and the customer is not prepared to discuss other solutions, consider not offering anything.

The second recommendation has a simple background: The public buyer is correct in assessing the risk of a discrimination case for the project and managing the risk. The buyer is not correct in disregarding the risks to the seller. This may be a general part of the organization's culture, and working with such a customer, who is unable to develop a partnership based on mutual understanding and empathy, is likely to become sour over time. A business partnership is like driving together in a car. You do not want to sit in a car that is dedicated to running people over—and certainly not to be in front of this car.

Depending on the project, the contract type, and the role of the future contractor, such customers may also be an easy source of a future stream of income with not much work to be done. Just consider that it is your job to place temporary fences around construction sites, and part of your calculation is to bring the fences to the sites, relocate them there from time to time as needed due to the progress of the project, and, at the end of the contract, take them home again. When this construction gets disrupted because of poor planning and—very commonly—poor change request management, your fences will remain there for a much longer time than originally expected. They will stand there for long periods, and as the project has no true progress, they will not require much work. All you need to do in such a situation is write your monthly invoices.

I consider this situation an exception rather than the rule. Mutual understanding and empathy are generally important to do a project with different organizations involved, and if one finds that the prospect lacks these fundamental characteristics, it is often best to not offer and leave the foreseeably difficult business to the competition.

2.12 Submission Deadlines

In my work as a trainer in proposal management, I was asked by a student some time ago, when an offer should be submitted to the customer. They had a case in which the deadline of the prospect lay shortly before Christmas day and had sent the proposal package so that it would arrive just on time. They forgot to account for the risk that, during this time of the year, the workload of parcel delivery companies is temporarily growing. It is typically highest in the days

---

[72] Some B2B online marketplaces have the functionality to provide an anonymized Q&A forum.

before Christmas, when the late gifts need to be delivered on time, and even with temporary staff hired for the holiday period, parcels may take them same days more to deliver. In the case of my student, the parcel was late, and the work of a proposal team of four over almost two months was made futile.

Submission planning should be part of any proposal development. In an online environment, the timing is rather simple, but with physical documents to be submitted, the risk is much higher.

Developing the offer can take a lot of time. Especially, some complex proposals and tenders against unclear customer requirements can consume a lot of time, energy, money, and professionalism invested by the people involved. Offers for very large projects are regularly done with consulting support from retired business professionals from the field—an approach one finds in defense, infrastructure, and energy projects.

On the other side of the business table sits the buyer, who also runs a kind of initiative—probably a project, possibly a customer project as well—that has deadlines too and a budget that needs protection from costly delays. The buyer has already booked resources for certain dates and made other kinds of commitments, and it is therefore in the buyer's interest to make the pre-contract business swift. Buyers therefore put a deadline on the submission, and missing the deadline may make all the work invested by the seller in the development of the offer lost.

It is also a signal to the customer that this vendor does not have a habit of meeting deadlines. On the supplier side, the departments winning the business and doing the project may be separate and their habits may be different, but a customer would not be aware of that. Approaching the customer early when one finds out that a submission deadline is impossible to achieve is at least a signal of respect. It may also be that the deadline of the buyer is much less fixed than it seems and that the vendor is given some extra time.

How does one justify asking for postponing of the deadline? There is a simple argument that most customers will accept: All resources are currently busy with ongoing business, and meeting customer requirements in running projects has the highest priority, more than winning new business. This message is given to the buyer, combined with the promise that, once the business has been won, the rule will also apply for it.

Most submission deadlines will not be open for re-definition, for the reasons described above, and the submission of physical documents comes with risks. A solution that I found working well with customers was to recommend a two-way submission:

- Submission as a digital package on a server provided by the seller or on a third-party internet server, and communicating the download address per e-mail. If the package is small enough, it may instead be just attached to the e-mail. This needs to be submitted in time before deadline.

- Simultaneous submission of the physical package per delivery service. If the e-mail is on time, this may take some additional days.

If the buyer allows such a two-pronged approach to all vendors, this is a fair offer. The vendor wins another two days for proposal development and the security that the offer will be submitted in a timely manner. The buyer benefits from having two versions of the offer, a paper version that is easier to read and a digital version that allows the use of a search function. Another benefit is that the submission date is documented with the e-mail.

In my previous book *Situational Project Management: The Dynamics of Success and Failure,*[73] I described that, in contrast to the assumptions found in most literature, projects can have more than one deadline. Figure 1.6 (page 16) describes the result of a survey that I made on the question of projects with no deadlines, one deadline, and multiple deadlines in September 2015. I received 466 responses, and projects with multiple deadlines were the most common among them.

During offer development and submission, managers are also faced with multiple deadlines. A timeline that I came across in 2016 during a request for proposal process that my customer attended is shown in Table 2.7. It displays how a buyer imposes the deadlines to coordinate the progress on the sellers' side with their own schedule.

Table 2.7 A Sequence of Deadlines Imposed by the Buyer to Coordinate the Work of the Sellers with the Buyer's Schedule

Timeframe	Buyer's Activity	Activity for Participating Sellers	Deadline
01–12 February	Seller prequalification and shortlisting	Submission of PQQ,[a] notice of seller's intention to send proposal	10 February
15–19 February	Submission of RfP[b] with SOW[c] to shortlisted sellers	Confirmation by seller to buyer of receipt of SOW	19 February
22 February–4 March	Completing and refining SOW	Submission of concerns and questions for clarification	26 February
7 March–29 April		Development and submission of proposal	29 April
2 May–6 May		Proposal presentations	6 May
10 May	Vendor selection		

[a] Prequalification questionnaire
[b] Request for proposal
[c] Statement of work

Multiple deadlines mean that there is a critical deadline. In the case above, it may be simple to meet the first deadline on 10 February, but the truly difficult deadline lurks further in the future, to be finished and have the proposal submitted to the prospect by 29 May. The danger with a critical deadline in the future is that teams tend to focus on the nearest deadline and ignore for the moment the later ones. When they then become aware of the schedule risk, it may be too late to make resources available to meet it. Looking at deadlines one after the other may be an inappropriate approach—one needs to assess them all right from the start.

2.13 Teaming Agreements

Teaming in this context stands for independent organizations that act together as one team to do projects that a single one would not be able to do alone. Teaming is a difficult task,

---

[73] (Lehmann 2016b)

developing a "Mission Success First" culture across the network of teaming companies without completely ignoring business needs—one does not want teaming companies to go bankrupt or be liquidated during the course of the project—or in the near future, when one may still need them. One should want to protect the project from corruption, one of the greatest obstacles of long-term success and of professional independence. One has to reconcile differing business interests and at the same time make sure that when the relation turns sour, one's own position is well-protected.

It is possible that the members of the teaming agreement have experience from joint business in the past, but often they are working together on the project for the first time, so to function well, the companies that participate in teaming will have to go through the Tuckman team development phases[74] similar to those that individuals go through when they come together for the first time to work as a team:

- **Forming.** The team roles, formal relationships, communication rules and channels, and other formal things are fixed. During this phase, good project managers organize teams for success.
- **Storming.** The team members now focus on things that separate them. In this phase, organizations that participate in the teaming agreements build cross-company processes and business rules, just like individuals developing their interpersonal interfaces. This process is often driven by conflicts about whose rules to apply, what interfaces regarding business and communications should look like, and how to deal with differences in processes, cultures, business style, and many more. Particularly if the teaming partners have not worked together previously, these differences may not be known yet, and discovering them can become a painful process.

  For a project manager, this phase allows the least influence, and the productivity during this time is also lowest. One should listen to concerns, make sure that conflicts do not escalate beyond an acceptable threshold, and hope that the phase will be ended soon. To avoid misunderstandings: This phase is often distressing and troublesome, but it is important for later reaching the performing phase.
- **Norming.** The focus of team members is changing from the separating to the common and connecting. It could be a favorite sport or even sports team, some common opinions in politics or rock music, common interests in children, or dogs, or even stamps—whatever it is. For organizations, this phase is often driven by the identification of common business interests, which can be better achieved together than by one organization alone. During this phase, the influence of the project manager is generally growing again, and so is the productivity.
- **Performing.** There is no guarantee that this phase will ever be reached. The team went through thick and thin together and has found modes to embrace and cultivate the common and to control the separating. Positive messages are communicated loudly, problems and sensitive issues are discussed in private. When teaming organizations have entered performing, they perceive and regard business benefits as outweighing the investments that the cooperation requires from them and as justifying the limitations that come with the cooperation.

---

[74] (Tuckman 1965)

Figure 2.21 describes the three most common types of teaming agreements. Each type of teaming agreement comes with specific benefits, but also specific risks, as follows.

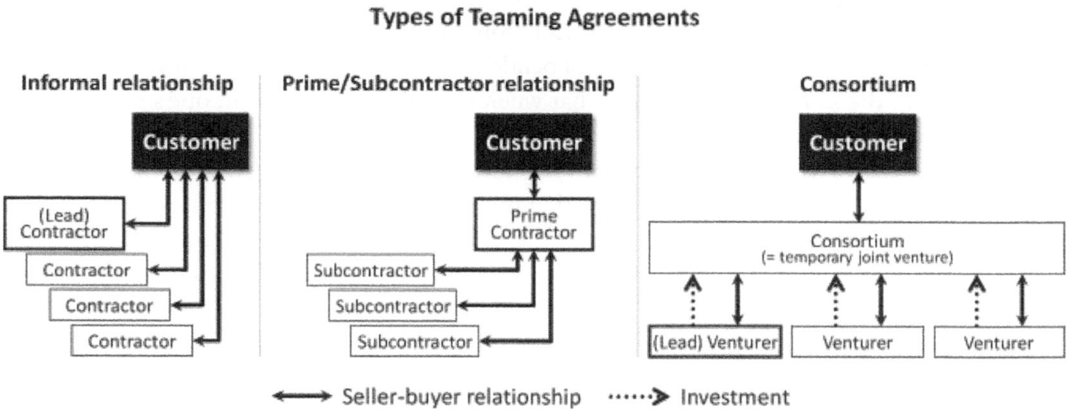

Types of Teaming Agreements

Figure 2.21   Different forms of teaming agreements for project work.

2.13.1 Informal Relationship

The customer selects and contracts each of the contractors and has a direct contractual relationship with each of them. It is possible that one contractor is in a lead role, informally or with a commission agreement paid by the other contractors for helping them to also enter the business.[75] The lead contractor may be the one who has the best relationship with the customer, has the largest part of the business, has licenses that the other contractors can use, or has some other advantages that are beneficial for the other contractors.

Benefits for the customer:

- It saves money. The customer manages and coordinates the contractors and does not have to pay a fee to a prime contractor for this work.
- Only one set of interests matters—that of the customer. Decisions are not influenced by other, often opposing interests.
- The customer is much more directly involved in the project and will learn many details that will be important later, during the operations of the result of the project.
- Implementing change requests will be easier, but the customer will have to take the full risks for their consequences, particularly if they are handled poorly.
- In case of default of a contractor, only a part of the project is jeopardized, and as soon as a replacement has been found, the project can go on as originally planned.

Disadvantages and risks for the customer:

- The customer may be inexperienced in the topic of the project and does not know to what degree one should trust contractors.

---

[75] In most cases, the customer will be kept unaware of such commissions.

- The customer is often not sufficiently aware of the special language and of caveats in the field of the project; this increases the risk of errors, whose costs will normally have to be born by either the contractor or the customer, and which can lead to delays, quality and security problems, etc.

2.13.2 Prime/Subcontractor Relationship

The customer has only one business partner—the prime contractor—who then selects subcontractors and outsources project work to them. In some contractual settings, the customer may reserve the right to name a subcontractor, with whom the prime contractor then needs to work. Instead of naming, the customer may nominate subcontractors, which means that the prime contractor is given a short list of potential subcontractors whom the customer considers acceptable, and the prime contractor can choose one of them.

Benefits for the customer:

- The customer has only one business partner, who coordinates the work and should be knowledgeable of the language and the specific caveats in the field.
- Depending on the contract type, risks that come with the use of subcontractors are fully or partially owned by the prime contractor.
- The prime contractor, like any other contractor, can be a vehicle for knowledge transfer based on the experiences made in other projects.

Disadvantages and risks for the customer:

- The prime contractor is an additional cost driver. The organization invests resources into the project for the customer and bears major project risks; they will want a fee to cover this and, in addition, make allowance for a profit.
- When the prime contractor defaults, the entire project is probably lost. In an almost worst-case scenario, and depending on applicable legislation, deliverables made for the customer but still not formally handed over may become part of the contractor's insolvency estate, and it may then become legally impossible for the contractor to get them handed over.

    To make the scenario *full* worst-case, the deliverables have been de-facto handed over, but without a document that verifies the legal transfer of property, and the customer may have to give the deliverables back for addition to the insolvency estate.
- The project is driven by two sets of business interests—those of the customer and those of the prime contractor. At times, these business interests are in alignment, but not always. The prime contractor may from time to time make decisions that are not in line with the interests of the customer.

    Sometimes, the consequences of such decisions are hidden from sight, becoming visible only after the end of the project and a potential warranty period, and sometimes taking the customer years to identify.

Benefits for the prime contractor:

- Business. The difference between the price paid by the customer and the costs to pay subcontractors is the prime contractor's contribution margin.

- The proximity to the customer opens new business opportunities without expensive marketing and often without competition. Many prime contractors are long-term incumbents with their customers.
- The ability to employ subcontractors temporarily as needed gives the prime contractor a high flexibility to scale teams up and down in relatively short term, mostly much more quickly than it would take to hire employees; it is easier to finish a business with them than to make employees redundant who rely on the long-term income from the job and expect job security from the employer. Employees organize themselves from time to time in unions; subcontractors don't do that.

Disadvantages and risks for the prime contractor:

- During the project, it may turn out that the contribution margin, essentially the difference between the price paid by the customer and the costs incurred by hiring subcontractors, is becoming too small or even negative, both leading the project to a loss for the prime contractor. It may then be difficult to step out of the business.
- The prime contractor must meet contractual deadlines as well as scope, quality, and safety requirements imposed by the customer. If the subcontractors do not perform as intended, the prime contractor cannot meet such requirements.

2.13.3 The Consortium

A consortium is a temporary joint venture. It acts as a prime contractor, but some or all of its subcontractors are also venturers, investors into the consortium, based on a mutually binding consortium agreement.[76] Consortia may be made just for a project. Often, consortia are developed for a combination of project and time-limited operation of the results. A common model is used in construction and infrastructure projects under public–private partnership (PPP) called *build-operate-transfer* (BOT) projects.

In this type of business, the consortium builds or expands infrastructure, operates it for a given time—allowing the consortium a profit to return the initial investment—and then transfers the deliverables to the actual customer, the licensing country, whose infrastructure has been developed this way.

An example was the Taiwan Highspeed Rail Corporation (THSRC), a consortium to build and operate a highspeed rail link from the city of Taipei in the north of the island Taiwan to Kaohsiung in the south.[77]

The lifecycle of the consortium began with its foundation and the preparation for the bid and ends when the infrastructure is handed over to the country. After the completion of the construction project, planned for 2003, THSRC was allowed 30 years to utilize the train system and make a profit, and then the system would be handed over to the state.[78] Unless the

---

[76] Sometimes, organizations that are neither temporary nor joint ventures also call themselves "consortia".
An example is the W3 Consortium, an association that defines the standards for the World Wide Web.
[77] (PMI 2005)
[78] (Chang and Chen 2001)

consortium has no other business to go on by the time the license period ends, it would then be liquidated.

Figure 2.22 shows another lifecycle of a BOT project, in which the contractor is given 35 years from the beginning of the license period until the final transfer to the country. This gives the contractor a strong incentive to do the project quickly, as delays not only affect the time when the cash flow turns positive, it also reduces the time that the consortium can cash in.

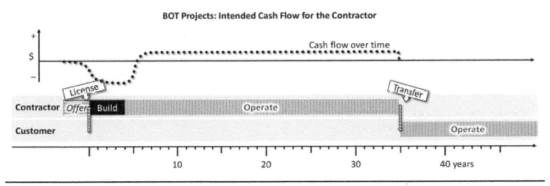

Figure 2.22 A typical build-operate-transfer (BOT) project has the intention that the contractor invests into the project (mostly for infrastructure) and later recovers the investment from the return made from operating the infrastructure. The example above has 35 years' total duration, including the build and the operation phase.

BOT projects promise self-funding infrastructure development and have therefore become very popular in the public sector. Some private companies have also discovered the BOT model for their internal infrastructure projects, which then are investments by a third party that get payed back through a fee paid for the usage of the infrastructure, until the contract for this use expires and the infrastructure is handed over to the customer. In a private setting, the contract duration is normally less that the 30–35 years in public projects; typical are rather 5–15 years, and the contractor may not be a consortium but a single company.

Except in the field of BOT projects, consortia are also used for political reasons. For the European military airlifter Airbus A400M, the consortium Europrop International was founded to develop and build the turboprop engines as a contractor to Airbus. Europrop is a joint venture by

- German MTU Aero Engines
- French Snecma
- British Rolls-Royce
- Spanish Industria de Turbo Propulsores

The workshare among the four venturers follows precisely the number of aircraft purchased by each of the four partner nations. It is no surprise that such a corporate construction, which follows more political wishes than technical, organizational, and business requirements, has not proven to be the most efficient and effective organization.

A consortium can also be just one contractor among others in a major project. The expansion of the Panama Canal performed in the years 2007 to 2016 was budgeted at $5.2 million, and roughly half that was planned for the construction of new locks, which was done by the consortium Grupo Unidos por el Canal, comprising:

- Sacyr Vallehermoso S.A. (Spain, lead venturer)
- Salini Impregilo S.p.A. (Italy)
- Jan de Nul n.v. (Belgium)
- Constructora Urbana, S.A. (Panama)

A consortium may have a leading venturer, which may be the investor with the largest share, the organization that has the best contact with the customer, or the organization that is entrusted by others to do the majority of the management tasks and allow others to focus on technical contributions.

Benefits for the customer:

- From the customer's perspective, the consortium is just another prime contractor. The customer has only one business partner, who coordinates the work and should be knowledgeable of the language and the specific caveats in the field.
- Depending on the contract type, risks that come with the use of subcontractors are fully or partially owned by the consortium.
- The consortium, like any other contractor, can be a vehicle for knowledge transfer based on the experiences made in other projects.

Disadvantages and risks for the customer:

- The consortium is an additional cost driver. The temporary organization invests resources into the project for the costumer, bears major project risks, and will want a fee to cover this and, in addition, make allowance for a profit.
- When the consortium defaults, the entire project is probably lost. All the consequences described above for a defaulting prime contractor also apply for a consortium, and because the consortium is intended to exist temporarily anyway, terminating it in a crisis situation is easier done by its venturers than doing it with an organization that has been set up with the intention to last.
- The project is driven by a multitude of business interests—those of the customer, and also those of the various members of the consortium. In addition to the risk that comes with a prime contractor, dysfunctionality of the consortium caused by incompatible interests of the companies involved pose an additional risk, which may be a major one.
- The consortium has no history of successes and failures. A customer can ask offerers to provide reference customers, but the consortium is new, founded just for the project.

  I recommend in such a case to get information on the venturing companies, but although this may be a good way to know more about the competency, reliability, and integrity of each of them, it does not provide information on how the companies can work together.

Benefits for the venturers:

- Together, the companies can work on projects that a single company alone could not do. The teaming allows the joining of free resources, skills, licenses, and so on.
- A teaming partner in a consortium may use the consortium to transfer knowledge from another one.

Disadvantages and risks for the consortium and its venturers:

- During the project, it may turn out that the contribution margin—the price paid by the customer minus the costs occurring from using the venturers also as subcontractors—is getting too small or even negative, both leading the project into a loss for the consortium and its venturers. It may then be difficult to step out of the business.
- A teaming partner in a consortium may use the consortium to transfer knowledge from another one. While this is a benefit for the receiving venturer, it may by a disadvantage for the company that gives the knowledge away. The partner in the consortium may turn up in another business setting as a competitor, strengthened with the knowledge gained.
- The consortium must meet contractual deadlines as well as scope, quality, and safety requirements imposed by the customer. If the contributing venturers do not perform as intended, the consortium cannot meet such requirements.

2.13.4 Mixed and Expanded Teaming Agreements

In reality, it is common to mix teaming agreements. A customer may have several prime contractors for different work packages in a project, who then have subcontractors. The teaming relationship between these prime contractors would then be informal. Or a consortium may have subcontractors who are not members and venturers but just selling their products and services.

Many projects today are built as complex PSNs.[79] They stretch over a multitude of tiers, and it is common that they are not engineered but have grown over time. It is also not rare that no one in the organizations involved has a complete picture of which organizations are actually involved in the project. The types of teaming agreements described above help gain an understanding of the complexity. Today, it is rarely sufficient to talk of just a customer and a contractor. Many parties are involved, and in large PSNs, the majority of participating organizations are at the same time both:

- Contractors to their customers
- Customers of the subcontractors

The subcontractors help these organizations meet their obligations, but they also require a fair share of the money that the customer is prepared to pay for the work. In a later chapter, I will discuss how to both engineer and grow the PSNs with the objectives of not losing control, deciding which organizations are actually involved in the project, and motivating those who are to build a strong teaming relationship.

---

[79] I consider the term "supply chain" no longer adequate for the intricate and dynamic networks used in project procurement today in many fields, and the complexity and the speed of change in them are growing further.

2.14 Pricing

2.14.1 The "Perfect Price" in a Non-Competitive Setting

"Customers always want to negotiate the offer made. This is so annoying. We were asked to give them our best price and our most generous payment conditions, and we did just that. Then, they come back, telling us that we are about to be selected, but open new discussions, in which they want to negotiate the price again. Why don't they simply accept what we offered?" I often hear statements like this from classroom attendees who deal with prospects in business development, and the frustration is commonly strong. I then try to convince them that this is actually good news. Here is why:

The competitive phase has been left. This means that the procurement done by the buyer uses non-competitive methods such as an invitation for quotation (IfQ) as defined above, or that a vendor selection has been finished, the vendor has been selected, but the contract is still not concluded and signed by the customer, and the customer wants to use the situation for renegotiations. There may still be many details that need to be agreed upon, and the customer may hope to bargain an even better price, a faster delivery, or other additional benefits.

Price is an important yardstick for a buyer when the vendor is picked from the options available. The survey on vendor selection criteria in Figure 2.15 (page 102) has shown that price came directly behind the understanding of the needs and wants of the customer on second place of the decision criteria used.

The complaint above on renegotiations is something I hear often from team members in bid and proposal teams and from sales staff. Pricing is a complex thing, and the complexity begins with three uncertainties (independent of the contract type, by the way):

- What will it actually cost us to perform the project for the customer?
- What is the customer prepared to pay?
- When we quote against competition, what numbers will they talk about, and by how much would we be allowed to quote higher or by how much would our price be lower?

As a reminder, competition is not necessarily another vendor; it can also be the *make* option. I will discuss pricing in a competitive setting in the next section.

In a non-competitive setting, there are different ways to look at a "perfect" price. The first is to compare the value of the business for both sides, as shown in Figure 2.23, or to be more accurate, the minimum price the seller organization is prepared to accept for its products and services, and the maximum price the buyer is prepared to pay for them.

In the upper situation, a price between US$3 and 4 million would be inside the corridor, where the value perceptions of buyer and seller overlap. In the bottom example, the business is almost impossible, because there is a gap between the maximum price that the buyer is prepared to pay and the minimum price that the seller wants. It is still not fully impossible, given the following additional factors:

- The project may be mandatory. The customer is not free to make the project or leave it, due to legal or contractual obligations.
- The seller applies benefit engineering and successfully increases the value perception by the buyer.

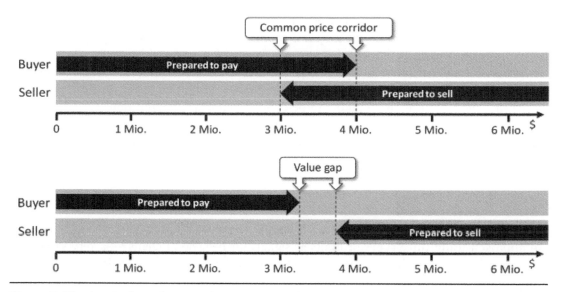

Figure 2.23   A swim-lane diagram showing acceptable price ranges by two parties. If these price ranges overlap, a contract is possible. A gap between the two price ranges makes it unlikely that the business can be developed.

- External or internal influences change the value perception by the buyer or the seller, with the effect that the value gap gets closed and a meeting-of-the-minds corridor is opened.

Value perceptions are indeed highly dynamic, as people experience in day-by-day life as much as in project management. A well-written article in a business magazine, for example, can change the value that a manager assigns to a business item and either increase its value or diminish it. Another strong factor is the unforeseeable behavior of competitors, to which an organization must swiftly respond.

As much as business would still be possible if a value gap exists, the business may also not take place in spite of a common price corridor among the parties. One reason is timing, combined with distrust: One party has to advance something, either the buyer a payment or the vendor work or a delivery, and a lot of trust is necessary for the initially outlaying party to meet the obligation. Trust can be based on the basic trustworthiness of the other party or on the assumption that the interest of the other party in a long-term business relationship is stronger than its interest in a quick win from cheating.

Experience with past business situations also influences a basic attitude on trust. When someone has bought goods repeatedly and business was conducted correctly, the person will be more trusting to a new vendor and assume that business will also be performed correctly. In contrast, a person who has just experienced being cheated will be more fearful when the next business requires advance payments, deliveries, or services. Distrust can make it impossible to gain benefits, but too much trust will inevitably lead into getting cheated. Trust is a situational task, depending on the people involved but also on the business situation.

Trust is about the preparation to advance payments, deliveries, and services. Another question is ability. Customer projects can diminish a vendor's liquidity quite strongly, when a lot of work needs to be done that will be paid for later. Excessive advances can drive a contractor into financial jeopardy, even if the project is profitable by the numbers. The project allows the vendor to bring money home, but the money comes too late to pay the rent.

As discussed above, profitability and liquidity are not the same, but they are in tight interaction. The customer is often in a similar situation, in that the burden to pay early for the project may cause the organization's liquidity to peter out. Even if the organizations trust each other, they will have to balance out payment terms as much as price.

The description of pricing using the concept of the common price corridor above is difficult to apply in practice for a simple reason: Seller and buyer do not tell each other where these limits actually are. They may give each other a number, a "best price" by the vendor or a "maximum budget" by the buyer, but it is unlikely that these are the true limitations of what the parties consider acceptable.

In my book *Situational Project Management: The Dynamics of Success and Failure,*[80] I describe an approach using three types of limits as signals for behavioral change—a very situational concept taken from quality management, where it has been in use for decades:

- **Specification limits.** Exceeding them likely leads to rejects or other kind of massive problems. In the context of this discussion on pricing, I will instead use the more appropriate term *walk-away limit,* but, in essence, this is the same thing.
- **Control limits.** Exceeding them is a signal that corrective action may be necessary.
- **Warning limits.** Exceeding these limits is an indicator that more attention is needed. In the context of this discussion on pricing, I will instead use the more appropriate term *just-buy limit,* but, in essence, this is the same thing.

An interesting approach on pricing occurs when the concept is applied to the buyer in the procurement process. As described above, pricing has two dimensions—the price quoted and the payment terms that come with this price—which define who will perform in advance and by what time.

Figure 2.24 illustrates an example of price limits by a customer for a project that is currently procured. The limits on the right-hand side are driven by the "get today, buy later" promise of protection of the credit balance. The limits on the left-hand side are driven by the desire to be low cost and make the organization profitable.

The limits delineate sectors that in turn will trigger behavior of the buyer, and depending on the business goal of the offer and the intended buyer behavior, the combination of price and payment terms should target for the right sector.

Under the buyer's just-buy limit:

- The buyer will just buy.
- Targeting this sector is favorable when:
  - A very quick and uncomplicated contract conclusion is desired without any discussions and negotiations.
  - Further communications with the buyer is impossible.
  - The seller wants to focus solely on technical aspects of the project and not on commercial ones (e.g., when the project is intended to become a pilot project for future business).

---

[80] (Lehmann 2016b, 212–214)

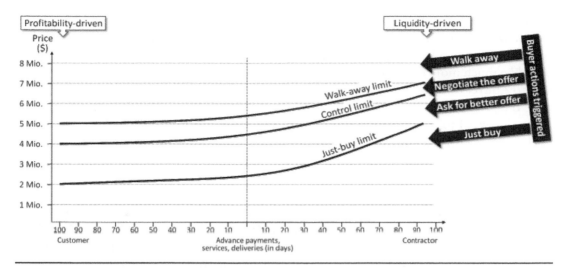

Figure 2.24 The three types of limits on the buyer side relating to price and advance performance, and what response by the buyer is to be expected for the sectors that are delineated by these limits. (Numbers shown are examples.)

Between just-buy and control limit:

- The buyer's attention to the price will increase, which leads to a request for a better price or more time to pay.
- Targeting this sector is favorable when:
  - A quick contract conclusion is desired without lengthy discussions and negotiations.
  - Further communications with the buyer is difficult.
  - The seller wants to focus on technical aspects of the project more than on commercial ones.

Between control and walk-away limit:

- The buyer takes action to reduce the price or delay the payment.
- Targeting this sector is favorable when:
  - The seller desires a contract at the highest price or with the least advance performance.
  - The seller has personnel capable of performing lengthy and often difficult negotiations.
  - The seller understands and accepts the risk that the negotiations may fail.

Above the walk-away limit:

- The buyer walks away from the negotiations.
- Targeting this sector is favorable when:
  - The seller desires to have no contract, but the buyer is too important to reject sending an offer.
  - The seller has no personnel capable of performing lengthy and often difficult negotiations and to satisfy the customer.

So, what is then the perfect price? I started this section with the common complaint of people involved in project offers that buyers so often do not simply accept the offers made to them but want to renegotiate. This is not necessarily a bad signal. If the prospect just accepts an offer, maybe with a careful request to reduce the price or allow for longer payment terms, the price may have been too low. After offer acceptance, the prospect has become the customer, and it could be much more difficult to get a better price.

If the buyer walks away, the price may have been too high. The prospect who wants to negotiate the price and the payment conditions essentially signals that the price offered is near to the walk-away limit, but still not exceeding it, and if the intention is to sell the project at the highest price, this is a good signal.

In practice, it is very hard to know where a prospect's limits truly are, and even a prospect who is prepared to answer such a question may not really know. These limits are mostly used in an intuitive fashion, not in full awareness.

I nevertheless recommend drawing these limit lines based on estimations and on the knowledge about the buyer organization that is accessible. The understanding of where these limits are will improve with the knowledge of the prospect, and one often overlooked aspect of knowing the prospect well is better pricing, which in turn leads to more profitability for vendors of project services and products. When a major number of these vendor companies are JAMs, companies that are "just about managing" financially, poor understanding of their buyers is among the major reasons for their difficulties.

2.14.2 The "Perfect Price" in a Competitive Setting

In a competitive setting, the desire is to not be outpriced by other vendors. But things are again far more complex. The cost that the buyer will need to pay will have different elements, and it may be interesting to look at them individually and then decide at which of them one considers oneself superior to competition. It is a kind of packaging, as my favorite Italian restaurant does, when the owner serves a glass of Grappa or Limoncello with the receipt to reduce the anger that some guests may have when they see the high price that they are expected to pay. Figure 2.25 shows how the price can be packaged or left just as it is.

It is advisable to have a detailed knowledge and a good understanding of the customer and the competitors to understand how a price can best be packaged. The knowledge is needed to have the correct numbers at hand, the understanding is needed to know to what arguments the customer will be prepared to listen and what arguments will be raised by the competitors. A price offer for a project can be packaged differently:

Just the "naked" price:

- If this what the seller is good at—being inexpensive for a mostly standardized product or service—and if this what the customer wants, one should focus on this number.

Costs and benefits of the usage of the project deliverable:

- The deliverables that the project is intended to make will be taken into operations by the customer and then generate running costs, possibly for years, or may become input to another project and save the costs there or bring benefits.
- Together with the benefits of the project deliverables, these costs may be more relevant for the decision.

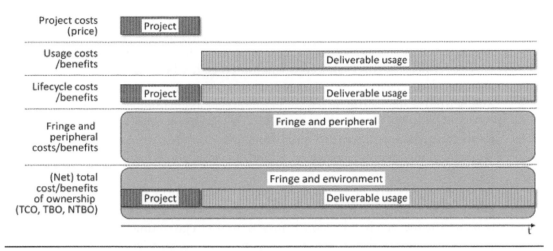

Figure 2.25 In a competitive setting, the "naked" price may be what matters most to the customer, but there are various options to package it and spruce it up with more beneficial combinations of numbers.

Lifecycle costs/benefits:

- Lifecycle costing means combining the project costs and the costs of operating the results of the project in one model.
- Project costs are one-time costs, whereas the operational costs are recurring, so one will have to model the lifecycle costs based on assumptions such as the operation time and intensity of the deliverables of the project.
- One can also add benefit calculations over the lifecycle, as shown later in the discussion of benefit engineering.

Fringe and peripheral costs and benefits:

- Fringe and peripheral costs may sound as if they are marginal, but often they are not. They can vastly exceed lifecycle costs:
  - The cost of implementing a new database with connectivity to lots of legacy systems may be high, but the much larger costs are those to make the old systems fit for this connection, particularly if their data structure is inconsistent and the quality of the data is poor.
  - It may be expensive to build a new chemical plant, but it may be much more expensive to also design and develop the surroundings of the plant in a way that avoids a glitch or an accident at the plant that leads to an environmental disaster.
  - Re-developing an organization may be an expensive and difficult activity, but managing the consequences of these internal changes in the local and organizational proximity of the organization may be even more difficult.
- The benefits for the buyer that come with a seller's offer may also lie outside the immediate realm of the project and its results:
  - A book publisher has paid a famous author a royalty for the rights to the person's book and earns money from the sales of the book. The publisher may profit even more

from the reputation of being this particular author's publisher, which then lures other authors and makes book buyers interested in trying out more of this publisher's books.

○ A chemical manufacturer may procure a plant to decontaminate and reduce waste water and benefits from reduced operation costs, tax benefits, and other incentives that come with such a plant. They may also be able to win new customers who prefer to buy from a company that feels responsibility for its environmental impact. Pointing to the investments they have made and to the effectiveness of these investments may be the final decisive arguments in a competitive field. Many managers are intelligent enough to understand that increasing profit is not the only responsibility of a corporation towards society.

○ A new business software program in a corporation reduces waiting times, when data are crunched internally before they can be presented to the user, as a lifecycle costing benefit: The user can handle more transactions over a given time. This in turn can lead to a fringe benefit: The organization becomes more responsive internally and in the marketplace, reducing frustration all around and leading to more satisfied customers and employees, with the effect of growing business. This fringe benefit can greatly exceed the direct operational benefit.

Total cost/benefits of ownership:

- *Total cost of ownership* (TCO) calculation includes all costs that are directly assignable to the project and the use of its results, with those fringe costs that are incurred elsewhere but must be paid by the same organization. *Total benefits of ownership* (TBO) calculation does the same for the benefits.
- These calculations can include many of the elements described above, such as:
  ○ The price—what it will cost the customer to do the project with the vendor.
  ○ The costs directly assignable to using the deliverables of the project, which may include further business with the vendor on consumables and service.
  ○ The directly assignable benefits from using the deliverables of the project benefits, such as additional income, cost savings, etc.
  ○ The costs and benefits that are not directly assignable to the project and its results, which the buyer would also own.

A problem for the seller is that the first number, price, is easily and reliably assessed. The seller just decides upon it. The following numbers are much more difficult to find out, as they are based on data that are hard to assess by the seller, and though they can be much more relevant to the buyer, they remain inaccessible to the seller, particularly when the buyer considers them confidential.

Net total benefits of ownership:

- The highest art in pricing is to package all these different kinds of costs and benefits into one calculation and include the price in a convincing number. *Net total benefits of ownership* (NTBO) calculation asks what net benefits the project and its deliverables will bring after deduction of all costs. If the project is mandatory, the number may actually be negative, because the project is not done to gain benefits but to comply with law or other binding regulations. If the project is discretionary and a seller would want to use this number in the offer, the seller would need a lot of quantitative data as well as qualitative information on the customer and the competitors, and then compute the numbers from that.

Figure 2.26 visualizes how simply quoting a price to the buyer is generally the easiest task for the seller, because it depends on the number in the offer that the seller generates. Moving to the right leads to cost and benefit forecasts that make the offer for the customer more relevant. These forecasts are based not only on actual data and facts, but also on estimates and other assumptions, and these all are much harder to collect.

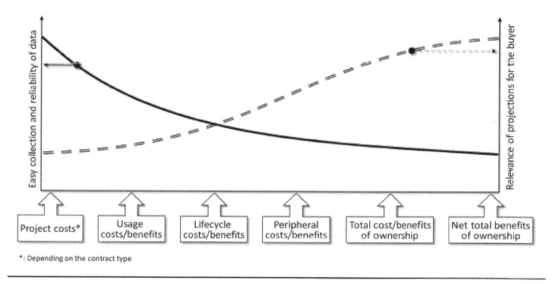

Figure 2.26   Quoting a naked price is the easiest task for the seller. Making the offer more relevant by adding further cost and benefit information is challenging, but may be the key to success.

Moving to the right may also mean an increased investment in time, effort, and costs to obtain the data, and this investment is limited by the resources that the seller is prepared to invest, in their availability, and in the business that is to be expected. One will probably be more prepared to drill very deep into data for a project with a value of $100 million than for another one with a business value of $100,000.

Estimating saves cost and time and may be necessary in cases for which data are inaccessible, but the more estimates are used, the less reliable and accurate the forecasts will be. A negative response from the buyer may then be that the forecasts are inaccurate or unreliable and that costs and benefits that they promise are not realistic.

Another negative reaction could be that the buyer considers the existence of too many and overly accurate forecast numbers intrusive. A perception that should be avoided would be that the seller used spy-like methods to tap data that the buyer considers private. It is therefore important to clearly mark all these forecasts as based on assumptions that have been derived from experience and basic market analysis and include uncertainties. If the numbers that underlie the assumptions are then found correct, it can then be taken as a signal that the assumptions were accurate. This approach also helps if the forecasts prove wrong.

As a seller, one will then ideally address which of these packages for the price are the most convincing for the offer, based on one's strengths and on the desires and wishes of the buyer, and then try to direct the buyer's attention to it. If one can just deliver cheaply, and the customer desires a low price, this is the perfect one, the one that should be offered. If the price is not the lowest but looks friendlier in a context with other cost and benefit data, such as in lifecycle costing or in TCO calculations, the seller will prefer to direct the buyer's attention to them.

Another limitation of the data and knowledge that a seller can develop about the buyer stems from the informal character of the relationship before a contract has been concluded. The buyer has no obligation to provide the seller with sufficient information, and as gathering, processing, and communicating this information on the buyer side binds resources that may be needed in other tasks as well, the seller is finally dependent on the good will and intentions on the buyer side to share information. One may argue that it is important for the buyer to make sure that sellers are fully informed when they are asked to submit offers, but the buyer's organization may have other, more urgent obligations, and the urgent is always the greatest enemy of the important.

I generally recommend for all binding offers to add a clause similar to the following: "The price, fees, deadlines, scope of work, commitment of resources, and all other considerations (obligations) in this offer are valid to the degree that the information given by Buyer is correct and complete. Information that Buyer makes available to Offeror only after submission of the offer and that is relevant for defining the mutual obligations invalidates this offer".[81]

2.14.3 When the Offer Is Too Low

Sometimes, buyers have another limit: when an offer seems too good to be true—when buyers have the impression that there is a catch in the offer and that a seller may finally rip them off, or that the offer is rather low-balling to get more than a foot in the door and then make the best out of the buyer's dependency on the seller. In Swabia in south-western Germany, the area around Stuttgart where I was born and have grown up, people often say, "Buying cheaply means having to buy twice". Some say instead, "I am too poor to buy cheap things".

A price that is far below a cluster of prices from other vendors could be taken by a buyer as a signal that something is wrong with this price. A low-balling vendor is a possibility, a lack of understanding of the requirements another. A vendor who does not understand what it takes to do the project will later become a problem. Either the vendor is unable to do the job and all consultations turn out after a while to be wasted time and money, or the vendor will come back later to renegotiate price, fees, and deadlines. Actually, I met purchasers who follow a strategy to not award the business to the vendor with the lowest price, but the second lowest, to avoid such risks. So, being the cheapest vendor may not be a promising strategy at all.

A good account manager or sales representative may be able to obtain from the buyer information about how numbers are perceived. Experience from earlier offers may be also helpful. It could also be that the person who is in charge of procurement is known from a previous job in another organization. Just when it comes to pricing, it is in the vested interest of the vendor to make as much information available as possible. A buyer may support the seller in the endeavor

---

[81] This is advice on contracting to the degree that it relates to project management under contract. For the correct wording of such a clause and the possible consequences in the applicable legal system, it is necessary to consult a lawyer.

to know more, or may prefer to keep it unclear to the seller, depending on the perception of the process as a kind of poker game or as an attempt to find a capable partner with whom one wants to develop a relationship based on mutual trust.

2.14.4 The Customer Dictates the Price

This is another purchaser's strategy. The buyer has a set of fixed numbers for price and fees, approaches a vendor who is considered trustworthy and capable, and if the vendor accepts the numbers, this company will get the job. The buyer's main desire in this approach is predictability of the costs of the project, more than finding the cheapest seller, and it saves the buyer from performing a costly and time-consuming competition. Vendor selection can actually be done in a fraction of the time that it takes to go through a competitive tender, bid, or proposal process, and the final price may even be cheaper.

For a vendor, the situation bears some risks when in internal discussions sales people clash with execution people over the decision to go for the business or not. The first want the business at any price, in order to win the customer and become an incumbent, and the second doubt both that the project can be done for that price and that the incumbency with this customer has so much value. The first—correctly—point to the business being a safe bet, because there is no competition; the second may—possibly also correctly—consider the business a loss and the buyer not worth the effort. The strategy of approaching just one seller with a pre-determined price can put this seller under tough pressure, and the risk of an unhealthy decision is high.

2.14.5 The Absolutely Last Price

Buyers tend to ask sellers to give them their lowest price right at the beginning of the negotiation process. Sellers expect that even when this number has been named, there will be further negotiations and will therefore state a higher price. Buyers assume that there is still some leeway to further reduce the "best price" given by the seller and start negotiations again. This may spiral the price further down as long as the seller allows or until the buyer considers the negotiations as finished, with or without a result.

A question that I am often asked by students is about the moment that a seller should select to quote the very last price: the one price (or fee, depending on the type of future contract) that is the absolute minimum for the vendor to quote, that the organization would avoid quoting at almost any price, and that describes the red line at which the business becomes uninteresting for the vendor. As a rule of thumb, I recommend to never state this number at the negotiation table. The negotiation table is a place for competitive behavior, and when the vendor quotes the very best price, this should be done at a moment when competitive behavior has failed and the buyer seeks cooperation.

Personally, I had good experience with the exit door. The negotiation was considered ended and failed by the buyer's purchasers, because it was their desire to come out of the meeting room with their job done, and instead they will have to undergo further negotiations. The lust for further negotiations is gone, and a feeling of realism creeps in. Just at the door, on the way out of the meeting, I turned around, seemingly contemplating over other options, and then told them: "I am just wondering . . . I have a little marketing budget that I could make available for our business".

I then observed their reaction: did they signal interest and relief, or did they consider the negotiation finally finished? If they signaled further interest, I offered them my very last price in exchange for something that costs the buyer not much money but has a high business value for the vendor. A recommendation letter is a good example of that, or a joint press release or article in a special-interest magazine. Something that helps the vendor sell in exchange for a very special price. Something that builds on the good will of both parties, that can often not develop at a negotiation table, where both parties are restricted by the constraints of their roles.

2.14.6 "Mission Success First" in Negotiations

Negotiations can be a stressful time, when each party tries to squeeze a maximum out of the needs of the other party. Negotiations are then like a tug of war, which is won through a combination of physical strength, stamina, concentration, grip of shoe soles on the ground, and—when larger groups pull on each end—also team function.

Negotiation can be done in a different spirit: A mutual "Mission Success First" attitude can shift the focus from the question, "What's in it for me?" to the question, how one can serve the other party best, in exchange for gains that one may get. This turns the parties into partners, putting completing over competing and taking the distress out of the negotiation situation.

One side in the negotiation will offer such a positive exchange of benefits by actively showing interest in the wants and needs of the other one and making offers based on them. If the other side responds in kind, the mutually beneficial negotiation can develop. It is then much less like a competitive power game and more like two persons building a team to achieve common goals. Developing a common "Mission Success First" attitude requires empathy and understanding between the future partners and the desire to get the maximum benefit, not simply the cheapest solution, out of the relationship.

This entire book is written around such a "Mission Success First" attitude, which helps ensure collaboration and communication in teaming organizations. Trying to implement it during negotiations has another benefit: Contract parties in projects often find it difficult to switch from the competitive behavior that mostly dominates business development to the collaborative behavior that is needed after contract award to perform a successful project. If the collaborative behavior is already applied before the contract is signed,[82] it is not necessary to switch behavior, and the joint working style can be implemented as a matter of course.

2.15 Writing Complex Proposals

2.15.1 Where Do You Place the Most Important Contents of the Complex Proposal?

Proposals and, to a lesser degree, tenders and bids can be long and complex documents. Several aspects must be balanced out, including the level of detail that the assessors of the proposal

---

[82] There may be legal matters to be considered, particularly in public contracting, where competition is regarded as a means to save costs for tax-payers.

on the side of the prospective customer need to make a decision in support of the proposal. The proposal may follow a statement of work from the customer and needs to show that this document has been fully read and understood and that the concerns in it are all addressed in the proposal.

On the other hand, the seller may want to make reading the document not too difficult and time consuming and will also consider the amount of information that is given away to the prospect. The major body of the proposal is dedicated to the buyer's employees who assess the details, but these may not be the people to finally make the decision.

Complex proposals are commonly introduced with an executive summary, which describes in a small number of paragraphs the central arguments for the proposal that will then follow in detail on the next pages. The generally correct assumption is that decision makers do not have the time to read the proposal in detail and need some quick help to understand its focal points. The approach is based on the assumption that they may delegate the analysis of the details of the offer to employees and will read in person only the summary. For some managers, this assumption may be true, not for all.

One problem may be that sometimes even the executive summary is too long, or that some executives may not want to read text that consists more of self-praise and marketing lingo ("Gobbledygook") of the seller than of the information that the buyer needs to make the decision. Observing this group when they assess proposals is interesting.

A common practice is to first look for the price, which is assumed to be near the end of the proposal. From there, they scroll backward through the document looking for something that catches their attention, and if they find nothing, they will then directly hand it over to their experts or place it on the stack with the competing proposals.

What may capture their attention? Text boxes with short text snippets in large fonts, of course, possibly also tables, but definitely images. This is related with the (simplified) model of the brain as made of two hemispheres. The left one deals mostly with words, texts, and other logical things, the right one with perceptions of images. Delivering information to the right hemisphere is quicker and less arduous for the recipient, but the information is less concrete. So the person may then read the caption of the picture to understand what it shows and how this is linked to the proposal.

Speaking captions, which repeat argumentation made in the text in brief form, so called *action captions,* may be the only pieces of text that decision makers read at all. Many images in this book also work as action captions, with which I try to convince prospective buyers of this book that it is worth buying. The opposite of the action caption is the *horse title,* which stands for an image of a horse with a caption saying something like: "Image_3.2.16_Horse".

2.15.2 Familiarity of the Audience

The public transport system of the municipality of Munich, the city where I live, has about 200 different fares, depending on distance that one wants to travel, including differences for adults, children, students, or seniors. It has fares for the trips, the number of trips, weekly or monthly cards, and so on. The tickets can be bought in person at kiosks and vending machines or digitally using smartphone apps.

Munich has a complex system with rings and zones that can make it difficult to find the right price for the trip that one wants to make. The system is run by a syndicate of local transport providers called *Münchener Verkehrsverbund GmbH* (MVV), which from time to time conducts surveys among its customers to determine how satisfied people are with their offers. The response is generally very positive, and the MVV commonly communicates these positive results as press releases and online. In particular, the vending machines are often praised by customers for their perceived usefulness, as they give access to all these 200 different fares so that every customer can find the best one.

In July 2014, in the context of a research project, I studied the application of the *Technology Acceptance Model*[83] for the ticket vending machines at Munich's light rail stations. The Technology Acceptance Model states that three factors influence the degree to which users will be prepared to accept an unknown technology:

- Perceived usefulness—can I achieve my complex objectives when I use the technology?
- Perceived ease of use—can I adopt the technology without major difficulties?
- Behavior intention—does the technology help me achieve the objectives and goals for which I am using it?

To better understand how they worked for different user groups, I defined three clusters of passengers and asked each passenger or passenger group to which they belonged:

- First-time users—travelers who have not used the Munich ticket vending machines before.[84]
- Rare users—regular passengers who use the vending machines less than once a month.
- Frequent passengers—those who use the vending machines at least once a month.

I measured for each user the time it took from the first physical contact to the moment when the machine dispensed the ticket and, possibly, small change, before I asked for further information. It took the frequent users on average 30 seconds to buy the ticket from the machines; most of the time was spent either sorting the coins needed or processing credit or debit cards. It took rare users on average 1:10 minutes, and first-timers on average 4:00 minutes. Three first-timers gave up; they could not figure out what the correct fare was and how to buy the ticket for that fare. I do not know whether they took a taxi instead or took the light rail without paying (Munich has no barrier system that prevents from that), but in any case, they were lost as paying customers for MVV, after having blocked the vending machine for a lengthy period of time.

Why does the city of Munich communicate high rates of satisfaction with the vending machines, when a group of customers has so many difficulties using them? People from this group do not turn up in the surveys conducted by MVV. The surveys focus on regular customers. Munich has many visitors, and understanding their needs better would probably further increase its reputation as a destination for city travelers. I consider it no good idea to frustrate them this way.

A similar situation can be found when we are writing complex proposals. Are we writing for those who are familiar with the topics addressed, or rather for novices? Can we expect readers to be experts, be accustomed to the terminology used and the concepts underlying the argumentation, and be able to understand the metrics applied and technical data given? Are

---

[83] (Davis 1989; Hess, McNab, and Basoglu 2014)
[84] This group would also have included locals who used the MVV ticketing machines for the first time, but this group did not show up.

we convincing the readers with a variety of options from which they can chose, so that they are finally able to select what they actually need or want and configure the project and product that suits them best? Or should we avoid overwhelming readers with details that they do not understand, numbers that they cannot put into context, incomprehensible special lingo, and choices that make them feel lost between equally attractive solutions?

Sometimes we have to write a separate section for each group, and may even have to create an "in-between" category, as my observations at the ticket vending machine show, and we may need to provide clear entry points for each of these groups, so that they understand which part of the offer is written for them. One reason may be that we do not know who will read the proposal, particularly if the process is done online; another reason may be the presence of a mixed group with both experts and laypeople. It may then be a good idea to mark text sections as dedicated to experts and others as moderately difficult or easy reads. The idea is always the same: not to frustrate the reader with text that is either incomprehensible or lacks the desired accuracy and depth of information.

I have found that organizations that make such a separation of groups of audiences are more successful. One of my favorite examples is a pizza home delivery service, which has two entry points in their online order service: One allows customers to order a "classic" pizza—that's the quick and easy way for the beginner, simply selecting a "preconfigured" pizza—or, for those who want to configure a personal pizza, there's an "expert's mode" which is open to all options and choices that the delivery service has on offer.

Aircraft manufacturers do precisely the same: An Airbus A380, the largest passenger aircraft in the world, can be individually configured, which is a project with seven engineers working for a year. An airline may instead choose to order a standard configuration, which is only special in branding, such as colors, logos used, and similar small adjustments. The preconfigured aircraft is cheaper, and the process of ordering and manufacturing is faster, but it will not have special attributes or features that set it apart from the aircraft of competitors.

As with the pizza vendor or the aircraft manufacturer, I recommend providing two or three entry points for readers of proposals: one with a focus on perceived ease of use, the other with a focus on perceived usefulness for the expert, and possibly a third in between. The topic here is not *technology* acceptance, it is *proposal* reception, but the underlying principles are the same.

2.15.3 Behavioral Expectations by the Buyer

When developing the offer, the seller should have a clear understanding of the expectations that the decision makers have on the buyer side on the kind of leadership behavior by the seller. Figure 2.27 describes how a seller is expected to propel change in a customer environment—transactional or transformational—and on what level this is expected to happen.

Without such a clear picture, misunderstandings can lead to losing the business opportunity, or to winning it and failing to meet customer expectations and finally losing the project.

2.15.4 The Friendly Dog Effect

Youth is the only problem that resolves itself over time. As a young project business manager, I made a mistake more than once, until I learned my lesson and made sure that I was talking

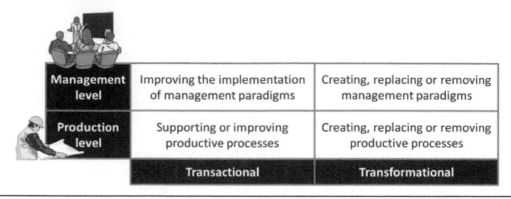

	Transactional	Transformational
Management level	Improving the implementation of management paradigms	Creating, replacing or removing management paradigms
Production level	Supporting or improving productive processes	Creating, replacing or removing productive processes

Figure 2.27   Four quadrants describe whether a seller in a customer environment is expected to transform the environment or to act within it, and on what level this should happen.

with the right person on the buyer side for development of the business. This is probably the person who is most difficult to arrange a meeting with—the person who has a lot of knowledge of the organization, its strengths and weaknesses, desires and fears, and either makes the critical decisions or strongly influences those who make them.

Families know the effect when they decide to get a dog as a pet for the children, and to get it from the place where pets are found that need a family—the animal rescue. When the family visits this animal sanctuary, they will be approached by some dogs that are exceptionally cute and friendly, and with every glance of their eyes, they seem to say "please, take me with you".

These may just be the wrong dogs. They may be the ones that chew on carpets and socks until they need replacement, playfully bite the children (who do not consider this fair play at all), and may be infested with parasites such as ringworm—which can be transmitted to humans; and while these skin infections are not dangerous, they are hard to get rid of, and one does not want to have them in the home.

So while the family was so much lured by the friendly and extravert behavior of this dog, another one may be far more suitable, but this dog does not approach the visitors but stays calmly in the background and remains unseen.

In large organizations, a similar effect can often be seen. The visitor from the vendor company is welcomed by a friendly person who listens to the full sales argumentation, watches the presentations shown, and promises all-over helpfulness.

But this may be the wrong person. The person is pleasant, likable, and an easy contact partner to talk with, but is not in a position to make the critical decisions. No one else in the company listens to this person, and this is why he or she talks to the stranger, who has an open ear and is grateful for any help offered. The person may have time because he or she does not get too many challenging tasks, based on the assumption that this work would not be in good hands. The person may promise to be a great door-opener for the vendor but does not have the keys to open the right door.

One of the main concerns for a seller when trying to find entry to a prospective customer organization is to separate the decision makers from those who are, in the end, not much more than a comforting waste of time.

2.16 Non-Disclosure Agreements and Non-Compete Clauses

Some years ago, I was invited by a major multinational corporation to offer my services for a qualification project with 80 project managers involved. The invitation had a document attached entitled "Non-Disclosure Agreement" (NDA), which I was required to sign and return. Only after that would I be allowed to phone the company contact given to schedule an appointment. The company's message made it very clear that without the valid NDA, there would be no initial meeting and no business. For a self-employed trainer, who needs to fill a schedule with training assignments and still has free slots, the decision to reject the NDA and with it the business with the customer is a tough one. The NDA had seven pages written in legalese that was hard to comprehend for a non-lawyer, like me and most people I know. It was made enforceable by an amount of roughly US $120,000, which I would have to pay to the company if I were to infringe the agreement.

Such NDAs come with a sequence of difficult decisions for a potential contractor. First, there is still no business with the buyer, but one is first required to accept risks that may be considered negligible for a major customer but are significant for a self-employed person. It may well be that the business will never occur—something more likely than unlikely, as described above—but the risks would nevertheless persist.

A second consideration was whether I should consult a lawyer to get a professional opinion. The fee that I would have to pay to the lawyer would be my first investment in this customer. I accepted that cost, and the opinion was worth the money. In essence, the clauses would have prohibited me from successfully defending myself legally if I were falsely accused of a breach of confidentiality, because such defense at court could be considered further breach of confidentiality. The applicable law of the agreement would not be German law, the law that I know best. It was a foreign legal system, which would cause additional uncertainties and might make it finally impossible to defend myself from false charges.

For a qualification project of the enquired size, I would have to subcontract assignments to trainer colleagues, which would add more risks, as I would have no control over their behavior. A competitor might fake allegations against me, and I would have difficulty defending myself. And so on. The risks that came from the NDA would jeopardize my professional existence dramatically, and my first thought was to simply reject it.

I then decided to respond in a different way: I changed some sections in the document, making the NDA a mutual agreement, in which the customer protects the confidentiality of the contractor as much as vice versa. I changed the applicable law to German law and reduced the liquidated damage amount to $60,000, which would also apply mutually. Then, I sent it to the customer as required. They had told me to return the document, but not to do so without changes, so I met the requirement. Some days later, I phoned the company. In this first conversation over the phone, the woman I talked with confirmed the receipt of the NDA, but it seemed that no one had so far opened the document. We then appointed a meeting. Two days before that happened, I was informed that someone had opened the NDA and was angry at me. The company still did not cancel the meeting but told me to be prepared for a difficult discussion. I took this as a signal that they were still interested in buying my services.

During the meeting, the person responsible for purchasing training services showed me how dissatisfied he was with the changes that I had made. The discussion then became a lesson in the "Golden Rule",[85] and on the reciprocity of obligations and rights that form the basis of a partnership among equals. We also discussed how a major company can easily work under different legal systems, but a self-employed trainer cannot. We ended with a document that had mutual obligations that were not enforceable with a liquidated damage threat or a penalty under German law. The business was easy to win, because all other vendors had dropped out due to the NDA. We then had a great staff development project, based on mutual trust and respect.

From a buyer's perspective, entering a process to find future contractors bears natural risks of mishandling of confidential information. This information may become accessible to third parties, or the seller may use the information to compete with the buyer. Descriptions of a buyer's wants and needs regarding the desired products and services allow identification and analyses of the company's immediate plans and long-term strategies, but also of its weaknesses and the threats it is encountering. Names and contact data of employees on the buyer side can become effective targets of head-hunting measures. Because many sellers also serve the buyer's competitors, forwarding such information may not happen due to bad will, but may be caused by negligence or coincidence.

The seller has the same risk: The buyer may use the information passed on in the offer to boot out the seller and use the knowledge to self-make the service or product that was originally enquired for purchase. The buyer may also transfer the knowledge to another, preferred seller to enhance this company's competitiveness. In an even worse case, the buyer may use the knowledge acquired from the offer to develop new offerings and directly compete with the seller.

To protect from such risks, organizations have such NDAs, also called *confidentiality disclosure agreements* (CDAs). There are various other names, but the purpose and the basic setup are mostly the same: One or two parties (if only two parties are involved) agree to keep secret the other party's or parties' confidential information, to which they get access during a developing or existing relationship and for a given time period after the end of the business relationship. Sometimes more parties are involved in a multilateral agreement; then several or all of them may guarantee this confidentiality to others. The idea of an NDA is to allow one partner in the business relationship to talk openly to the other about confidential issues without having to be afraid that this confidentiality will be broken by the other party. Such issues may include:

- The business relationship as such. The future, current, or past business partner may, for instance, not want the other party to use the name and logo on a reference list or in a cloud of logos used on a presentation slide.
- The broad or specific contents of the relationship; for example, the products or services that are to be delivered.
- The names of people involved on the customer side—they may be approached by head hunters.
- Business secrets, technical secrets, and similar confidential information.

To be fully binding, effective, and enforceable, non-disclosure agreements must be contracts by nature. They are not project contracts that will only be established later in the process, but instead preliminary contracts with often much lower value; but while they are intended to reduce risk for one party, they may increase risk to the other one, as the case story at the beginning of this section showed.

---

[85] "Treat others as you want to be treated".

My recommendation is definitely to either have no NDA at all; have it as a non-binding, diplomatic document with mutual responsibilities; or have it as a mutually obligating contract. On top of the "Golden Rule" argument, the mutuality is necessary in some jurisdictions to consider it enforceable, if indeed the NDA is considered a contract. An agreement that obliges only one party is in such countries considered "gratuitous" and therefore invalid, or at least unenforceable, because it lacks *consideration*—obligations on both sides.

Often, parties leave unclear which information is considered confidential and which is not. The document should clearly state the conditions that make information confidential. The clearest way to achieve this goal would be to refer just to documents that are marked as secret, confidential, etc. For some parties, such a clause may be too restrictive, so a wider reaching definition may be found, but this should nevertheless be clear when defining the protected information.

Having signed an enforceable NDA as a seller, I recommend documenting everything that may help you defend against allegations in a possible court case, including:

- Information published by the buyer and the buyer's employees in magazines, social networks, and other media. This documentation may help you make your point in case you have to verify that certain information was not treated as confidential by the buyer.
- Breaches of confidentiality by the buyer to your disadvantage. In case the buyer raises claims, this will allow you to defend yourself by raising counterclaims.
- Documentation of infringement of the buyer's confidentiality by a third party. In case of claims by the buyer against you, this may show that the claim is unjust, because other parties after similar actions have not been sued.[86]

2.17 Submitting and Presenting the Offer

2.17.1 The Most Fundamental Consideration for the Presentation

My customer Chicken Flea GmbH[87] was perfectly prepared to present their proposal for the customization and implementation of an issue management and tracking system at Earthworm S.A. The company had spent a significant effort in the short time of two weeks analyzing the needs of the customer for such a system, and on what would be necessary in regard to customization of functions and features to adapt the software to the specific needs of the buyer. They had interviews with the key people at Earthworm to ensure that they would support the software implementation actively by communicating the change to employees and by providing resources.

Chicken Flea's presentation team members made sure that they knew the details of the presentation environment, so that they had the right materials with them, including their own data projector, which was brand new and better than those in Earthworm's meeting rooms. In addition, they provided Earthworm with some expensively printed and bound copies of the proposal to make a positive impression and had an A1-sized poster made that included the basic concepts of the proposal as an easy-to-understand infographic. Because the pricing of

---

[86] This defense will not work in all jurisdictions.
[87] I changed the name and some details of the case story to protect my customer.

Chicken Flea also seemed reasonable, they considered the chances good to win the business with Earthworm.

I was invited to attend the meeting in order to give me some insights on Chicken Flea's business that would help me adapt seminar contents to their needs. During the meeting, Earthworm's executives raised many questions that related to their own uncertainty as to whether the proposal addressed their actual needs. This was not meant as criticism against Chicken Flea's offer, they were actually uncertain. They had never used an issue tracker before and were not sure how to make best use of such a product and how to integrate it into their processes and working style.

In response to this uncertainty, the Chicken Flea people pointed to their vast experience in implementing and adapting such software for customers and insisted that they had already investigated what the customer needed and would respond to these needs with a standard approach that had been found to work with other customers. Chicken Flea also emphasized the need to ensure the integrity of their internal processes and how important it was that their methodology would be implemented (the one on which I was about to train the company's project managers).

During the meeting, I had the impression that it was it was turning adversely for Chicken Flea, and this impression was right. I later had an opportunity to talk with Earthworm's executives, and they confirmed my impression that Chicken Flea's presentation came over as arrogant and disinterested in their concerns and worries. Chicken Flea (my customer, with whom I had experience) was actually great at resolving such customer problems, but at least during the presentation that I had attended, its employees were unable to give the customer the peace of mind that they would do that.

The decision a buyer has to make in hiring a project contractor is different from a standard product order at an online web shop or from a physical supplier. There, a product is ordered, and once the ordering and payment process has been completed, the business relationship is over. There may be some long-term commitments, such as complaints, warranty, and service needs, or software updates from time to time, but the essential business in most such cases would be considered finished.

The decision of the buyer in a contract developing process for a project, and particularly during and after an offer presentation, is based on several objective and subjective questions, such as understanding of customer needs, price, and the capability record of a seller. On top of this is a more delicate, often unspoken, and even unconscious decision criterion—the question: "Do we want to bind ourselves with these people in an agreement for long-term cooperation?"

Many factors may influence the answer to this question:

- Rapport, sympathy, and chemistry
- Togetherness
- Shown openness of the seller's people to the concerns and objections of the buyer's employees
- Business risks for the buyer that come with the seller

The question for the seller should therefore be: "How will we make the buyer's staff want to work with us?"

I would like to add to this contemplation that as a seller, one should ask the same question: "Do we want to work for this buyer in a long-term cooperation, under a contract that will not allow us to just walk away when we find the relationship more damaging than satisfying and profitable?" The offer presentation is a moment at which both parties should decide if they want to work with the other party, and if they decide that they do, should the do their best to also win the sympathy of the other party.

2.17.2 Preparing the Offer Presentation

The success of offers and associated presentations are too often risked by seemingly minor blunders that devalue all the time and effort invested. A case that I know was a group of young professionals who wanted to present a software solution for team communications and collaboration in an organization that was more traditional and whose meeting room equipment was not the very newest. The meeting room had an older but still powerful data projector that worked only via a VGA cable, and the presenters had a tiny modern laptop PC without a VGA connection. They had no adapter available, and the prospect did not have a suitable one either; their corporate PCs were older, larger, and all had the VGA socket installed. The version in which the presentation file was made was then too recent for the software installed on the PCs of the buyer, so it could not be simply shown from a data stick. After a while, a solution was finally found, but a lot of time was lost that could have been used for dialog about the offer, and the impression on the buyer side was that the vendor was poorly prepared.

There are many questions that one should ask if one wants to use the buyer's location and equipment for presentation, including:

- Aspect ratio and screen resolution of the buyer's presentation equipment.
- Connectors used at the location, such as RGB, HDMI, or even wireless.
- Are you expected to use a projector, an LED display, or will you do this online? You cannot use a laser pointer on an LED screen, and it also will not work on a silver lenticular screen used in some presentation rooms.
- Should you bring your own presentation equipment to the session?[88]
- If one has a video with sound as part of the presentation, what is the availability of audio equipment, and how does the PC connect to this equipment—over a separate audio cable or via an inbuilt channel in the video cable?

For presentations with audio, such as interviews, movie clips, or a welcome message from the chairman", I generally recommend having a wireless active speaker at hand that can be placed inside the meeting room near the audience, if the customer has no audio equipment at the location of the presentation. PC speakers are rarely loud enough, and during the presentation, they generally point in the direction of the presenter, not the audience.

To make the presentation stand out among the other presentations that the buyer will receive, and to make it memorable for the audience until the contract is awarded, it is necessary

---

[88] For such opportunities, don't save money on the quality of your equipment, but consider mobility and noise generation of the equipment. Good projectors commonly come with good speakers too. They are a practical solution to reduce the number of items to carry to the session and the clutter of cables.

to almost "hijack" the presentation rooms temporarily. Here are some examples of items that are useful for such a purpose:

- Expensive paper blocks and pens for notes that the listeners can use later use as memory helpers.[89]
- Large posters or roll-up displays with the core arguments of the offer packed with information that is of general interest.
- Door hangers, saying "Presentation, do not disturb" or similar that are placed outside the presentation room. They are offered by many inexpensive digital printers and can be used to protect the meeting from disruptions. Attendees often reuse them after the session.
- Handouts with core information of the offer in packaging that conveys a sense of value with a register that makes details easy to find. The offer, the handout, and the presentation should be aligned in structure and appearance.
- A USB stick or a similar item containing the data, which the audience can reuse later for other data. Alternatively, a download option from a publicly accessible website serves the same purpose.

Selecting such items must be balanced. They should communicate the worthiness of the vendor and the offer, and at the same time elevate the presentation experience, but one must avoid the perception of bribery. If the items are useful for the presentation, there should be no discussion on them.

When you have video sequences in the presentation, make sure that they are technically professional. Loud wind noises during the greetings from the seller's CEO make the message hard to understand and the presentation appear unprofessional. Video professionals are expensive people to hire, but they arrange light, background, sound, and many other details that amateurs may miss and that separate a professional video from a dilettantish one.

A boilerplate of images and videos for use in such purposes can be helpful, but a membership in a photo archive is also helpful. They have great functions such as keyword search and light boxes that make selection easier, and the photo material is definitely of professional quality.

The selection of the speaker for the presentation is also important. The presenter must be able to communicate to a group, balancing professional self-assurance against the perception of uncertainty or arrogance. The person must be able to speak free from stage-fright, as a nervous voice can damage the credibility of the seller and the offer. Depending on the will of the seller to invest in the presentation, it may be beneficial to have a line-up of experts who can answer other, more detailed technical, commercial, and organizational questions.

The seller's representatives should have a good sleep before the presentation. Recent research has shown that people who have been deprived of sleep are still able to sense raw emotions in others, such as fear or rage, but facial expressions that signal sadness or happiness may no longer be interpreted correctly.[90] To ensure responsiveness of the presenters, this ability is

---

[89] An alternative option is a footer section on each handout page with space for notes. These have the benefit that the notes are written at the most useful place, directly under the text to which they relate.

[90] (Killgore et al. 2017)

essential. After a long discussion, it may be also difficult to remember what was discussed and agreed upon, even if minutes have been taken during the meeting. It may in this context also be interesting that lack of sleep reduces the ability to memorize details.[91]

Most presentations take place at the buyer's location. A better solution for the presentation may be to have a presentation room at the seller's facilities, which should then be well equipped for the purpose of convincingly selling a solution. If the buyer comes visiting the seller, this will allow the seller show the buyer their premises and has the advantage that the event takes place on the seller's home turf.

The presentation should generally be developed by a professional who has mastered the presentation software, not by a layman. It is often annoying to see presentations with great content devalued by poor layout, unaligned bullets and indents in lists, a lack of images that help the audience conceive the message, and slides cluttered with vast amounts of text in small font size. Another common mistake is illegible text—for instance, black text on dark blue background. One must also consider that text that is easy to read on the PC screen may change color and luminance on the buyer's presentation equipment, and what is easy to read at home will no longer be that easy during the presentation.

It is advisable to limit the amount of text presented to the audience. People cannot listen and read at the same time, so if there is too much text, the listeners will need to decide if they want to listen to the spoken explanations or read the presented text. Spoken text and photos or graphics are a better combination.

Before using accessory items such as laser pointers, one should make sure that the presenter is practiced in using them, so as not to disrupt the presentation by uncertainty about how to use the equipment.

It is generally recommended to have a separate computer at hand for presentation purposes that does not have business data on its disc. This helps protect the data of the presenter's organization and its business partners. It further avoids nasty situations in which, by coincidence and neglect, such data become temporarily visible to the audience over the projector. On this PC, all disruptive functions should be switched off, such as desktop notifications of news feeds or e-mails.

It is recommended to check the presentation at least twice for spelling and grammar errors. Using an external copyeditor, ideally a competent professional from outside the business, is preferable, as some errors are commonly overlooked by reviewers from the subject matter. Among the grammar blunders commonly found in presentations are false apostrophes[92] or multiple exclamation marks.[93] One should remove them from the presentation before they can spoil an otherwise positive presentation.

---

[91] (Kuhn et al. 2016)

[92] Sometimes called the Greengrocer's apostrophe, such as when "Granny Smith apple's" and other goods are offered, using a false apostrophe in "apples", which indicates possession instead of a plural. Someone may be in the audience who considers this unprofessional.

[93] British author Terry Pratchett wrote in his novel *Eric*: "'Multiple exclamation marks', he went on, shaking his head, 'are a sure sign of a diseased mind'". Someone in the presentation audience may know the text passage.

2.17.3 Preparing for Q&As

Good project management should always be a good combination of preparation and improvisation. An essential part of such preparation is to prepare for improvisation, which means to have the resources in place that allow immediate adaptation to new tasks and requirements that turn up. In a presentation, these may occur at any time, but they are most likely to come up during a question and answer session.

A great backup is to have additional slides in the presentation that would normally not be shown, unless a question is asked that they can help answer. They may include numbers or graphics that help explain a difficult point. The slides may be placed following the final slide of the presentation and would only be opened if needed, or have a setting "hidden" in the presentation software and would be made visible when needed.

It may also be helpful to ask for a flip-chart, whiteboard, or similar to help visualize concepts when answers need to be developed ad hoc. Certain questions are generally likely to be raised, and one should be prepared to answer them:

1. **Operational disruptions.** To what degree will the seller disrupt the buyer's operations, and what will the seller do to reduce the impact of these disruptions?
2. **Need for resources provided by the buyer.** What resources will the buyer need, and how long will these be blocked from doing their normal work?
3. **Management attention.** What seller-side management attention will the project have, and how will this help the project meet the buyer's requirements?
4. **Cancellation terms.** What are the various "points of no return" for the buyer that make it expensive, difficult, or impossible to step back and terminate the project.
5. **Assurance.** What guarantees will the seller give that the project will be performed as promised?
6. **People.** Who are the professionals that the vendor intends to use for the project? A brief CV with experience, credentials, and a photo are what most buyers expect. A profile in LinkedIn, Xing, Viadeo, or a similar professional network could also be beneficial.

Depending on the offer and the wishes and needs of the buyer, more questions are predictable for the presentation team, and even if they are addressed in the prepared presentation slides, the buyer's employees may bring them up again and ask for more details or simply for another confirmation. The better the team is prepared for questions, the smoother the presentation will take place and the higher the likeliness of success will be.

2.17.4 Some More Don'ts of Offer Development

The offer may be a bid, a proposal, a pitch, or any other kind of offer; the following rules should apply to avoid misunderstandings.

- When you are making an offer in a foreign language, don't use translation software. The results will probably be too erroneous to reflect the energy, time, and professionalism that you put into offer development. The impression given to the buyer will be that of a blunder, independent of the technical, or otherwise, quality that your offer has in your own language.
- Take care with measurement systems. In a car, 40° Fahrenheit (4.5° Celsius) means that one needs heating; at 40°C (104° F) one would want to have the air conditioning

to cool the cabin. Misunderstandings on different measurement systems cost NASA a $125-million mission to Mars in 1999, and can also lead to losses in offer management and in project delivery.

- Acronyms can be dangerous. In UK, the acronym BS commonly stands for British Standard, in the US, it has a more derogatory meaning. Never assume an acronym is understood and that the buyer and the seller have the same understanding.

2.18 The Contract

The contract is the ultimate trophy of business development. It turns the buyer into a customer and makes a contractor from the seller. It is the starting point of the customer project on the contractor side, at least in theory; many projects are already active at this point of time.

Contract signature fundamentally changes the relationship between customer and contractor. The parties have stepped into a legally enforceable bond with mutual dependencies. Not meeting obligations by either party may have significant consequences for the other, and both are facing the possibility of the other party becoming insolvent, which means that the contractor can no longer deliver services and goods, and the customer can no longer ensure provisions and enabling services for the contractor to use and can no longer pay the contractor.

2.18.1 Binding and Nonbinding Agreements

Agreements between a buyer and a seller may be legally binding or not and may be "complete" to the degree that they can be considered a valid vehicle on which to base the business relationship.[94] Figure 2.28 shows different types of agreements between buyers and sellers.

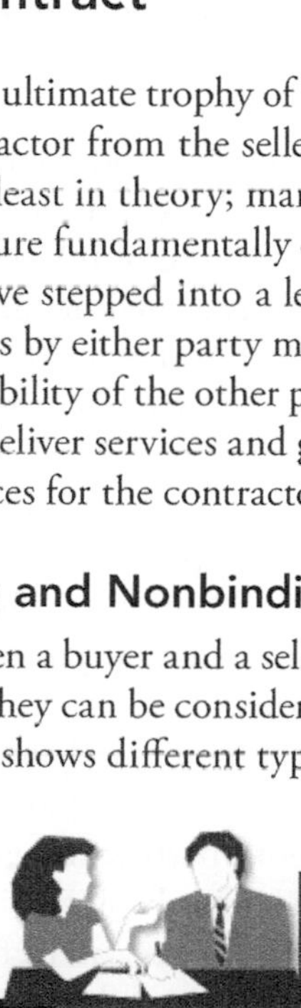

		Commitment	
		Diplomatic	Legally enforceable
Completeness	Preliminary	Memorandum of understanding (MOU)	Letter of intent (LOI)
	Final	Gentlemen's agreement	Letter contract (LC), contract

Figure 2.28 Different forms of agreements.

Assuming that people are mostly not accurate in the use of terminology, it is helpful to recommend distinctions among these types, generally based on the plain wording of these descriptors:[95]

---

[94] Nobel prize laureate Oliver Hart has convincingly stated that there is no such thing as a complete contract. A contract can be accurate (and he recommends contracts to be that), but cannot be fully complete (Hart & Moore, 1998). This will be discussed later in more detail.

[95] These definitions may not be applicable in certain jurisdictions, and parties may use these terms differently. I strongly recommend clarifying terminology at the beginning of the negotiations among the parties and gain legal advice before applying them in agreement practice.

- **Memorandum of understanding (MOU).** The term *memorandum* originated in the diplomatic context, in contrast to a legal context. An MOU is a diplomatic document that describes the status of an ongoing negotiation as a baseline to ease further discussions. Jointly writing an MOU may also help identify areas of understanding or disagreement that have not been clear to the parties before.

  An MOU may also be used by the negotiators as a report to their mandating managers, which has the benefit over unilateral reports that the points of view of both parties are addressed. The agreement does still not include enforceable consideration (obligations on both sides), and either party can withdraw from the negotiations without further obligations to the other party.

- **Letter of intent (LOI).** "Letter" can have the meaning of a letter sent to a person as a written message, or of a document that formally certifies and guarantees entitlements of a party. The meaning here is definitely the second, as in other legally enforceable documents such as *letter of credit, letter of exchange* (a kind of IOU), and similar legally binding documents. In project contracting, an LOI is a small temporary contract that eases developing the actually intended contract, which is much larger.

  I had project managers in my classes of airlines and manufacturers who develop aircraft configurations. When one configures a pizza on the website of a delivery service, one is normally done in two or three minutes, and one can do it alone. Configuring a new car before it can be purchased takes at least two people, the buyer and a salesperson, and they may work on the configuration for up to 30 minutes. Developing a custom configuration for a large passenger aircraft may be a one-year project, and it involves a team of up to seven engineers from the manufacturer.

  The three examples have something in common: One can only order the pizza, or the car, or the aircraft after the configuration has been developed and agreed upon. In the case of the airline and the aircraft manufacturer, both parties may enter an LOI, a contract for the configuration development, which describes liabilities if any of the parties steps back from the development project before the purchase can be finally closed. These LOIs are still not sales contracts over the aircraft, despite the observation that some manufacturers communicate them as if they were.

  In the case of aircraft configuration, the LOI often also includes reservation of production capacities on the manufacturer side to avoid long waiting times when the configuration has been developed and the airline has finally ordered a batch. The LOI is a contract that creates the legally protected environment that allows development of the actually intended major contract.

- **Gentlemen's agreement.**[96] A gentlemen's agreement, if it is documented at all, is a diplomatic document that prescribes obligations that will not be enforced by legal action. The parties rely on the mutual trustworthiness and long-term interests of both parties in a mutually beneficial business partnership. It may sound illogical, but many agreements on large project businesses are actually gentlemen's agreements, which is great if the

---

[96] Some people hold an opinion that a gentlemen's agreement cannot be written down. Others allow for written gentlemen's agreements, as long as these are agreements that would not be taken to court. One should further note that today it may be more appropriate to talk of "Gentlepersons' agreements". I have met many gentlewomen in Project Business Management.

business goes as expected: Gentlemen's agreements are quick and easy to enter and easy to update and change if everything works out right. They can become a nightmare for both parties if it does not.

- **Letter contract (LC).** The LC is in literature often considered a synonym to the LOI. From my experience and observation, particularly with US contracting, I think they are not the same. A letter contract does not prepare the field for the development of the actual contract but is a temporary legal substitute for the actual contract, which has been developed to a status that is final enough to make it valid through management signatures. Obtaining such signatures as a final approval of the contract may be time consuming. It may need signatures by managers who spend much time travelling; in a public environment, parliaments and other entities may need to be involved, which may take a lot of time. An LC is an agreement to take the full contract as if it were already signed and valid and start working according to it. The idea is to win time for the project and use resources that the contractor has available right now. It is commonly the buyer's desire to block these resources, while the contractor wishes to keep them productive and generate income with them.

- **Contract.** There are actually two definitions of the word, and both can matter in Project Business Management:
  - *The legal definition.* Any agreement that is legally binding. There are exceptions in some legal systems, but the validity generally comes with the characteristic that it can be enforced at court in case of violation. Depending on the legal system, the characteristics that define which agreements are legally valid and which are not can vary substantially, but according to this definition, an oral agreement can be a full contract if it has these characteristics.
  - *The commercial definition.* For a business person, a contract is generally a document. Together with applicable law, it sets out the rule book for the business partnership. Depending on the business culture, this rule book may be considered sacred or as a rough guideline. It is also a baseline when change requests are decided upon that come with contractual implications and require amendments (alterations, deletions) or addenda (additional rulings). It further provides information on applicable law, place of court, and how to treat the document if a part of it may not be valid or enforceable.

A good contract is written in a way that provides a clearly elaborated delineation of what should be considered in compliance with the agreed-upon terms and what is in conflict with them. This sounds simple, but in project business reality, this delineation is often blurred owing to a lack of time to develop the contract when deadlines are pressing, or to a lack of competence in the project, when the budget does not allow hiring experts in contract development who are competent in both commercial and legal matters. In a worst-case scenario, it may be up to a judge to construe a set of criteria from the incomplete contract and the actions of the parties to the contract, and then decide, based upon these criteria, whether an infringement of the contract has been committed or not. Depending on the legal system and the personality of this judge, the results can be different and are highly unpredictable.[97]

---

[97] Research indicates that it my even matter whether a ruling is made before or after lunchtime (Kleiner, 2011).

In the description above, it was assumed that the agreement was made between two parties. Reality may be far more complex. In the aircraft business for example, it is quite normal that one or more aircraft are bought by a leasing company, which then leases them to the airline. The leasing firm sits like a proxy between the two parties, who do not have a contract with each other, at least not for the sales and lending of the aircraft.

For configuration management, the direct contact between the airline and the manufacturer will nevertheless be necessary to make sure that the manufacturer knows the airline's requirements first hand and that the airline knows the options and constraints that the aircraft has in place for configuration development. Project contracts can become complex treaties with many parties involved, and with the growth of the number of these parties, the complexity that the project manager and the team will grow.

Again, the terminology above is a recommendation, based on experience, observations, and common sense, written by a business trainer, not a lawyer. It is a good idea to use this terminology to avoid misunderstandings, but if these terms are used in the context of Project Business Management, make sure that all people involved have the same understanding of what they mean and that this understanding is in compliance with applicable law. The final liability remains with you.

2.18.2 Signing the Contract

Contract signature is a very special moment in the development of the relationship between the seller, who now becomes the contractor, and the buyer, who becomes the customer. It takes a lot of uncertainty from the seller—particularly the uncertainty as to whether the offer made will be accepted—but brings new uncertainties.

- Will the seller's organization be able to meet all obligations so that the customer will not be in a position to reasonably refuse payments?
- Will the customer meet obligations, particularly provisions, enabling services, and payments over the entire lifecycle of the project?

Uncertainties also exist on the customer side:

- Will the customer's organization be able to meet all obligations so that the contractor will not be in a position to reasonably claim impossibility to do what was ordered by the customer?
- Will the contractor meet obligations, particularly functions and deadlines, over the entire lifecycle of the project?

The moment of contract signature with one seller is ideally also the moment for the buyer to inform other sellers that their offers have been rejected. This may be a short and otherwise meaningless note or an elaborate explanation of the reasons. The latter would take some time in a moment when time is particularly tight. One should consider it an investment in a future procurement process, when this seller will be asked for another offer and may use the lesson from this procurement to improve understanding and skills.

Figure 2.29 shows how contract signature is embedded in the typical process.

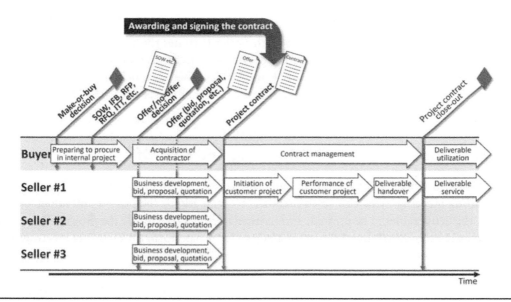

Figure 2.29 The competition for the contract with the buyer and with the award and the signature of the contract.

The description in Figure 2.29 may look idealistic to many practitioners:

- Even relatively large projects are often performed for customers without a written contract. They are based on the assumption by each party that the other party is sufficiently trustworthy and that the combination of mutual dependency and shared long-term interest is strong enough to replace the document. If things go well in the business relationship between the parties, this is the most efficient approach that the parties can have. Developing and negotiating the contract costs a lot of time and energy, and different opinions and interests during that time can sour the negotiations before the business relationship actually begins. Sometimes the relationship does not go well, and the approach can then lead to massive problems and losses for at least one party.
- It is also not rare that project work is begun before the contract is finally signed. It takes at least one of the two parties too long to finish the processes necessary for signature, or the cause may be just a manager who is reluctant to sign or badly organized. The risk is the same as with the business that has no contract at all. The business relationship is based on mutual trust and interests, and the risks for the parties involved can be enormous.
- The description here assumes that the contract is made between two parties. There may be more, and timing may then become even more difficult, when some parties have been quick in signing and others have not.

From a legal perspective, one may consider a business relationship without a written contract a *verbal* contract. Depending on the legal environment, a verbal contract may be considered valid or not, but in all systems, it will be hard or even impossible to enforce such a contract. One should therefore always be aware of the risks that come with the easiness and acceleration of the business development process under a verbal contract.

When the contract is signed, Project Business Management for both sides enters a new phase: The buyer becomes the customer and the seller becomes the contractor. The contractor is now in the position to consider the activities necessary to bring money home with the project, but also enters the obligations that the contract describes, as much as the customer enters obligations, which include payments, provisions, and enabling services. This will be the core topic of the next chapter.

2.18.3 Startup Meetings: On-Boarding and Kick-Off

On-boarding meetings and kick-off meetings are often confused, but they are not the same. On-boarding meetings help new team members to get a grip of the project and understand its mission, the deliverables to be created, and particularly their role inside the project. These team members may be internal, but a project manager can also perform on-boarding meetings with staff from contractors.

The kick-off meeting is different. The project manager invites supervising managers to present a plan of the project that is sufficiently mature to be shown and discussed and to get the approval for the plan and the promise of support that the project will need. The kick-off meeting is commonly done with members of the core team present to help answer questions from the managers that go into more detail than the project manager has, as a preparatory measure to allow managers to make a well-funded decision. A project can have a multitude of on-boarding meetings, depending on the number of new people in the project and on the need to on-board them. I also recommend off-boarding meetings for team members who have finished their work and whose information on that work and observations on the cooperation in the team may be valuable for the future management of the project.

There should normally be just one kick-off meeting in an internal project. In a customer project, there is often the need to have two kick-offs: One as a meeting together with management from the customer's and contractor's management, in which cooperation over the project will be discussed, as well as how both organizations will act together to make the project successful and the customer happy. On top of matters such as costs, workload, and key dates, organizational disruptions may also be discussed, particularly if it is the customer whose business will be the one that suffers, and who therefore has to agree with predictions on timing and severity of the disruptions. A second meeting, an internal kick-off without the customer, can then be held, in which the project manager discusses with his or her own management how the project intends to bring money home and affirms the necessary internal support. This second meeting will focus on all issues that are not the business of the customer.

Both kick-off meetings in the customer project may include core-team members, just like the internal kick-off. Who constitutes the core team? I recommend the definition that the core team includes all team members who are intended to stay with the project during the entire lifecycle and who are supposed to regard the project's success as a personal success, its failure as a personal failure. A number of documents will be presented to managers in a kick-off meeting, such as project scope statement, WBS, schedule, human resource plan, the various management plans, and more. It may be appropriate to have some of them in two editions, one that is shared with the customer and ideally formally accepted there, and a second that is an internal edition and includes all the business information that is not the customer's business—not as an act of distrust, but to focus communications on what is relevant and appropriate.

Chapter 3

Contracting

3.1 Contracting as a Process

I have found that the term *contracting* is used differently within the profession:

- Contracting to refer only to the process of contract development. In this understanding, contracting ends with the award of and signature on the contract.
- Contracting to describe the process of contract development, contract administration, and final contract close-down. Procurement in this understanding is then the sum of the different contracting lifecycles in a project from the buyer's perspective, each of them beginning with the first contact between seller and buyer and ending when all contractual obligations have been met. On the contractor side, where project business is mostly focused on one contractual relationship, contracting includes all activities that relate to selling, performing, and delivering.

In this book, I will use the term *contracting* in the second sense and assume that both buyer and seller perform the process as described. In projects performed under contract, the contract types and clauses agreed upon massively influence the project by assigning obligations, risks, and control levels to the parties. Contracts connect organizations that may be located in different geographic regions, countries, cultures, time zones, and also in different legal systems. This adds organizational and legal complexity to the common technical and interpersonal aspects of project management, and project success may depend on how these various aspects of complexity are managed.

3.2 Introductory Questions

The following questions are written in the style of a certification test. They are intended to give you an understanding of the contents of the following text section and the questions that will be discussed in it. It may be interesting for you to answer these questions before you read the section, and then again once you've finished it.

1. What does one need to take care of in international contracts?
   a) International contract law follows other rules than national contract law.
   b) The contract is valid under a legal system that is probably unfamiliar to at least one party.
   c) International business works best without an international contract.
   d) The contract works is a momentary snapshot of a power relation. It becomes less important as this relation changes over time.

2. Which types of project contracts are commonly designated in codified law in civil law jurisdictions?
   a) Cost reimbursable contract with fixed fee
   b) Fixed price contract, time and materials (T&M) contract
   c) Product contract, service contract
   d) Rental contract, purchase contract

3. What is true for contracts?
   a) To be valid, a contract must be complete and without areas that are open for interpretation.
   b) Project contracts cannot be fully complete; there will always be areas that need change and refinement later.
   c) It is generally better to perform project work for a customer without a written contract.
   d) To be valid, a contract must be in writing.

4. A customer's contractual obligation is generally to pay. What other obligations are common in projects under contract?
   a) Contractual provisions and enabling services by the customer
   b) A fixed schedule by the contractor, including all work of the contractor
   c) Buying a share in the contractor company by the customer
   d) A scope statement written by the customer

5. A customer has a supply network of contractors working under a capped target cost contract (TCC). What is this?
   a) A T&M contract with cost/benefit sharing and effort ceiling
   b) A cost reimbursable contract with cost/benefit sharing and cost ceiling
   c) A fixed price contract with cost incentive
   d) A cost reimbursable contract with cost/benefit sharing and price ceiling

6. A rolling award fee contract uses a monetary incentive to motivate what?
   a) Meeting precisely specified contractual obligations
   b) Improving the project and continuously saving costs for the customer
   c) Improving the project and saving costs for the customer in a special moment
   d) Getting additional functions from the contractor free of charge

3.3 Good Faith and Mutual Obligations

Descriptions of contracts between customers and contractors often assume that the parties have rather simple obligations. In Project Business Management, these obligations can become quite complex. Table 3.1 describes a selection from them; the system of mutual obligations can include many more than what is described here.

Table 3.1 Some Typical Obligations Contract Parties Have Toward Each Other[a]

Obligation	Customer Side	Contractor Side
Basic obligations	Payments	Products and services
Deliverables	Provisions	Project deliverables
Services	Enabling services	Project services
Information	As necessary for the contractor to do the job	Progress data, problems, performance, projections, possibly work and costs
Guarantees and insurances	Insurance of contractor staff and deposits for contractor's property on customer's premises	Bid bonds, performance bonds, insurance against liabilities
Monetary considerations	Payments	Outlays
Organizational	Project management	Project management
Schedule	Timeliness of provisions and enabling services	Timeliness of deliverables handover
Disruptions of project work	Protection from disruptions	Support of disruptions

[a] Of course, depending on the contract and applicable law.

The term *provisions* in this context describes deliveries of goods by the customer that the contractor needs to have in order to perform the business. Examples may be technical drawings of items or site layouts that the contractor needs to do the work. It may also be data structures of existing database systems, interfaces, process descriptions, sample data that the contractor can use for development, and many other items. For a project to translate and localize literature or software, this original literature or software must be provided to allow the contractor to start translating.

Enabling services can include the addition of the contractor's staff to the customer's electronic communication systems by arranging internet and intranet access, accounts on the customer's e-mail server, corporate phone extensions, access to the customer's internal call centers and technical services, and possibly access to internal social network systems. Enabling services may further include having a team site in place that the contractor can use together with the customer for document exchange, online conferences, task assignment and tracking, and other forms of team communications. Access to the corporate restaurant and to the coffee break zones for contractor staff working on the customer's premises is another common example. There are organizations in which providing provisions and enabling services is uncomplicated and mostly ad hoc; in others, complex and often tedious processes need to be followed.

I remember a customer for whom I had a two-years' qualification program on-site, and it took a full week until everything was in place so that I could start working. I could not complain about payment—the week was compensated by the customer—but the time that I had left to meet some challenging deadlines was further shortened due to the delay. Provisions and enabling services will be discussed in more detail later in this book.

Customers are often very clear and challenging when it comes to defining the obligations of the contractor, whereas the contractors in turn are much less insistent when the customer does not meet commitments timely and in full. This can lead to problems with costs, deadlines, and other challenges for the contractor.

I recommend that contractors ensure that the correctness, completeness, and timeliness of the customer in meeting obligations is documented in detail. This is of particular importance when the customer's performance is incorrect, incomplete, or late, and the contractor organization needs to defend itself in disputes over the correctness, completeness, and timeliness of its services and deliveries. The defense of incorrect, incomplete, or untimely fulfilling of obligations by the other party concurrent to one's own errors and delays may be a strong argument in negotiations on settlements, during alternative dispute resolution, or in the worst case, at court, especially if the failures of the other party were among the causes for one's own failures.

3.3.1 The Doctrine of "Good Faith" in International Project Contracting

The intention of this book is to give guidance to project managers who are dealing with Project Business Management, either on the contractor side to bring money home and make the customer happy, or on the customer side to manage often complex and dynamic *project supply networks* (PSNs). This book is not intended as legal advice, and no guarantee is given when it is applied that the project, the project manager, the performing organizations, and other stakeholders are protected from legal challenges. In case any kind of statements, documents, or actions can have legal implications, I recommend seeking professional advice by legal experts trained in the specific legal system.

The following describes the concept of *good faith* in the two great legal environments, *common law* and *civil law*. The legal aspect is important, but the relational aspect is even more important in the context of this book, and I will discuss this also.

Common law has its roots in the British mediæval times, beginning in the 12th century. It was applied later in the various British colonies, which mostly adhered to it after their independence but then developed some differences in their application and interpretation. Its central element is a system of precedent called *case law,* in which current decisions and opinions by judges form a binding body of law for future cases with similar contents, a doctrine called *stare decisis.*

Civil law is based on old Roman law, but its modern expression is based on the requirement to separate the three powers that are present in a state—legislation, executive, and judiciary— as was postulated by the French philosopher Montesquieu.[1] Because a judge is part of the judiciary, he or she cannot make laws. The laws must be present in such a system before the judge can interpret them. The strongest signal that a country has jurisdiction is therefore a universal book of laws, a so-called *civil code,* which the judges then interpret and apply.

---

[1] (Montesquieu 1748)

Originating in France, it was during the short time that Emperor Napoleon Bonaparte ruled over major parts of Europe that the concept was taken over by European countries, further developed, and from there exported to other countries and continents. To my knowledge, there is only one place in common law where such a civil code exists, and this is California.

It may be interesting to see the distribution of common law and civil law worldwide, as shown in Figure 3.1.

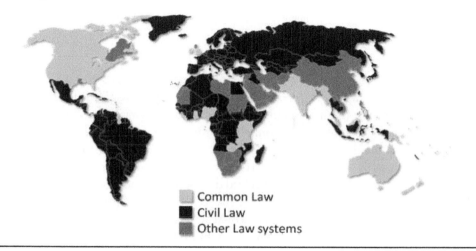

Common Law
Civil Law
Other Law systems

Figure 3.1   The global distribution of legal systems.[2]

Other legal systems in the figure include areas with mixed legal systems combining common law and civil law principles, with Islamic law, and with party law, where the law reports to the ruling political party, which excludes the separation of powers.

The purpose of this book is not a fundamental discussion on legal systems with their benefits and disadvantages, but in developing an understanding of the impacts that these differences can have on Project Business Management in a legal environment. Cultural differences between parties can impact a project strongly, but legal differences can damage a project even more, and to complicate things further, law is a major factor influencing culture: People commonly behave in a way that they do not get into conflict with the law, and as laws are different, so will then be the behavior of people.

In essence, good faith means that one party does not intend to benefit from a contract by causing damage to the other party. It is different from breach of contract, in that it includes acts that are in contradiction not with the *words* of the contract but with its *spirit,* and with the intentions that led to its creation. It deals with the asymmetry of knowledge and influence of the contract parties and also of their ability to act to gain the benefits from the contract. Good faith requires a party to consider the fair interests of the other party in statements and actions and not gain undue benefits to the detriment of these interests.

Good faith is a requirement in human behavior in contractual situations. It is imposed by certain jurisdictions to various degrees, but not in all of them.[3] It can, for example, impact the

---

[2]   Some assignments of states are unclear and cause disputes as to which systems they actually belong.
[3]   (Moss 2007, Reeves and Murphy 2014)

application of contractual terms if these are too unfair for one party, and a court may wish to seek fairness among the parties; or when one party enforces its business interests against the other in an aggressive way, so that the common goals of the business cannot be achieved.

The meaning of good faith in a contractual context is that one party assumes a degree of responsibility for the success that the other party gets out of the project. It is primarily based on the understanding that the contract parties are, first of all, not opponents but partners in a joint endeavor.

3.3.2 Good Faith in Common Law

The Anglo-American legal system, also called common law, is rather hesitant to apply a concept of good faith in contract law. The degree of such hesitance varies across the different common law countries but seems visible in all of them. Common law assumption in a contract is that two or more parties are in a partially or fully competitive situation, and the contract defines the rules of this competition more than its outcomes. *Caveat emptor*—literally, the recipient of a benefit must take care—means in essence that all parties "stand on their own feet" and have to take care not to get deceived, at least as long as the perceived deception does not constitute criminal action.

The legal focus is on the interests of each of the parties and the degree to which the implementation of such interests can be limited by the interests of the other party. In common law, there is mostly no obligation for a party to subordinate its own commercial interests to the common interests shared by both parties, unless this is clearly stated in the contract. The benefit of this approach is predictability: A judge will rather not assign responsibilities to contract parties in hindsight that they did not describe in the contract.[4] The disadvantage of the approach is that a judge is not in a position to protect a contract party from unfair treatment. Another disadvantage is that contract documents become very long in order to prescribe every aspect of the agreement in exhaustive detail. These contracts are hard to read and generally even harder to understand for laypersons in law—people who are involved in the execution of the contract and should have an understanding of which actions or non-actions can be regarded as compliant with the contract and which cannot.

3.3.3 Good Faith in Civil Law

In civil law environments such as Continental Europe, and also in major parts of Latin America, in Japan, and in other countries, it is a basic principle that obligations must be met under the application of good faith. In German, this is called *Treu und Glauben,* meaning *trueness and confidence.* In France, it is called *bonne foi, buena fe* in Spanish, and so on. The concepts are generally written into the civil codes that characterize these legal systems, and the precise meaning is then found in actual court cases that required their interpretation in specific cases. The good faith principle assumes that each party in a contract has certain basic obligations against the other party or parties in addition to the plain wording in the contract. To varying

---

[4] This is emphasized in a statement in the *PMBOK® Guide,* 6th Edition, under Project Procurement Management: "Anything not in the contract cannot be legally enforced" (PMI 2017, p. 461). This statement is correct for a US standard; it is not true for an international standard.

degrees in different civil law countries, these duties particularly include that the actions (or inactions) of a party must show an attitude of loyalty toward the other party, which includes reliability, honesty, and thoughtfulness, and that the other party can in turn base actions on the confidence that the first party acts in such good faith.

In a civil law system, contractual agreements are rather focusing on completing than competing, and a judge will decide on a contractual disagreement based not only on the words of the contract but also on its apparent spirit and on the principle of fairness and just expectations of the parties. A benefit of this approach is protection of the parties in a contract from dishonest and disloyal behavior. Another one is that contracts are much shorter and generally easier to read for the untrained person. A disadvantage is that the interpretations of an arbitrator or a judge add uncertainty and unpredictability to the business relationship, because the understanding of what constitutes fairness may be different from person to person, and decisions of judges may even be different before and after lunch.[5]

3.3.4 Good Faith and Basic Trust

A contract in a project consists of at least two parties, one as the customer and another one as the contractor. In addition to the legal requirements, there will also be a relational connection that develops concurrently with the contractual relationship. In essence, the quality of this relational connection not only affects the degree of mutual trust between the parties but is also affected by this mutual trust. A project needs both relational connection and trust to be successful for two reasons:

1. I have already mentioned Conway's law, which in essence says that teams working together building systems must have well-functioning communications structures in order to make the system function, which requires that its components are working well together. Interfaces between teaming partners (and actually people inside their inner teams as well) that are not fully functional will build components that add up to systems that will not be fully functional, and each party will spend a lot of time proving that the dysfunctionality is not its fault and that another party must be blamed.

2. A project manager who wants to know what is going on in a project needs people around who trust each other and also trust the project manager. Distrust leads to communication failure, to prettification of issues that should be addressed urgently and vigorously, and to people wasting energy and time with mutual blaming when it is instead necessary to find solutions. Overly protective behavior disrupts any "Mission Success First" culture—something we need to situationally manage projects in complex and dynamic environments and finally lead them to success.

As the project manager on the contractor side, one commonly has two goals: Bringing money home with the project and making the customer happy. The first task relates to the survival of the contractor organization, its ability to respond to unforeseen challenges, and its intrinsic power to grow. The second task is needed to turn the contractor into a (or even "the") incumbent supplier, which makes winning future business much easier. The second

---

[5] (Levav 2011)

task includes building a trustful relational connection with the customer, so that the customer desires to keep the connection alive.

The first task may at first glance seem to contradict the second, because the profit will come from the bills paid by the customer. Indeed, the first task may instead lead to destroying this trust. To give an example, many project contractor organizations establish a position of a *claim manager,* whose job it is to identify claims—constructive changes that allow for additional billing by the contractor on top of what was agreed upon originally. Constructive changes are changes that are understood to be covered under the contract only in hindsight. An example may be over-time work done by employees of a contractor that was necessary to finish some contract work on the premises of the customer, and because the customer has not sent these employees home at the end of the regular working time, one may construe a change against the original contract and bill the additional working time.

Intensive claim management can earn significant amounts of money for the contractor but puts the customer's employees under pressure. They will have to explain to their managers the causes of the increased costs, and the reaction of these managers may not be driven by under-standing but by anger. Customer organizations respond to this threat by employing claim managers too, either to place counter-claims against the claims of contractors or to reduce the bills that they have to pay to the contractor. Claim management may also be outsourced to specialized contractors. It is highly competitive, and the desire to compete is often incompatible with the need to complete.

On top of the communication failures and the incapacity to build complex systems by parties who should work together, lack of trust and relational connectivity has many more negative consequences:

- **Communications with potential lawsuits in mind.** Communications bear the risk that the things communicated today may be used against oneself later in a conflict situation. Without intensive communications, the ability to build a working system in which different components act toward each other gets diminished.
- **Lack of perceived affirmative action.** Humans need affirmation to go on with things they do right. In an environment of distrust, praise will not be regarded as affirmation and encouragement but as deception and flattering.
- **Burnout of team members and contributors.** It is existing knowledge in psychology that burnout syndrome is commonly caused by the combination of two elements:
  - A perception of effort–reward imbalance (ERI, a sensed discrepancy between what a person or a group invests in a job and what is returned to that person or group), and
  - Exhaustion. When people do not have enough confidence to talk timely about their problem, and when the people they talk to do not have trust in them and take the notion seriously, burnout will be a common result.
- **Reduced error tolerance.** It takes a lot of trust in people to assume that they will not repeat an error they have made and will voluntarily fix the consequences. In an environ-ment in which management behavior is not regulated by trust and by the desire to sustain this trust, people who have made errors will feel that they should not talk freely about these errors, which often increases the damage from such failures—solutions are not searched for, too much time and energy is spent finding culprits, and those who are

pushed into the role will spend the same energy to defend themselves. The driver for this behavior may be interpersonal, but it also has a business intention, when errors have costly consequences and parties wish to shove these costs as liabilities on other parties.

- **Need for micromanagement.** Micromanagement apparently becomes necessary when one believes that one cannot trust in the abilities, the sincerity, and the good will of subordinates or contractors. Micromanagement adds a massive workload on the micro-manager, binding time that should be used for actual management tasks by the person and letting employees burn out.

- **Misunderstandings.** There are many causes of misunderstanding. Distrust is one of them. A trusted boss who communicates the need for rework on an item to an employee is considered a supervising person giving direction. The same communication from a distrusted boss will rather be understood as criticism and possibly disrespect regarding the work and the attitudes of the person.

- **Sophistry.** People will stick in a literalist fashion to the words of agreements and ignore their spirit. The result will be a go-slow and work-to-rule attitude rather than one that puts the "Mission Success First" and supports this with proactivity, quick responses, and the preparation to go the extra mile when questions are raised and issues become visible whose swift resolution is critical for success.

- **Stress.** Lack of trust and relational connectivity puts people and businesses under stress, but, as the forensic psychiatrist Charles Morgan said, "No one becomes smarter under stress".[6] When people's mental resources are consumed by stress, their ability to act as a problem-solving team diminishes, and the desire to have a fast, effective, and efficient project remains unsatisfied.

One may argue that these *effects* of a lack of trust (and others) are also its *origins,* and this is actually true. Distrust is a self-confirming vicious circle, and once the project finds itself deeply entangled in this circle, it is hard to get out of it again. Once the project is locked in this circle, it is hard to know what is actually going on. When the project manager asks for estimates, the answer will be political estimates; when the project manager asks for opinions, he or she will be given those opinions that people use in self protection; and if they ask for facts and data, these facts and data will either be cherrypicked and communicated selectively or completely replaced by opinions. Without trust inside the project, a project manager does not know what is going on in the project. Analyses of so-called "melon projects"—projects that are green on the outside, but the deeper one drills into them, the redder they get—have repeatedly shown that a lack of trust was the basic cause of why the project manager did not know what was actually going on in the project.

Too much trust, and particularly too much trust in the wrong people, is also a cause for project failure. All project managers, one may presume, have had the experience in their professional lives that someone to whom they had given the present of trust had then forsaken them.

The difficulty for project managers is to achieve a balance between the monsters of mutual distrust on one side and gullibility and credulity on the other. In addition, project managers must find out who deserves their trust and to what degree, and who does not. Trust has two layers:

---

[6] (Bond 2017)

- **Basic trust.** Trust in the environment and the people around one, based on one's own experience in life and also hardwired in people's personality. When this trust exists, self-protecting measures will be kept at a minimum.
- **Individual trust.** Trust in a specific person, based on his or her past behavior and current living situation. When this trust exists, the person is considered trustworthy and can be entrusted with confidential tasks and knowledge.

The principle of good faith builds on the desire of basic trust, which generally makes life easier. In an environment in which basic trust applies and is justified, one does not need to invest much time and money in protecting property. In such environments, one can just leave a locked car parked on a road and trust that the car will still stand there. Where basic trust goes very far, people even leave their cars unlocked and trust that no one will steal something out of the car or even the entire car. In environments in which this basic trust is missing, cars are not only locked but additionally protected with alarm systems and a steering wheel lock. Owners will prefer to leave the car in a guarded garage for additional protection and have GPS trackers attached secretly to the car that allow finding and reclaiming it if it gets stolen. In an environment with basic trust, life is much easier and simpler, and less money and time is invested into protective measures. Basic trust in an environment where this is inappropriate leads to losses.

Basic trust is the basis of any true faith approach. In a business environment, basic trust can be developed by people who are surrounded by persons and organizations with whom they have had successful long-term business relationships, and who must be more interested in a common long-term future than in quick gains from competitive and hostile behavior. Functioning true faith implementation, legally but also in relational connections, in turn leads to increased basic trust. Legal systems are a commonly ignored influencing factor of cultures.

Another factor is long-term orientation. The principle is comparable to restaurants in cities with many tourists who casually visit them. In the center of the city, where many tourists gather to see the famous attractions of the town, restaurants are often expensive and their quality of both food and service is below average. Restaurant owners have to pay expensive rental costs, and they know that the unhappy tourist family does not matter that much—the next family is already waiting for a free table. Restaurant owners at the outskirts of the city have different business situations. Their guests are regulars, and if they stay away, they cannot be easily replaced with casual customers.

This rule does not always hold: Sometimes restaurant owners in a city center follow their passion to have a great restaurant as much as their desire to make profit; meanwhile, on the outskirts, a restaurant with poor quality may open up, and while it may not survive for long, this is just the time that one visits it. Another game-changer is restaurant reviews on popular websites, written by normal visitors (and unfortunately sometimes polluted by friends or enemies of the restaurant owner). These give a long-term quality motivation for the locations that are more frequented by casuals as well, at least when they find that tourists read these reviews before they decide where to go for a meal.

In general, *long-term orientation* is a great motivator for people to develop trustworthiness, which in turn is the basis for sound trust. I cited Stephen Covey above, who said correctly: "If you want to be trusted, be trustworthy".[7] Long-term–oriented people in business are generally

---

[7] (Covey 2004, 51)

more likely to develop trustworthiness, because their interests are more in mission success than in quick wins. In project management, and here particularly in Project Business Management, where contractors and customers meet, as well as people and organizations in other roles, this leads to a dilemma: We do not have much time to unhurriedly develop trust, we must function quickly in order to deliver quick wins and meet the project's deadlines.

In the complex PSNs in Project Business Management, people often have to deal with thus far unknown organizations and individuals, and time for the development of rapport and trust is scarce. Team members across the diverse organizations must get to know each other, develop interpersonal interfaces as much as technical and organizational ones, and learn to manage the little conflicts that turn up several times a day over marginal things in a way that allows them to be successful together and to complete their work and, with that, finally the project.

These techniques can help build rapport rapidly with business entities for which you are working as a contractor or that work for you in such a role.

- **Identifying common interests and opinions.** There are many areas of interest in subjects such as hobbies, family, lifestyle, sport, pets, politics, etc. Although diverse business interests can be disruptive for a relationship, the commonalities of interests drive bonding between humans.
- **Mirroring and synchronization.** As with all primates, mirroring creates a perception of togetherness. Raising glasses, drinking, and placing them back synchronously is an example of how one person mirrors another.
- **Spontaneous helpfulness.** When a person drops a coin and another one stoops down to pick it up, a positive relationship develops. This is just momentary and no big thing, but if such situations reoccur, the bond gets a little bit stronger with every little service.
- **Adjusting one's tongue.** When fast speakers slow down to make sure the other person understands, or slow speakers speed up to match the quick thinking and impatience of the other person, a signal is sent that the speaker desires effective communications. The same happens when people who normally would use a strong dialect or accent turn to standard language, or when people who normally use special terms explain issues in layman's terms.
- **On-boarding of people with cultural congruence.** In areas in which people speak with a strong dialect, it may actually be helpful to have people on board who can speak the same dialect. My own dialect, for instance, is Swabian, which can help build rapport much more rapidly when I am dealing with people from companies such as Daimler, Bosch, Porsche, among others, in which many people have grown up with this regionalism. This is particularly helpful given the traditional reluctance of Swabians to develop trust with people they do not know. Cultural congruence may also mean having an industry veteran on the team who knows the peculiarities of the specific trade.[8]
- **Maintaining the right extent of eye contact.** This is again culturally, but also individually, different. With some people, one should apply the three-second rule and keep eye contacts brief but frequent to avoid a perception of being intrusive or aggressive. With others, it may be more appropriate to keep the eye contact much longer to show

---

[8] In offer management for projects in a military environment, getting support from military veterans may also be helpful.

interpersonal interest. It takes some sensitivity to find the intensity of eye contact that the other person is comfortable with.

- **Questions.** As with eye contact, questions can be a sign of interest and an essential element of active listening. They can also come across as intrusive and annoying. Balancing between too few and too many questions helps build rapport swiftly.

- **Allowing others some time.** This may sound counterintuitive, given that the task is rapid rapport building, but it is much easier to build rapport with an exhausted person once they have been given sufficient time to recover. A person that just came out of a conflict will need time to relax and digest the experience. A person on steroids after a great achievement may need a break to come back to normal reasoning.

- **Saying "Thank you" more often than normal.** There are few expressions that act more universally as bonds than an honest, heart-felt "Thank you". It signals attention and interest. Consider person A telling person B that B has done a great job. B could say "Yes, I know, and I am very proud of it. I think I can do even better, next time"; B's attention is obviously consumed with his or her own achievement. A simple "Thank you very much" signals that B's attention is directed to person A, building rapport instead of seeking admiration.

All these techniques must be used with care. Each of them can create the impression of flattering or of being intrusive, turning the good intentions into the opposite. The assumption that rapport building is generally a well-controlled process also ignores the influence of "chemistry" between humans—sometimes, it seems impossible that certain people will ever develop a constructive relationship with each other. It may even be that such conflicts are very old, going back to struggles from earlier projects in which the collaboration has not worked, and it may then be difficult to overcome such vendettas—at least to do it timely before they hurt the project. There are actually handbooks for rapid rapport building, mostly used by spies and insurance salespeople, but their basic weakness is that human relations are not built by following a handbook like a cooking recipe, but by applying the interpersonal and social skills that most of us have as humans, by observing the people we are in interaction with, and by applying basic common sense.

When companies in a project under contract strive for mutual rapport, trust, and a mission-oriented relationship, another dilemma turns up. Many of the vendor selection methods discussed in the previous section are deeply competitive on price and/or attractiveness of solutions. They are meant to be competitive, assuming that this helps the customer get the best offer for the task. Now these vendors must finish competing and focus on completing the work in a collaborative fashion toward the customer and other vendors that work for the project. The qualities that made them win the contract are no longer helpful to fulfil it. To make things worse, only now, when the vendor has become a contractor, will the company be granted access to data and people on the customer side, and there is now a strong business case to invest time and go into these details.

During the offer development phase, this business case was much weaker, given the commonly low hit rates in new-customer project business. It is not uncommon that, during the planning of the freshly won business, it turns out that the price offered was too low, that deadlines agreed upon are not realistic, and that technical solutions desired are not feasible, at least not with the resources that the project can use. A first approach would then be to find ways to reduce costs, effort, and time that the contractor would need to invest in the customer project, which is in essence competitive behavior. There are two limitations—one is the contract with the customer and the requirements it specifies on the contractor's work and results; the other one is the rival objective of ensuring a happy customer during and after the project to ease winning future business. Depending on the project and its legal environment, there may also be regulatory or legal requirements that the project must meet and that restrain the freedom of the contractor to descope or otherwise downgrade the project.

The need to transform from a player in a competitive setting to a fundamentally collaborative affiliate, from a party in a rivalry-based contest to a teaming partner, can become a major issue in a business situation with just one customer and one contractor. It gets much more difficult for the large PSNs that we find in more and more modern projects, and the complexity and dynamics of these networks increase the potential for conflicts that can no longer be settled among the parties and will need to be resolved in arbitration or even at court.

Another risk for the development and sustaining of inclusive PSNs are loose cannons. People with a lack of self-control in challenging situations are common, and with the right support, they can climb to high and influential ranks. An inconsiderate statement or action can be enough to frustrate people and organizations involved and to disintegrate the PSN partially or in total. Such a statement may be made in a moment of anger, and although this anger may have cooled down after a short time, the damage created may be lasting.

The tension between divisive and competitive dynamics on one side and the need for an enduring, inclusive relationship among teaming partners on the other will impact the project during most of its lifecycle. The intrinsic dynamics of a business system with two or more parties generally tend toward competitive behavior. It is like marriage. It takes two to keep it alive over the years; one partner could be enough to end it. Game theory is very helpful in understanding the inner forces of PSNs, but for most cases, educated common sense is sufficient.

What can be done to overcome these disruptive tendencies? One needs some kind of glue. In teams, one often finds specific persons who act as adhesives. They have a calming and integrating effect on co-workers and can make a team from a loose group of people. When these people leave the team, divisive trends will often prevail again, and the performance of the team will suffer. New people may increase these divisive effects, and a team that worked well in one moment will fail in the next.

Organizations often behave similarly. The presence of one organization can glue the different organizations together, helping them collaborate even while their business interests may diverge to some degree. When this glue organization leaves the PSN or when another organization joins it that brings a more competitive approach with it, the PSN may lose both effectiveness and efficiency, replacing alignment to the common mission with blaming, finger-pointing, poor communications, and other protective behavior. Adhesive people or organizations (and in them specific people who define the organizations' attitudes and aspirations) should be kept with the project, or if they must leave the team, efforts should be taken to find or on-board other adhesive people or organizations that can effectively replace them.

This section discusses Project Business Management from the perspective of the contractor. Contractors sometimes forget that in most situations, they have as much interest in a well-working business relationship as the customer does. Such relationships allow the contractor to contribute with pride to a common success story, to overcome difficult challenges much more easily, and to finish the project with a reference story that helps win future business. In addition, it is the basis for becoming the incumbent seller for future business, making it easier to win such business against competition.

Contractors can participate in sustaining the relational aspects of a PSN by upholding the principle of good faith. In addition to the legal quality of good faith, which is described above, good faith also has a behavioral and a relational quality. In civil law countries, particularly those of the "Germanic" legal realm, including German-speaking countries, Scandinavian countries, and even Japan,[9] ignoring the principle of good faith can lead to successful damage claims by the other party. The principle is enshrined in civil codes[10] or constitutions[11] and upheld at courts. A signal that a jurisdiction values good faith is the common use of the term "contract partner".

This reflects the concept that a business contract is first of all the foundation of a partnership in which the parties join assets to achieve a certain goal. In most project environments, the majority of assets provided by the customer are of a financial nature, and the assets provided by the contractor are technical, human, and organizational. Reality will be more complex, because customers often also provide non-financial assets (provisions and enabling services) and contractors prefinance a lot of project work and goods in advance that the customer is expected to pay later. In such a legal environment, implementing good faith principles is required by both contract-oriented law and relation-oriented common sense.

In common law countries, the principle is rather nonexistent in jurisprudence, statutes, and regulations. If it is present, much less emphasis it is given to it. But for the contract parties, there is no restriction on applying the rule in the business relationship to ascertain mutual respect and thoughtfulness and create an environment with resilience against the ever-luring divisive forces threatening project success.

The key behavior is consistent care for the contract partner:

- **Clear mission goal.** To put "Mission Success First", the mission success criteria must be identified and agreed upon. Changes in the mission success criteria will occur; they will also be decided upon in mutual agreement.
- **Communications.** The contract partner gets informed early of all incidents that may impact the party's success and its ability to meet obligations.
- **Helpfulness.** Other contract partners are offered help to increase their business success, as long as this does not put their own business success from the contract work for the project at risk. This may include financial help if a party is in liquidity troubles, technical help if a task is found overwhelming, or any other action that supports the other party and thus the common goal.
- **Interfaces.** Interfaces are defined in a way that all partners can contribute their best to achieve the common mission, not in a way that benefits one party to the detriment of another.

---

[9]   The Japanese Civil Code was modeled after the German Civil Code and enacted in 1896. With some modifications, particularly after World War II, it is still in effect today.

[10]   For example, in Germany in §242 Bürgerliches Gesetzbuch (BGB), the Civil Code in Germany (Juris GmbH 2013).

[11]   For example, in Switzerland in Article 9 of the Constitution (Admin.ch 2017).

- **Error tolerance.** It is accepted that errors will be made by contract parties. Accountability is assumed, but solutions are searched for jointly.
- **Self-restraint.** Opportunities to gain an advantage over another party are dismissed. "Loose cannons" are restrained or removed from the team. Decisions are made in the light of their effects on the other parties and on the success of the entire mission.
- **Mutuality.** It is made clear that all parties adhere to the good faith principles to avoid one party benefitting by going competitive while the others remain cooperative.
- **Observation.** A major risk for a "Mission Success First" culture are changes in ownership and management structures of one or more parties involved. Even a bank as a creditor may change an organization's behavior by requiring a more aggressive business style in exchange for new credits. Such new decision makers and influencers may no longer accept the restraints that come with such a culture and make decisions for the benefit of their own organization only. Such changes must be observed diligently and measures taken early to avoid damages to the project.
- **Continuity and consistency of purpose.** I generally recommend being situational both in the selection of practices that one applies—including approaches, behaviors, tools, and techniques—as well as in how far one plans the future and how much independence or interdependence one should establish for the project. The deep, underlying purpose should be maintained consistently and continuously. If one develops a machine or a software solution, the purpose of making it effective, efficient, easy to use, and possibly fun to work with remains constant, and this must be communicated repeatedly.
- **Owning shame.** Overly competitive people do not feel any shame about their actions. They are driven by appetites that are generally not held in high esteem, at least not among people whose job it is to complete a project: joy of conflict, lust for power over others, desire to hurt others, greed. Shame is the understanding that one is observed and judged based on one's behaviors and their results, and the desire to be judged in a positive light, not in contempt. Shame performs a vital function in group endeavors, and people who do not feel it strongly and therefore act shamelessly can disintegrate a PSN and the entire project.
- **Praising by megaphone, criticizing by telephone.** It is generally a good rule to spread good news loudly but to communicate disapproval in private. This keeps up the team spirit without sweeping issues under the carpet. Human nature is different. Most people's first reflex when they want to complain about something is to do it loudly, and there are situations when this is justified and the best thing people can do. When the task is to build a "Mission Success First" culture, it is rather detrimental to undercut teaming partners in front of others involved in the project. If a teaming partner behaves unlawfully, the way to deal with this should be to go to the police. In most other situations, I recommend sticking to the rule to praise loudly but criticize in private.
- **Joy from joint achievements.** This may be the strongest driver of group success. Experiencing what teaming partners can achieve together, results that one of them would not be able to achieve alone, creates strong bonds among these partners and confidence in a common future. Planning frequent quick wins on the way to the final result—intermediate achievements that give evidence of how well the partners cooperate and show the need for corrective action in areas in which they do not harmonize sufficiently—can be helpful in creating strong bonds among these partners.

- **Respecting teaming partners.** Disrespect toward other companies prohibits any "Mission Success First" approach. Such disrespect damages the fine balance of technical, financial, and interpersonal commonality that is necessary to uphold the teaming partnership. Disrespect blurs the perception that the joint achievements are truly joint. Hurting teaming partners will create a sense of shame if it later turns out that these partners actually deserve respect, but then, with teaming partners who do not deserve such respect, it may cause a lot of schadenfreude. In any case, all the behaviors and attitudes mentioned above will fail if there is no respect among the teaming partners, and this respect must be mutual in feelings and actions. Completing over competing is an attitude that must be shared by all. If this attitude is present only among some of the project business partners, they may be able to teach the others by being a role model and having some patience. If the other parties are not prepared to learn or if the time is too short to be patient, it is better to apply diligence in selecting these partners and observing the signals from a company that it is trustworthy—or not.

3.3.5 Concurrent Sourcing

There is another factor that often makes it difficult to build a cross-organizational team from the customer and the vendors that is based more on the desire to complete than on competing.

Figure 3.2 shows a project team that is active in seller acquisition and selection for two procurement items (#4 and #5) at a date n, which requires a high degree of competitive behavior. At the same time, they are collaborating with other vendors who have already been selected and are now under contract to complete procurement items #1 to #3. Changing social behaviors between competitive and collaborative several times a day can be very difficult for these teams. It can also be difficult for vendors who are in working mode for one procurement item and are at the same time offering for another one.[12]

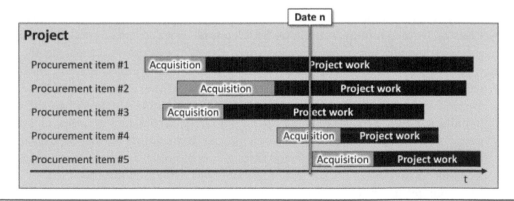

Figure 3.2   At date n the project team and the vendors have to act in a competitive way in the procurement items that are in the acquisition phase, whereas they have to work in a collaborative style in the items that are already worked on.

---

[12] A customer of mine actually mentioned in a discussion that he considers this the greatest source of problems between a customer and contractors: the inability to adjust behavior situationally.

Non-competitive behavior during source selection and contract development is considered corruption in many jurisdictions, particularly in public projects. The borderline is thin that separates the justifiable desire of a procuring project manager to have trustworthy partners under contract on one side from favoritism and nepotism on the other. The borderline is also thin between the reasonable need of a vendor to protect its own company from damaging competition and possibly criminal bid-rigging of competitive sourcing.

The basic problem shown in Figure 3.2 may occur even in a project with just one procurement item. At the moment of contract award, the project manager and the project's procurement team must change their behavior from acting more or less competitively and switch to much more collaborative conduct. During the acquisition phase, their collaborative behavior may have been very successful. They were able to reduce prices, rates, and fees; forced vendors to accept challenging deadlines; and obtained guarantees for the availability of people, technology, and other resources. The confirmation that applying pressure helped them do a good job is then likely to encourage the team to carry on the competitive behavior into the phase when the project work needs to be done.

Before the contract was signed, the buyer may have driven the seller into dilemmas, such as deadlines that are impossible to meet, a price that would not cover the contractors' own costs, and agreement on technical specifications that the contractor is unable to meet. During this time, the buyer did not have to care about these dilemmas too much—they were someone else's problems. Sellers during this time are also mostly parsimonious with the information that they give to sellers: These sellers are not yet contract partners of the buyer, just parties in negotiations, and the actual effectiveness of non-disclosure agreements (NDAs) is also often questionable. After contract signature, the customer should put more consideration into them: The impossibility of meeting a deadline will delay the project, missed technical specifications will harm the benefit that the customer expects, and a contractor on the way into bankruptcy is among the greatest nightmares for a project manager. The same is true on the side of the seller, who considers the risk of giving the buyer too much information too early—information that the customer employees may pass on to favored competitors or use to rethink their make-or-buy decision.

The contractor may be in a similar business situation as the customer during the time of contractual work. The relationship with the customer requires a collaborative attitude and behavior, but the contractor may concurrently bid for other work items in the project. The contractor at that time is also likely to be in bidding processes with other prospective customers, and it may well be that the contractor has the role of a prime contractor who needs to bring subcontractors into the project, possibly using competitive processes as well. In these scenarios, the contractor should also situationally separate those work streams that favor a more competitive approach from others for which a partnership attitude is required.

At the moment of contract signature, parties become partners—at least until the partnership turns sour and the contract requires dispute resolution. Independent of legal requirements, there will be an organizational and social requirement to develop a mutual good faith attitude and consider the interests of contract partners when decisions are being made in order to ensure completion of the project deliverables.

The time before the contract award is also very much characterized by mutual distrust, at least in new business. There is no experience between the parties that tells them how far they can trust the other. Giving away too much information and other assets and making oneself dependent on the other party may hit back later, but not doing that makes it impossible to build the business relationship.

After the contract signature, except for the perception of success on both sides, the lack of mutual experience still remains, but a credit of trust will be necessary—given the need to work together—and there will be not much time to build mutual trust through common experiences, because time pressure is already mounting in many projects at the beginning of the contractual relationship.

After contract signature, when the project actually will be performed, one should also keep Conway's law in mind, which, as already mentioned, states that to build a working system as a project deliverable, one needs a working system of all the teams involved. If the communication between these teams is insufficient, errors between the system components become likely to obstruct the performance of the final system.

3.4 International Contracts

Two of the most fundamental questions in contract development between two parties in different countries are:

- Selection of the applicable law
- Selection of the place of court

Often, this comes during the discussions of whose *terms of business* will be applied, but this may also be a separate discussion. It is common practice that, right from the start of negotiations, both parties try to make their own legal system applicable and to select the place of court near to their home location. To allow the conclusion of the project contract, it is inevitable that one party will have to give in. The other party will then be the winner in this discussion and have the benefit that, from the legal perspective, the project will be a kind of home match. The "winning" party may celebrate this success. It has avoided all the risks that come with having to act in an unknown legal system and, in a worst-case scenario, the costs and difficulties of managing a lawsuit over a distance, and finally win the case. If the focus is just competitive contract law, having won that argument is a clear success, and the considerations on applicable law can be considered finished. The losing party will have to make sure that they manage their risks that come from the agreement. This is *caveat emptor* on steroids.

If the reflection gets expanded to relational connectivity, includes good faith considerations, and follows a "Mission Success First" attitude, things become more complicated. One of the two parties must act under an unknown legal system. Actions that are acceptable and lawful in the party's own legal system may be rejected and illegal in the system under which it has to meet its contractual obligations, and the organization may not even be aware of the differences. Most people do not fully understand their own legal system—how should they understand a foreign one?

One may argue that this matters for lawyers only. It is their job to look at legal matters and resolve them accordingly. Things are unfortunately not that simple. The language often used in law and, even worse, in legally relevant documents such as contracts seems to be directed to experts only, and a lot of it is incomprehensible to the normal citizen. But this law applies to the normal citizen and is actually written for this person. In our daily actions as project managers, on a private level as much as professionally, we create precedents that will influence the future of the project in all aspects, and the legal aspect is one of them. Most lawyers who represent parties in negotiations, in alternative dispute resolution, or during legal action confirm that their success very much relies on the actions of the parties long before the actual case, and how these actions have been documented.

Compare this with public street traffic. As participants in this traffic—that is, when people drive cars—they have to adhere to the laws that regulate this traffic, and when the red light in front of them turns to green, they will not ask a lawyer if they are now allowed to drive off. If participants in traffic have a small dispute with another driver, or with police, they will generally sort this out ad hoc and in private. Only if the dispute over the legality of actions in traffic becomes very demanding and the potential consequences, such as damages or penalties, are high, will people call a lawyer to help them to follow formal processes. If I travel to a foreign country and will drive there by car, my own one or a rental car, I will inform myself about the rules of traffic that apply there with a special focus on those rules that are different from what I know from my home country. Not knowing a law will not be an accepted excuse if I break it, so it is my obligation to make myself familiar with it.

The same is true for project managers, whose actions or inactions will inevitably create legal precedent. They have to understand what actions are appropriate in the home countries of their teaming partners.

Project managers in international contexts are given valuable seminars today that help them navigate through *cultural* diversity, but in my observation, no help is given to them to understand and cope with *legal* diversity. Cultures and law interact intensively, but they are not identical. Culture influences law, because the people who make these laws embue them with their cultural understanding of what is right and wrong. Law influences cultures, because the people whose collective behaviors constitute this culture behave mostly in a way that does not bring them into conflict with the law.

On another level, cultures also interact with the behavior of those people who must enforce the law. In the USA, where many people own guns in many states without legal restrictions, police staff must take into account that the person they stop for speeding or running a red light may be armed to the teeth and may be prepared to use those arms against them. In Europe, where gun ownership is very much restricted and requires proof of reliability in character and lawfulness, policemen and -women are allowed to be far more relaxed, just doing their job without having to fear for their lives at every incident.

The person who gets stopped should understand the difference and behave accordingly: In the US, one should rather show submission and avoid quick gestures that could be taken as threats by the police person. In Europe, one is better relaxed and polite in order to avoid making the conflict over the traffic affair a personal conflict. Most European police people are experts in de-escalation and will show with a smile that they are thankful for self-constrained behavior. US-American cops must rather be experts in self-defense. The mutual influences of law and culture in a country are strongly amplified by the people whose job it is to ensure that law is complied with in daily life.

My clear recommendation is to teach project managers who interact with customers, contractors, and other business partners in foreign countries what the differences of their respective legal systems are. For many project managers, it may even be necessary to learn first the fundamentals of their own home law and how they relate to Project Business Management.

Again, one may argue that this is only a problem if one has lost the discussion on the applicable legal system and has to accept working under the rules of a foreign system. All one needs to do is to ensure that one's home law is the applicable one, and there will be no problem. In a competitive understanding, this is right. In a "Mission Success First" culture implementing good faith, one would also consider the problem that, in international contracting, there is always one teaming partner who has to work under foreign law, and both parties should discuss and implement measures that are necessary to make understanding and complying with this law easier for this company, particularly in consideration of the contract and the desired cooperation of the partners. In this approach, one does not desire to have any party involuntarily breaching the contract or the law.

One should also consider: If all parties insist that only their own law is accepted, there will no longer be international cooperation in project management. It is an inevitable element of international project business that at least one party works in a legal environment that is not its home environment.

3.5 Incomplete Contracts

I mentioned above the work of British–American Professor of Contract Theory Oliver Hart and others on incomplete contracts. They refer to a situation which they describe as follows:

> *Imagine a buyer, B, who requires a good (or service) from a seller, S. Suppose that the exact nature of the good is uncertain; more precisely, it depends on a state of nature which is yet to be realized. In an ideal world, the parties would write a contingent contract specifying exactly which good is to be delivered in each state. However, if the number of states is very large, such a contract would be prohibitively expensive. So instead the parties will write an incomplete contract. Then, when the state of nature is realized, they will renegotiate the contract, since at this stage they know what kind of good should be traded.*[13]

The statement was not dedicated to Project Business Management, but it applies to our field in a compelling way for two reasons:

1. Customers and other requesting project stakeholders are often not able to describe their needs. In my classes, I commonly ask a student if he or she sits comfortably. Because I select someone for the question who is visibly sitting at ease, the answer is typically, "Yes". Then I ask this person to describe the chair they are sitting on. I then ask them to focus on the cushion and describe the perfect softness of the cushion, between hard and extremely soft. In most cases, the student will finally describe this as "middle soft" or something similar.

   Then I ask a second student whether, if asked to make the chair cushion, would know what "middle soft" actually means, and how to make the cushion precisely "middle soft". Then, the answer is normally, "No". We probably spend most of our time sitting.

---

[13] (Hart and Moore 1998, p. 3, quoted with permission)

At work, in the car, aircraft, train, when we eat, and so on, we spend a lot of time in a sitting position; however, we are usually unable to accurately describe the perfect seat upholstery. In normal life, this is not a problem, but for someone who has the job to make the perfect seat, it does. This limitation applies to many projects: Requesters are typically not able to say exactly what they want until they see results. Then, they can easily describe the attributes of the deliverables that make them unhappy.

2. Requirements on the project are subject to change. These changes have many sources— internal as well as external—and have become an essential element of consideration and methodological development in project management. Being open for such change and adjusting practices accordingly was among the core topics of my first book, which focuses on "The Dynamics of Success and Failure".[14]

Figure 3.3 is a repetition from Chapter 1. It shows the result of a survey that I made in 2012, and for which I received 140 responses from a globally dispersed group of project managers.

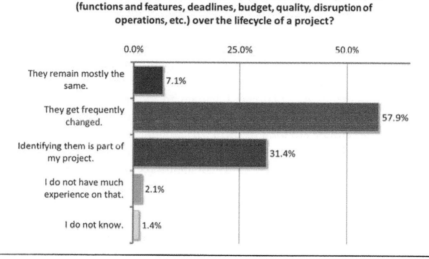

What is your experience regarding original stakeholder requirements (functions and features, deadlines, budget, quality, disruption of operations, etc.) over the lifecycle of a project?

Figure 3.3   The dynamics of stakeholder requirements on projects.

The results show how projects with changing requirements as well as the inability of stakeholders to describe requirements are far more common than projects with well-described and mostly static requirements. The situations described in the survey require different planning approaches, as Figure 3.4 shows.

Predictive approaches are good for situations in which the stakeholder requirements are static, allowing for long-time predictions. The agile approach is appropriate for projects that require exploration into the actual requirements and adaptation when these requirements are changed frequently. Between the two extremes is the *rolling wave* approach, which combines planning over a longer period than the typical one to four weeks in agile methods, but also allows for changes when new information becomes available, environmental conditions of the

---

[14] (Lehmann 2016b)

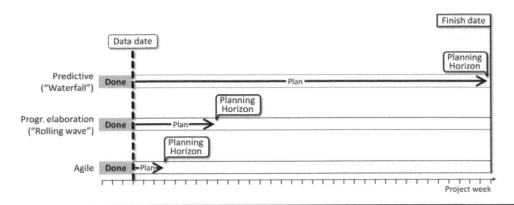

Figure 3.4   Differences in planning approaches in project management.

project change, or stakeholders communicate new requirements. In a rolling wave approach, a project manager asked what the planning horizon is and why it is exactly there should always be able to give a well-considered answer.

Linked with these different planning horizons are diverse granularities in planning and management approaches. Predictive planning is generally much more detailed, often micro-managed, to address all aspects that can make the investment in development fail and the plan fail. To avoid such failure, a strong top-down approach is then chosen in order to ensure that the project will be executed as planned. Changes are a major threat to this approach: The investment is valuable for projects with fairly predictable futures. Changes pose challenges, because when plans need to be adjusted to them, the amount of time and effort for replanning is also increased. Waterfall projects are mostly successful when the team is able to keep changes and other disruptions out of the project. Waterfall approaches are not appropriate for projects that include a high degree of discovery of requirements by exploration or creative work.

On the opposite extreme are agile approaches. They come with a much less detailed and granular planning approach. Often, they have no plan at all. Teams are rather self-managed in a bottom-up fashion. Scrum, for example, as the most popular agile method, has several roles defined, but project manager is not among them. Agile methods are best for projects for which the line applies, "Wanderer, there is no way, the way is made by walking".[15] Agile methods have difficulties in projects that require long-term predictions—for example, in order to book resources with a long lead time to have them available when they are needed, or to place orders to contractors for long-term work that (1) needs to be procured early to allow for a detailed vendor selection process, and (2) allows the contractor sufficient time to make what is needed.

From a project management rather than a legal perspective, a project contract is a type of plan, developed to govern the joint activities of the parties involved. It is different from other plans in that it binds not only the project: A work breakdown structure (WBS), a schedule, a human resource plan, a communications plan, and other project plans are binding for the project. The contract is binding for the entire organization. A contract is made of at least two parties, possibly more, and this rule is true for all of them. The binding nature goes beyond the limitations of the project and applies to the entire organizations. Another difference is the

---

[15] (Machado 2012)

legal nature: In a worst case, when conflicts cannot be remedied elsewhere, the parties will seek solution at court. Conflicts over plans are commonly managed inside the organization.

Besides this difference, similar rules apply to the contract as apply to those other plans. The contract can have a planning horizon designating the end of the period in the project up to which forecasts are made, resources are booked, and decisions are made. Figure 3.4 describes different planning horizons on the continuum between agile and predictive approaches. The planning horizon may be at the end of the work of the contractor, or it may exceed this date when subsequent warranty and service agreements are included in the contract.

The contract may also be valid until a certain deadline or milestone, by which time it may be renewed or replaced with a different one, or the relationship will then be terminated. The contract may be seen as a sacrosanct document or as a loose guideline, and it may include clauses for the processes that are used if the need for refinement or change arises. This would be similar to the schedule management plan that some project managers use to describe the processes to update or change the schedule. Oliver Hart and his colleagues called contracts of such an intentionally incomplete nature "agreements to agree".[16] The incompleteness is an adaptation to the uncertainty relating to the future—and sometimes even the present—that comes naturally with most projects.

In complex PSNs, refinements and changes become even more complex, as they can trickle down the network, or up, and even laterally. A change at one place in the project may make changes at other places necessary, and the more sophisticated the solutions used in the project, the more difficult it may be to identify and manage all consequential other changes, and the more important it gets. To make things even more difficult, many contractor organizations have professional claim managers, as mentioned before, who try to find constructive changes that allow for additional billing to their customer. Changes in complex PSNs open up many opportunities to finding such implicit changes, with the effect that the "Mission Success First" culture gets disintegrated and project costs rise massively.

3.6 Project-Related Contract Types

The following discussion will focus on contracts that relate to future work and results. If one goes into a shop and buys an existing product off the shelf, or orders a product that will be made as a one-off result in an existing production line, this purchase would also constitute a contract, but this would not instigate a customer project on the side of the supplier and is therefore not what I wish to discuss here.

The following contract types are the common types under which project contractors work for customers. The official definition of contracts is another difference between civil law and common law systems. In non-legal practice, the two typologies described in the following paragraphs are often used interchangeably. I often hear project managers in Germany refer to T&M contracts, which is in a legal sense not a contract type in the country, but a German project manager is nevertheless free to use this typological description in the country when it is correct for the contract.

---

[16] (Hart and Moore 1998)

3.6.1 Contract Types in Civil Law Systems

In civil law, project-related contracts are mostly defined based on the obligation of the seller, which could be either delivering a product or another kind of more or less tangible set of results, or providing a service, which means in essence to make resources available to the customer. A significant difference between the two types of contract is that a result contract[17] typically establishes the legal obligation on the seller to provide warranty against hidden defects for a defined period of time, which can, for example, last two years in the EU,[18] one year in Japan.[19] For a service contract,[20] there is generally no such legal obligation, but the contract can have such clauses, for example in a service level agreement (SLA) for software services, when the agreed-upon service level is not achieved by the contractor.

Incentives and other kinds of bonuses are rarely used in civil law countries, but price deductions or partial returns of upfront payments in the form of *contractual penalties* are common, and the word *penalty* is used here in its original meaning—as a punishment for not meeting contractual obligations independent of a detrimental effect to the customer's business, something that one would do better to avoid in common law countries.[21]

3.6.2 Contract Types in Common Law Systems

In common law countries, projects are rather typified by the obligation of the customer to pay. There are two groups of contract types:

- Fixed price contracts (with/without price adjustments)
- Flexible price contracts, based on the work or the costs incurred in the project

Fixed price contracts can have different forms. The most common are:

- **Firm fixed price.** A price has been agreed upon, and whatever happens, the price will not be adjusted. This contract protects the customer from cost risks until the first major change request occurs or until a judge may find that it conflicts with the doctrine of good faith (depending on the national law and the wording of the contract). There are also examples in which contractors have found ways to blackmail the customer, who is strongly dependent on the contractor.
- **Fixed with economic price adjustments.** These adjustments place some of the economic risks that may impact project costs on the customer, such as changing costs of raw materials and commercial off-the-shelf (COTS) goods, fees for preparatory work, transportation, licenses and other intangible cost items, and more. The base price as such remains unaltered.
- **Fixed price with motivational price adjustments.** These are adjustments to the price based on the performance of the contractor in the form of liquidated damages (LDs) or

---

[17] German: *Werkvertrag;* French: *Contrat d'entreprise;* Spanish: *Contrato de obra.*

[18] Implemented in partially different national laws, for example in Germany, Bürgerliches Gesetzbuch §438, §479, §634a.

[19] Japanese Civil Code, Articles 566, 570.

[20] German: *Dienstvertrag;* French: *Contrat de service;* Spanish: *Contrato de servicio.*

[21] "The ban on penalties [. . .] is one of the oldest rules in Anglo-American law". (George 2007, 52)

incentives. They are mostly linked with schedule dates, but they may also be linked with delivering special items or functions that are not mission critical but are "nice to have". Another application that one comes across from time to time is the linkage to operational disruptions, which a customer wishes to be limited, and for which a monetary incentivization may be considered. Another form is award fees, which are paid for the subjective performance of the contractor and are not subject to appeals at court. I will discuss them in detail below. The mechanics of motivational price adjustments will be discussed later.

- **Unit price contract.** In this form of fixed price contract, the price is not agreed upon for the entire project scope but for certain units that occur within it. An example is roll-out projects that plan to implement software in a number of countries, and a price tag is agreed per country.

Other contracts have a variable price, depending on costs incurred, resources made available for the project, or on the amount of work done for the project. They are a form of "lean contracting", in that changes in scope rarely require changes in the contract, saving time for negotiations, rewriting, and re-signing.

The forms of variable contracts most commonly found are:

- **Cost (reimbursable) plus percentage fee.** For contracts of this type, the contractor needs invoices from subcontractors and other vendors that can be re-invoiced to and reimbursed by the customer, with an agreed-upon percentage as an add-on to cover the prime contractor's general and administrative costs, account for the risks that come with the business, and allow for a profit. The price to the customer is then the sum of original costs plus the fee.
- **Cost plus fixed fee.** A disadvantage of the percentage fee contract is that it gives the prime contractor an incentive to generate cost overruns, because these would also increase the fee. A fixed fee avoids this; it remains static when costs for subcontractors rise. Here, the price is also calculated as the original costs plus the fee, but the fee will remain the same independent of the prime contractor's costs.
- **Time and materials.** In these contracts, the price is calculated as an hourly or daily rate, multiplied by the hours or days that human resources and equipment work for the customer—or, alternatively, are available for the project—plus agreed-upon prices for materials consumed by the project. These prices are agreed upon independent of the original costs that incur for the prime contractor.
- **Target cost contract.** A cost reimbursable contract with a cost target agreed upon. Cost overruns and underruns against the cost target are distributed in a shared ratio between customer and contractor. Target cost contracts often have a price ceiling—a form of mixed contract described in detail below.
- **Variable contracts with motivational adjustments.** These contract types can also be combined with penalties, LDs, or incentives, as described further below.

3.6.3 Assigning Cost Risks to Contract Parties

The two definitions of contract types can be combined in four ways. Commonly, a result contract is coupled with a fixed price for just this result—for example, a product needs to be made and delivered, and a service contract often has a variable price to allow the customer flexibility when more or less of the service is found to be needed during the project. This does not disallow selling a service at a fixed price. A common day-by-day example is a contract for mobile phone and data services, which is based on a flat rate, a form of monthly fixed price, for access to the provider's mobile network.

Figure 3.5 shows how the assignment of cost risk varies between the different types of contract.

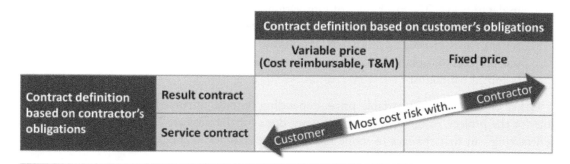

Figure 3.5   Assignment of cost risks depending on the project type.

It is worth repeating that the protection from cost risks that a customer may seek by insisting on a fixed-price result contract can vanish with the first change request.

3.6.4 Motivational Price Adjustments

When a certain delivery or completion date for a contractor is mission critical for the project, the customer will probably impose a deadline. In addition to this critical date, there may be dates that are desirable for the customer, but not critical. Such dates are often linked with motivational price adjustments to incentivize or penalize the contractor to meet the date. Motivational price adjustments may also be linked to other performance criteria, such as special functions or features, limitation of operational disruptions during the project, and other aspects of the project's performance that have a value for the customer. Motivational price adjustments can be agreed upon with every contract type described above. The handling of such adjustment clauses is different between the legal systems and also between countries.[22] Three types are commonly used, and different countries prefer different clauses. I will describe a fourth type—award fees—further below:

- **Contractual penalties.** These are commonly used in civil law contracts. Independent of an actual damage that missing a deadline or another criterion has for the aggrieved party, a price deduction (or a partial return of payments made in advance) is agreed upon if the damage case occurs.

    The criterion is defined (mostly) by the customer to enforce that the contractor meets the contractually agreed-upon criterion, a deadline, certain functionality or product

---

[22] (Smith 2008)

performance, or a maximum duration that the project is allowed to disrupt the customer's operations.

In most civil law jurisdictions, prohibiting parties from including a penalty clause in a contract would be regarded as an undue constraint of the parties' freedom to enter contracts as they desire, but the penalty will have to stand the test of good faith. A penalty clause, for example, included in a contract by party A with the objective of unfair enrichment to the disadvantage of the party B, will probably not pass this test and be either reduced by a court to what it considers "reasonable" or not enforceable at all.

- **Liquidated damages.** As written above, the term *penalty* should be avoided in common law, where it will be considered highly suspicious when it comes to legal action. A common-law judge considers the parties as free to enter a contract, but considers them as generally being on eye level.

  Each party has applied *consideration*—that is, whether the benefits that they expect from the contract trump the disadvantages that they will have to bear, such as costs, obligations to do or omit something, or risks that they will assume. The parties were equally free to enter into this legally enforceable agreement or not.

  Penalty signals a top and bottom situation: A country can penalize citizens, parents can penalize children, and so on. A penalty that is in essence independent of an actual damage and is not intended to compensate for such a damage would be a breach of this eye-level principle. To the common-law judge, a penalty is rather a means used for assault and battery than for an appropriate agreement between two or more parties that at least have to respect each other's free will.

  The court will also want to make sure that the clause does not lead to unjust enrichment by the aggrieved party. A solution used in most common-law jurisdictions is therefore to talk of LDs instead, for which the aggrieved party receives compensation. "Liquidated" in this context means that the damage has been assigned a monetary value that is fixed in the contract, and whose amount the aggrieved party therefore does not need to verify when the damaging situation has occurred.

  Technically, LDs function identically to penalties, but their justification is different. They are regarded as compensation of a loss that is hard to anticipate, calculate, and verify, more than as an enforcement or deterrence.

- **Incentives.** The enforceability of LDs to litigate breach-of-contract situations has limitations in most common law countries: If the actual monetary value of a damage suffered by the aggrieved party is much less than the amount specified in the contract, and if this amount is easy to forecast and to verify, a court will probably regard the claim from the contractual LDs as inappropriate overcompensation and as a kind of hidden penalty, which makes it unenforceable.

  Contract parties may therefore seek for an alternative solution to such a *malus* system by providing a bonus instead, in the form of an incentive for adhering to specified agreed-upon criteria.

The most common criterion used for such adjustment clauses is probably linked to meeting delivery dates. Table 3.2 shows how they work.

Table 3.2 Motivational Price Adjustments Used to Motivate a Contractor to Deliver by a Certain Date Under Certain Fixed-Price Contracts

Examples for Motivational Price Adjustments	
1. Frequently Used in Civil Law: Contractual Penalties	
Price for product xyz:	€1,000,000
Penalty for delivery after 31 Dec:	–€100,000
Net price in case of delivery after 31 Dec:	€900,000
2. Frequently Used in Common Law: Liquidated Damages (LDs)	
Fixed Price:	£1,000,000
LD for delivery after 31 Dec:	–£100,000
Net price in case of delivery after 31 Dec:	£900,000
3. Frequently Used in Common Law: Incentives[a]	
Fixed Price:	$900,000
Incentive for delivery until 31 Dec:	$100,000
Net price in case of delivery until 31 Dec:	$1,000,000

[a] In my observation, most frequently used in the USA.

Incentives communicate a more positive attitude than do penalties or LDs, but they have limitations. They do not protect a party from complete non-performance, from only partial performance, or from other forms of breach of contract by the aggrieving party. Contracts that involve incentives therefore often have an additional LD section in place to account for a party missing minimum requirements that have been contractually agreed upon.

3.6.5 The Capped Target Cost Contract

This is a mixed contract type that I have found quite common in the USA, often inaccurately named as "fixed-price plus incentive contract". It is widely unknown in Europe and in other countries. As we will see, it is not a fixed price contract at all, but it can turn into one when cost overruns exceed a certain limit, called the *point of total assumption* (PTA). I will explain the contract in a step-by-step approach.

It is always preferred for this model to assume that the contractor is a prime contractor, who hires subcontractors. These subcontractors will send invoices to the contractor, who will (1) pay them, and (2) re-invoice their amounts to the customer with the fixed fee added on top. Often, a calculation of internal cost is used, which makes this contract type effectively a T&M contract, in that these costs are a model rather than actual costs.

The explanation begins with a simple cost reimbursable contract with fixed fee, which the customer is prepared to pay the contractor for general and administrative costs, cover the risks of being the prime contractor, and allow for a reasonable profit. In the example, a cost target has been agreed upon at $1,000,000, the fee is at $250,000. This adds up to a price target of $1,250,000. We are looking at costs, price, and fees for three scenarios, one with a saving of $500,000 against the target, one at target, and a third scenario in which the price exceeds the target by $500,000. These three scenarios are shown in Table 3.3.

Table 3.3 Step 1: Cost Reimbursable Contract with Fixed Fee[a]

Uncertainty over scenarios (±)	500,000
Cost target	1,000,000
Fixed fee	250,000

Scenario	Cost of the contractor	Variance	Fixed fee	Price to the customer	Margin for the contractor
Low-cost	500,000	−500,000	250,000	750,000	250,000
Target	1,000,000	0	250,000	1,250,000	250,000
High-cost	1,500,000	500,000	250,000	1,750,000	250,000
Cost/price risk	*1,000,000*			*1,000,000*	*0*

[a] It begins with a simple cost reimbursable contract with a fixed fee.

Following this model, the customer assumes the entire cost risk, $1,000,000 in the example. The customer may complain that there is no incentive for the contractor to save these costs, and therefore a cost/benefit sharing is introduced to the contract to give the contractor a benefit if the costs to be paid by the customer are kept at a minimum by the contractor. The sharing ratio of 80/20 for customer/contractor shown in Table 3.4 is quite common:

Table 3.4 Step 2: Cost Reimbursable Contract with Fixed Fee and Cost/Benefit Sharing (Target Cost Contract)[a]

Uncertainty over scenarios (±)	500,000	
Cost target	1,000,000	
Fixed fee	250,000	
Cost/Benefit sharing	80/20	(Customer/Contractor)

Scenario	Cost of the contractor	Variance	Customer share of variance	Contractor share of variance	Fixed fee	Price to the customer	Margin for the contractor
Low-cost	500,000	−500,000	−400,000	100,000	250,000	850,000	350,000
Target	1,000,000	0	0	0	250,000	1,250,000	250,000
High-cost	1,500,000	500,000	400,000	−100,000	250,000	1,650,000	150,000
Cost/price risk	*1,000,000*				*0*	*800,000*	*200,000*

[a] The cost risk is now shared between the customer and the contractor. The margin that the contractor makes is the fixed fee ± the contractor's share of the cost deviation from the cost target of $1,000,000.

The margin for the contractor can be calculated price − cost or fixed fee ± contractor's share of the cost variance over the scenarios. It now entails a risk of $200,000. The customer is left with a risk of $800,000. This mirrors the 80/20 sharing ratio.

The customer is still unhappy. It may be taxpayer's money that is being spent, or there may be other reasons to cap the price. The price ceiling in the example has been fixed at $1,450,000, which means that this is the maximum price that the customer will have to pay, according to the contract. Table 3.5 shows the numbers.

Table 3.5 Step 3: Target Cost Contract with Price Ceiling[a]

Uncertainty over scenarios (±)	500,000	
Cost target	1,000,000	
Fixed fee	250,000	
Cost/Benefit sharing	80/20	(Customer/contractor)
Price Ceiling	1,450,000	

Scenario	Cost of the contractor	Variance	Customer share of variance	Contractor share of variance	Fixed fee	Price to the customer	Margin for the contractor
Low-cost	500,000	–500,000	–400,000	100,000	250,000	850,000	350,000
Target	1,000,000	0	0	0	250,000	1,250,000	250,000
High-cost	1,500,000					1,450,000	–50,000
Cost/price risk	*1,000,000*					*600,000*	*400,000*

[a] The cost risk is now shared between the customer and the contractor. The margin that the contractor makes is the fixed fee ± the contractor's share of the cost deviation.

Compared to Step 2, the low-cost and target scenarios remain the same, but the high-cost scenario changes very much. The invoices from subcontractors in the example are adding up to $1,500,000, but the payment by the customer has been capped at $1,450,000. The contractor has therefore a negative margin—a loss—of $50,000. The contractor's risk over the three scenarios has increased to $400,000, but at this point, the cost risk is already fully assumed by the contractor. The contract that was originally a cost reimbursable contract with cost/benefit sharing around a cost target has turned into a fixed-price contract, and the customer has shifted the full cost risk onto the contractor.

It is interesting to note that the cost point at which the contract changes its character must be somewhere between $1,000,000 and $1,500,000. This so-called *point of total assumption* (PTA), meaning the total assumption of the cost risk by the contractor, is calculated following the formula:

$$\text{PTA} = (\text{price ceiling} - \text{price target}) / \text{client share} + \text{cost target}$$

For the example, this computes to:

$$\text{PTA} = (1{,}450{,}000 - 1{,}250{,}000) / 0.8 + 1{,}000{,}000$$
$$= 200{,}000 / 0.8 + 1{,}000{,}000 = 1{,}250{,}000$$

Table 3.6 shows what happens with the contract when the PTA is reached. It is the cost number at which the contract becomes a fixed-price contract.

Table 3.6 Step 4: Calculating Costs at the Point of Total Assumption (PTA)[a]

Uncertainty over scenarios (±)	500,000	
Cost target	1,000,000	
Fixed fee	250,000	
Cost/Benefit sharing	80/20	(Customer/contractor)
Price Ceiling	1,450,000	

Scenario	Cost of the contractor	Variance	Customer share of variance	Contractor share of variance	Fixed fee	Price to the customer	Margin for the contractor
Low-cost	500,000	–500,000	–400,000	100,000	250,000	850,000	350,000
Target	1,000,000	0	0	0	250,000	1,250,000	250,000
At PTA	1,250,000	250,000	200,000	–50,000	250,000	1,450,000	200,000
High-cost	1,500,000	500,000				1,450,000	–50,000

[a] The contract changes its type at the point of total assumption. Up to this cost number, the contract is a target cost contract with cost/benefit sharing, but when the costs exceed the PTA, the price can no longer go up, and the contract turns into a fixed price contract.

3.6.6 The Rolling Award Fee Contract

The previous contract type may not seem easy to understand, but another group of contracts cause even more confusion: award fee contracts.

The award fee is an incentive that is not linked to objective data such as a delivery date or the achievement of certain measurable performance criteria by the project's deliverables, but rather to more subjective criteria that culminate in the question of the degree of excitement that the contractor could cause for the customer. If the customer is not sufficiently excited with the contractor, this incentive will not be paid to the contractor, and the contractor cannot appeal against this decision at court. One cannot sue someone for not being excited.

When I introduce the award fee contract to my students in project management classes, the common reaction is amazement combined with incredulity. Can it be that customers decide on their own discretion whether they are going to pay this kind of bonus to the contractor? And if yes, why should they do that?

Award fee contracts can be a strong means to develop a system of relational contracts in a complex PSN, which should combine good faith among its members with agility and intensive communication. In order to achieve that, the award fees should be paid in a rolling fashion.

There are some prerequisites. The first one is a budget to pay the fee when the contractor deserves it. Developing this budget is rather simple. One estimates the costs that poor communications by contractors, protective and competing behavior, and general distrust will cause to the project in the form of costs for delayed benefit realization resulting from late delivery, rework, lost incentives, overtime work of project staff members, and more. One then takes a part of that sum, I recommend 20 percent, and makes this the budget for the reward fee. When the budget is created this way, the contractors are essentially asked to support the customer in saving costs, and a part of these cost savings are then shared with the contractors. I recommend communicating it precisely as such a share to the contractors.

Then a simple scoring system should be installed, which both gives immediate feedback to the contractor for the cases that the award fee is paid or not paid and helps the contractor understand the causes. I recommend a monthly rhythm, but bi-weekly or bi-monthly may sometimes be more appropriate for specific situations. Figure 3.6 gives an example of such a score sheet.

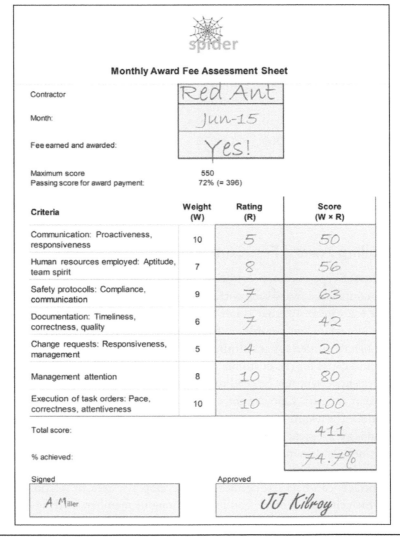

Figure 3.6 A score sheet indicates the results achieved by the contractor and how these are computed from a rating and a weight.

The score sheet should have between five and eight, maximum ten criteria to make it swift and easy to fill in and to understand. I recommend for the weighting column numbers between 1 and 10 for simplicity, but any other number range can do as well. For the rating, numbers between 0 and 10 can be used, or other numbers that are preferred. I saw one case using percentage weights that added up to 100 percent. This has a disadvantage in additional calculation effort needed when one wants to change the number of criteria or the weights assigned, but otherwise, this is also OK. The scores are then calculated from multiplying the rating with the weight for each criterion. The total score is then calculated as the sum of the individual scores, and this number will lead to the payment of the award fee when it equals or exceeds a predefined passing score, or the fee will not be paid when the score is less than the passing score. The passing score in the example has been calculated as 396, 72 percent from the maximum score possible, which here is 550. Because the contractor has exceeded this number in the example for June 2015, the award fee will be paid for this month.

The rolling award fee contract based on a weighting system has some advantages for relational contracting:

- It can be combined with any other contract type—such as fixed price, cost reimbursable, or time and materials—and also with the civil law contract types.
- The contractor knows how to contribute to the project to make the customer happy.
- The feedback to the contractor is short term, which makes it more effective.
- Every time the award fee is paid, mutual trust is increased.
- The contractor understands the value that the customer assigns to the different criteria and knows on which criteria more emphasis should be put, because these bring higher effect on the total score.
- The contractor gets immediate feedback on the performance of the last month and can make adjustments in the current month to better delight the customer.
- The contractor's project manager gets a business case to invest more money, time, and energy in the customer; if this is done well, it will pay back.
- The contractor gets additional monetary resources that can be used to the benefit of the project—for example, to hire better people and acquire better machinery.
- The contractor can use some of the fee paid and share it with subcontractors. In such a way, a cascade of award fees can ensure meeting customer needs and building great deliverables across complex PSNs.
- Award fees are a strong endorsement for the contractor's ability to perform. Documented in a way that protects the privacy needs of the parties involved, they can help win better future business.
- They are an appropriate addition to contracts that include the use of agile methods, in that they can make the successful implementation of agile practices one of the criteria.
- A score sheet is a great basis for a meeting between the parties to discuss successes, issues, and areas in which improvements can be made.

There are some reasons that can make the rolling award fee contract fail:

- **High-pressure environments.** Under high pressure, people learn less the lessons of effective collaboration, but more how to hide poor work or to blame others when this is no longer possible. The rolling award fee contract puts completing over competing; high-pressure environments bring competing back.

- **The intention not to pay the award fee.** The rolling award fee contract works best when the customer's intention is to pay the fee, not to save it. The fee is a signal that the customer has a great project that, among other benefits, saves cost, and the customer gives some of the savings back to those who help realize these benefits. The payment of the award fee should be considered good news on the side of the customer, not a liability.

- **New areas of conflict among contractors.** A contractor may find that poor collaboration by another contractor leads to dissatisfying results and costs him an award fee payment. A contractor may also find that the good work delivered led to an award fee payment to another contractor, who took the laurels for the results. It may be a challenge in a complex PSN to always correctly assign the successes to the right parties, and I recommend considering this question early, before work is assigned to the contractors, to avoid new blame games.

3.7 Protective and Relational Contracting

One of the most difficult things in project management is knowing where the project actually stands and where it is going. High-pressure environments are common in project management, and I know many practitioners and experts who speak proudly of them. In an e-mail, I was told some time ago, by a person I would consider a kind of bedrock in project management, that pressure in a project "separates the men from the boys". My response was that I would consider a project successful when it delivers what is needed and wanted, and do not care about such a separation. The promoters of pressure are very noisy, and their arguments seem convincing at first glance.

This book is more focused on helping project managers to complete, but there are situations that require our ability to adopt a protective and sometimes very confrontational attitude and to compete. It may be a good idea to reduce the number of such situations to an unavoidable minimum and focus most of the time on building a "Mission Success First" culture in the project and around it, but sometimes too many stakeholders are involved, and too often others bring up conflicts that we cannot avoid.

The rolling award fee contract is an example of a contract type that attempts to reduce pressure on the parties involved and to build an environment based on good faith, mutual trust, and fairness—the objectives of relational contracting. The capped target cost contract discussed previously is rather unfair by design: Cost savings are shared in mutual partnership, but cost overruns will be the liability of only one party—the contractor—from the moment that the PTA has been exceeded, as Figure 3.7 shows.

The contract intends to ensure that both customer and contractor benefit from cost savings—the customer by a lower price and the contractor by a higher margin. The same happens when costs increase, which also increases the price for the customer and reduces the margin for the contractor, but only until the price ceiling is reached, at which point the contractor's cost are at the PTA. At this point, the sharing agreement gets broken, and further cost increases will directly reduce the contractor's margin, as the price does not grow any further.

The PTA as a Breakpoint in the Cost-Price Development

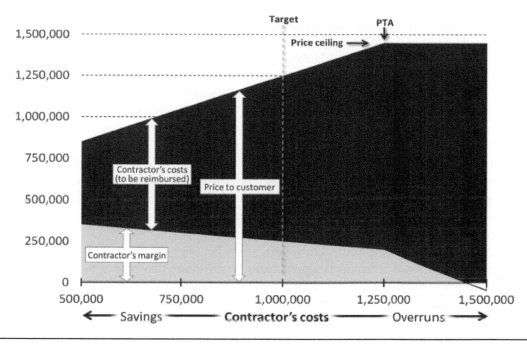

Figure 3.7 Cost savings to the left of the target line are shared between customer and contractor. Sharing cost overruns ends at the point of total assumption, from which point on the contractor assumes the total cost risks.

The attitude of the price ceiling is obviously to protect the customer from cost increases beyond the PTA by putting the full cost risk on the contractor from this point on. One may argue in defense of this contract type that it is solely the contractor who is responsible for meeting the cost target and for keeping the project costs under the PTA; there are of course projects for which this statement is true.

For many projects under contract however, achievement of cost objectives relies as much on the customer's discipline and self-control to meet cost targets as on the contractor's skillfulness. The customer can make a project less or more expensive with the quality and timeliness that buyer-side obligations and communication needs are met. A customer can also influence project costs with the way change requests are brought up and with the speed that upcoming problems that need the customer's input, responsiveness, and collaboration for resolution are sorted out.

Protective contracting is based on confrontational behaviors: The central question is, "How can we shield ourselves from liabilities and damage claims when things turn nasty, making sure that these claims will be directed toward the other party?" Relational contracting rather asks, "How can we protect all parties involved from things turning nasty?" Both approaches may be appropriate in specific situations; the problem is to foresee what a business relationship will look like in the future and then find the best balance between the two extremes.

3.8 Refinements and Changes

Swallowtail, LLC.,[23] is a construction company that had the order to build a major section of a new light rail line in a US city. The work was done under a combination of cost reimbursable and time and material (T&M) contracts, the former for work done with subcontractors, the latter for work done with their own staff and equipment. The customer used the freedom that the selected contracts offered to exert a lot of influence in the project, which was considered mostly beneficial to the contractor, because it allowed them to invoice additional work to the customer and keep resources busy and billable for a much longer time than was originally agreed upon.

Under the circumstances, the project got delayed by several months. One day, Swallowtail received an unwelcome surprise: a threat to open a lawsuit by the customer over a significant amount of US dollars for delays in the project. The delays added costs to the project that the customer needed to bear and also prevented the city from gaining earnings from ticket sales. The customer wanted the contractor to pay damages for these losses, which the contractor was not prepared to pay.

The case was then taken to court. During the court hearings, Swallowtail's project manager reported the existence of 50 change requests from the customer that needed to be assessed for impacts and risks, and that needed decisions on their acceptance or rejection. The large number of change requests, Swallowtail claimed, was the single most important cause for the delays. The customer rejected this statement, saying that they essentially had five change requests, a number that one would consider normal in a project of such a magnitude and duration. The case was finally settled in a compromise agreement, which granted the customer about 50 percent of the original claim. For the contractor, the settlement removed all profit from the project but prevented it from becoming a loss maker.

The most fundamental problem in the project was the unclear distinction between *refinement* and *change*. The customer considered certain changes (in the eyes of the contractor) to be refinements. Under the contracts selected, the city would pay the additional costs anyway, which gave it the feeling of freedom that many great ideas could be implemented and disliked work results were reworked. The contract also gave the contractor a feeling of certainty that the changes were only beneficial to the business with the customer. The contractor missed the fact that the customer might not understand how the requests would lead to delays and finally seek reimbursement for them. The damage from the lawsuit to the business of the contractor was indeed significant; the year in which it happened turned out to be the worst years in the history of Swallowtail.

I have repeatedly used the words *refinement* and *change*. In a contractual environment, it can be important to have a clear understanding of their common aspects, but also, importantly, their differences. The assumption by a customer that a new requirement or specification is a refinement, while the contractor considers it a change, is among the most classic causes of conflicts in customer projects. So what is what?

Refinement is a normal and mostly predictable element of progressive elaboration. The rolling wave approach used in the project uses only rough estimates and descriptions of work and results early in the project. Later, with more knowledge, these requirements and specifications will be made more precise and accurate.

---

[23] Name changed.

Table 3.7 The Distinction Between Refinement and Change

	Refinement	Change
Generation	A result of an iteration cycle during progressive elaboration	Requested by a stakeholder
Generally predictable?	Yes	No
First contact for request	Project manager	Preferably: Project sponsor
Changes requirements on scope, time, etc.?	Mostly: No	Mostly: Yes
Should impacts and risks over knowledge areas be analyzed?	Often	Always
Requires a written change request?	No	Preferably: Yes
Customer projects only: Should amendments to the contract be considered?	No	Yes
Should the change decision be escalated to the project sponsor or CCB[a] or an SC[b]?	Possibly	Possibly
Should the process be documented?	Yes	Yes
Can it turn a project from crisis to success and vice versa?	Unlikely	Yes
Fixed price contract: Can the price be renegotiated?	Probably no	Probably yes
All contracts: Can deadlines be renegotiated?	Probably no	Probably yes
Will the projects benefit from it?	If done at the right time, yes	If done at the right time, yes
Can it drive the project into problems or even crisis?	Unlikely	If poorly managed, yes

[a] Change Control Board
[b] Steering Committee

Change requests come as surprises. If one would have known about them in advance, one could have planned differently. Processes for change requests should be in place for all projects unless agile methods are used, which are basically designed to cope with frequent changes. These change request management processes become essential when projects are performed by companies working together in customer–contractor relationships. Here, two or more process worlds, one in each organization involved, must be coordinated to manage these changes.

Often, refinements and changes are hard to separate from each other. A predictable refinement may be used by a party to include some changes. Now that the books that include the specifications, forecasts, and plans are opened to process and document refinements, there may be some things that could be done differently. It can also work the other way: Assessing the impacts of a change request may make it necessary to plan things in more detail; otherwise, the understanding of the impacts gets lost in the overall ambiguity and vagueness of the plans. This overlapping may make it difficult to clearly separate change and refinements, but a brief glance at Table 3.83.7 reveals how important this distinction is.

The case story shows how ambiguity in the question of what constitutes a change versus a refinement can lead to conflicts that damage the project at least for one party in the contract. I recommend paying particular attention to clarifying the borderline between the two forms of project management activity to avoid misunderstandings and quarrels.

The intention to replace *competing* with *completing* in the *project supply network* and to build and uphold a "Mission Success First" culture jointly with our contract partners forces us to be acutely aware of the moments when this intention gets challenged.

The management of change requests and refinements are indeed among the most vicious challenges to this intention. These are times when it will not be sufficient for our contract partners to do their work as ordered; They have to pay particular attention to business interests, their own and those of their customers, contractors, and other partners in the *project supply network.* Moments of change and refinement are the moments when projects can be improved, but also when the foundations can be laid for future troubles—in the worst case leading to lawsuits and other forms of crises in the PSN.

Chapter 4

Managing Complex and Dynamic PSNs

The book so far has focused to a major degree on sellers. Some aspects, such as contract selection and the entire contracting process, would be interesting for both sellers and buyers. The following contents are more of concern for buyers who need to manage complex and often highly dynamic project support networks (PSNs). They should nevertheless also be of interest for sellers. One of the major success obstacles and profit destroyers is poor project management on the customer side. Another reason is that a project manager working for a prime contractor is simultaneously a contractor to the customer and a customer of the subcontractors. In project management under contract, it is often impossible to say: "This is not my business".

4.1 Change Requests in Complex Project Supply Networks

Crane Fly Corp. is a manufacturing and service company with several large production facilities distributed all over the world. In 2016, they initiated a project to expand one of their existing production plants in order to increase output and modernize production. A major element of the project was an innovative production management system, with thousands of sensors distributed over the plants to measure the utilization and condition of equipment and tools and to identify areas for ongoing optimization during production. Data from these sensors would be collected in a data processing center, which would analyze it and make decisions for actions. As a learning system, the data processing center would be able to make these decisions in interaction with humans first and later turn to an autonomous modus operandi. The company called that "artificial intelligence in manufacturing".

Another major goal was reduction of setup times, when production batches needed to be changed, and cutback of idle times of machinery and personnel, which was in some of their productions at over 90 percent. They considered this number highly inefficient and costly and desired to reduce it by bringing higher flexibility and adaptivity into production workflows. The data processing center was expected to help on this task as well—it was expected to cut back *maintenance, repair, and overhaul* (MRO[1]) durations—and to also reduce the *mean time between overhauls* (MTBO) by calling for services based not on equipment working schedule but on equipment condition and on the health of the allover manufacturing process.

A further option was to use highly flexible and mobile production equipment whose idle time during one production workflow, in which it was not involved, could be utilized to support another one. Humans can support such a task to some degree, but the complexity of a production facility, in which a major number of production workstreams runs concurrently, might prove too complex to achieve such flexibility without computer help. The business case was clear and easy to achieve, assuming that the project would run smoothly (which it obviously did not, otherwise I would not tell its story here).

The principal element of the new production management system was the data processing center, whose development and implementation were outsourced to Earwig Ltd. Earwig was the main contractor to a number of subcontractors. The company was expected to manage these companies, but also to cooperate with other direct contractors of the customer, which provided services and delivered additional infrastructure for the manufacturing plant. The plant was huge, and so was the number of contractors involved, with an even larger number of interfaces between them that needed to be taken care of so as to not negatively affect contractual work assignments by doing work twice, leaving work half-done, having workers and equipment stand in the way of others, and allowing conflicts between contractors' workforces to develop.

In the end, the project was successful, but with major delays and at about twice the costs that were originally budgeted. Post-mortem analysis of the project showed that this was caused by the frequent change requests by management on the customer side that needed to be implemented by the complex PSN. Every change request trickled down the PSN, and even seemingly minor changes at one point of the plant not only needed to be communicated to the contractors but also resulted in changes to their prices, fees, and delivery dates. Management was not aware of the major claims from contractors that would follow these changes and were not prepared to listen when they raised the changes. Change request management in a complex PSN is difficult, because it can change the contractual relationship as much as the interpersonal.

4.2 Introductory Questions

The following questions are written in the style of a certification test. They are intended to give you an understanding of the contents of the following text section and the questions that will be discussed in it. It may be interesting for you to answer these questions before you read the section, and then again once you have finished it.

> 1. Which of the following observations is NOT among Lencioni's five dysfunctions of a team?
>   a) Avoidance of accountability
>   b) Absence of trust
>   c) Fear of conflict
>   d) Inattention to processes

---

[1] Another commonly used explanation for the acronym is *maintenance, repair, and operations.*

2. A project supply network is developed, consisting of organizations that act together to deliver the mission of the project. Each of the companies in the supply network works under contract for a company that took over this project as a prime contractor from a customer. What is probably true for the business situation?

   a) Subcontractors will do work for the project, and the customer has no contractual relationships with them.
   b) Subcontractors will do work for the project, and the customer has contractual relationships with the prime contractor and also with them.
   c) The prime contractor will not be responsible for delays, quality problems, and other issues caused by subcontractors.
   d) The customer will have to accept the prime contractor's prohibiting direct communications between customer and subcontractors.

3. A customer has subcontractors nominated to the prime contractor. What does this mean?

   a) For a specific work item, a subcontractor has been named by the customer. The prime contractor must subcontract this company for the item.
   b) The customer gave the prime contractor a list with companies that are approved as subcontractors for a certain work item, and the prime contractor selects one of them.
   c) The customer gave the prime contractor a recommendation list with potential subcontractors, but the prime contractor is free to subcontract to someone else.
   d) The prime contractor is given a blacklist with companies that are not acceptable as subcontractors, and while the prime contractor is free to subcontract, listed companies are excluded.

4. A customer project is approaching handover and acceptance. What is true for that?

   a) Final handover and acceptance must always be done in one process, whose completion includes proof that both are formally done.
   b) Although an orderly acceptance is important to ensure the success of the project, the value of the final handover is rather negligible.
   c) Although an orderly final handover is important to ensure the success of the project, the value of acceptance is rather negligible.
   d) Final handover and acceptance can be done in one process or separately, possibly with weeks between the dates.

5. Alternative dispute resolution (ADR) includes which of the following?

   a) Mediation and arbitration
   b) Coaching and mentoring
   c) Distributing boxing gloves
   d) Smoothing and avoiding

6. A project manager on the customer side deals with a prime contractor but does not know who the subcontractors in the project are, and she is not interested in knowing. She is assuming that the prime contractor has the full responsibility, and that everything is covered in the contract with that company. However, why should it matter to her?

a) The subcontractor organizations are potential employers that could recruit her when the project is failing and she gets fired.

b) The customer has legal and contractual obligations against the subcontractor and vice versa, and she must ensure that these are fully met to avoid legal action.

c) The contract regulates what happens when the subcontractors do not perform. It does not protect the project from such malperformance.

d) The subcontractors' employees are potential objects for recruitment. They are experts and know the project and its products.

4.3 Teaming Agreements

Teaming here is used to describe how two or more organizations work together to achieve results that each of them alone would not be able to achieve. As discussed above, they do that by tapping the assets of other organizations. In the simplest teaming structure, which consists of just one customer and one contractor, the customer taps into those assets of the contractor that can be used as project resources, such as people, equipment, licenses, and management attention, and the contractor taps into the customer's financial resources by sending invoices. Even such a simple structure may become more complex—for instance, considering provisions and enabling services that the customer must provide to facilitate the contractor's work.

What does teaming look like when there are more organizations involved? There are in essence three types of teaming agreements that are used for projects, as Figure 4.1 shows. Each model has advantages and disadvantages, which will be discussed below.

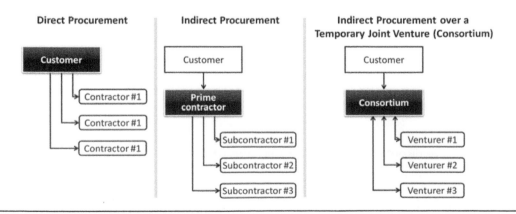

Figure 4.1   Different structures of teaming agreements. The parties in the black boxes are in charge of managing the project. The arrows depict the general flow of money.

4.3.1 Direct Procurement

In this model, the customer manages the project and all contractors directly. The customer hires the contractors and has independent contracts with each of them. The contractors have, in most cases, no contracts among them, except possibly for commission when one contractor brings another one into the business.

Often, one of the contractors is in a lead role. This contractor may have the best relationship with the customer, may have the largest share of the work, and may have critical licenses or any other assets that are essential to completing the project. Direct procurement comes with benefits and disadvantages:

- **Benefits**
  - Generally lower costs, because no prime contractor is paid for organizational work.
  - The project manager considers only the interests of his or her own organization in decision making.[2]
  - Default of a contractor damages the project only partially.
- **Disadvantages**
  - The customer may not have sufficient experience and knowledge to perform the project and manage the contractors.
  - The customer bears the risk of failure alone, particularly failure from poor performance of the contractors and from a lack of coordination of their work.

I will discuss these options again later in this section.

4.3.2 Indirect Procurement

The customer contracts the organizational tasks to the prime contractor, who in turn contracts this work partially or in full to one or more subcontractors.

The lead contractor role in this model is naturally filled by the prime contractor. The customer may be aware of some or all subcontractors assigned or may not know at all that the direct contractor has given work to subcontractors.

In some cases, the customer is involved in the selection of subcontractors by *naming* them, which means that the customer tells the prime contractor who to work with; by *nominating* them, which means that the prime contractor gets a list of accepted subcontractors from the customer and makes a selection from this list; or by an *acceptance process,* in which the prime contractor must introduce a potential subcontractor to the customer and can only close the contract with the company when the customer has approved the subcontractor. It may also be that the customer leaves the prime contractor the freedom to select and appoint subcontractors. In all these cases, the prime contractor will be held accountable for errors made by the subcontractor(s), because the prime contractor is the only direct business partner of the customer.

Indirect procurement also has benefits and disadvantages:

---

[2] I still recommend applying true faith considerations on the interests of the contractors, as discussed before.

- **Benefits**
  - The prime contractor has the overall responsibility and is probably more knowledgeable, experienced, and skilled in performing this type of project successfully.
  - The prime contractor gives the project the management attention to a degree that the customer may not be able to give.
- **Disadvantages**
  - Generally higher costs, because the prime contractor must be paid for organizational work and for the risk of managing the subcontractors, and the prime contractor will also need to make some profit to have an interest in doing the project for the customer.
  - The project manager in the organization of the prime contractor must consider at least two groups of interests in decision making: those of their own organization and those of the customer.
  - Default of the prime contractor can kill the entire project.

4.3.3 Indirect Procurement over a Consortium

This model is similar to the previous, with the exception that the prime contractor is not an independent organization but a temporary joint venture—a *consortium*—founded by two or more companies for the project, as discussed earlier. In most cases, these venturing companies will not only invest in the consortium but will also work for it as subcontractors. Alone, they would not be able to offer to work for the project, but together, they think they can make it.

The customer in this model has again only one contract with the consortium, which will then distribute the work among its venturers and may still give some more work to other subcontractors. The venturers/subcontractors are also free to subcontract their work to other companies, which can make the PSNs that arise out of such structures very confusing.

The term *consortium* is used here for joint ventures that have been founded temporarily for a project or a major program; there are other uses of the term. The consortium may have some more tasks on top of the project. I have already mentioned build-operate-transfer (BOT) projects, in which a consortium has been tasked to develop some kind of infrastructure arrangement, such as a railway line or a toll road, and to operate it for a limited time, after which the arrangement is handed over to the state that tasked the consortium.

If the consortium will not have a new project or operation by that time, it is likely to be liquidated subsequently. Consortia are a common example for *coopetition,* cooperation of companies that are otherwise in competition. This unclear nature of the business relation is among the ambiguities that can make teaming in projects difficult, not only in consortia. Another challenge is the double nature of the participants, who want to make the consortium successful as venturers, but also desire to make profitable business with it as contractors. As venturers, they want cheap and reliable suppliers.

In the supplier role, their business interest is achieving high prices and not investing more in the customer than what is unavoidable. Add the business needs of the customer to that, and the multi-objective nature of consortia becomes obvious.

Working with a consortium as a prime contractor has also its specific benefits and disadvantages for the customer:

- **Benefits**
  - ○ The consortium has the overall responsibility to perform the project for the customer.
  - ○ Although the consortium is a new organization, in most cases founded specifically for the project—and therefore cannot be expected to have experience in the business (and reference customers)—the venturers are expected to be experienced and skilled in performing this type of project successfully and are expected to transfer this competency to the joint venture.
  - ○ The consortium focuses on the project and therefore gives the project more management attention than would a contractor who performs several customer projects concurrently.
- **Disadvantages**
  - ○ Conflicts between the venturers can endanger the project.
  - ○ Different business interests can add further risk, when some venturers see the project as strategic and long-term, while others strive for quick win.
  - ○ The venturers can threaten in a crisis situation to liquidate the consortium, something they would not do with their own company.

4.3.4 Mixed Structures

In real life, one often sees mixed forms of these teaming structures. A customer may get one work package done with a consortium, hire contractors directly for a second one, and do a third with its own resources. One should also not underestimate the dynamics in PSNs, mostly driven by changing requirements on the project and changing business situations of the contractors and subcontractors involved. One can also add further tiers to the PSN with sub-subcontractors and so on. Managing customer projects stretches from a simple customer–contractor scenario to highly complex structures, and the more complex they become, the greater the demand on project managers will be to manage them. The challenge to meet this demand also goes up with the frequency of changes, and project managers will be needed with expertise in contracting and cross-corporate relationship building as well as in purely technical matters.

4.3.5 Customer's Involvement in Subcontracting

Customers often want to be involved in the selection of subcontractors. There are some options that they can use:

- **Naming of subcontractors**. The customer tells the prime contractor who to work with.
- **Nominating of subcontractors**. The prime contractor receives a list of accepted subcontractors (often three companies) from the customer and makes a selection from this list.
- **Acceptance process**. The prime contractor must introduce any potential subcontractor to the customer and can only close the contract with the supplier when the customer has approved it.

In many projects, customers leave the prime contractor the freedom to select and appoint subcontractors at their own discretion. They trust in the contractor's ability to find the best partners for the project, and they do not want to spend time and energy for any involvement in the selection process.

In all these cases, the prime contractor will be held accountable for errors made by the subcontractor, because the prime contractor is the only direct business partner of the customer. The doctrine of *privity of contracts* makes it clear that from a legal perspective, the customer has no direct connection with subcontractors, but in a relational understanding, it will nevertheless exist.

4.4 Managing PSNs Is Managing Interfaces

4.4.1 The Sentiments of Industries

In March 2017, I had the opportunity to speak at a conference in London dedicated to the oil and gas industry. It was an interesting event, dedicated to an industry that had lost, in just one year between 2014 and 2015, about half of its income stream, which was necessary to keep its operations as well as its projects alive. Figure 4.2 shows the price development of Brent Crude Oil. There was a similar development of the oil prices, particularly when one looks at the net price that the industry takes in, which is street price minus taxes and other fees.

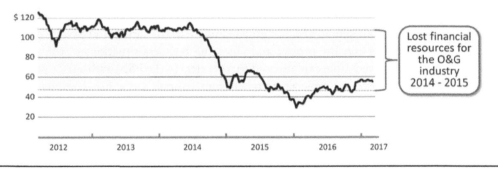

Figure 4.2 Development of Oil Price 2012–2017, Brent Crude, price development per March 13, 2017.

The content of most speeches in the conference, including mine, was that the industry should respond to this difficult situation by closing ranks, increasing professionalism, and building on strong collaboration to avoid further losses from dysfunctionality inside operations and projects. Although attendees officially agreed, in private they noted that the industry is ridden with struggles for shrinking resources and a general distribution conflict.

Money was not the only resource that the industry was short of—talent was another one. As an example, it is interesting to look at where the industry stands in employee attractiveness rankings such as *Fortune 100 Best Companies to Work For*,[3] which focuses on corporations active in the USA (see Table 4.1).

---

[3] (Fortune 2017)

Table 4.1 Fortune 100 Top Ten Companies to Work For[a]

Top Ten Companies	Oil & Gas Companies
1. Google	37. NuStar Energy
2. Wegmans Food Markets	41. HilCorp
3. Boston Consulting	
4. Baird	
5. Edward Jones	
6. Genentech	
7. Ultimate Software	
8. Salesforce	
9. Acuity	
10. Quicken Loans	

[a] Among the 100 Best Companies to Work For, one finds only two oil and gas companies, and none of the big players is listed there.

When I ask young people working in the oil and gas industry how their friends react, the response is rarely that these friends see this career decision as "cool" and "awesome". Instead, the decision to seek a career there is questioned, and instead of admiration, they rather sense a lot of negative sentiments. Young people are rather lured by technology companies than by oil and gas companies.

Indispensable for such an industry is an influx of young talents with new ideas, different lifestyles and success models, and with a different interest in matters of environment and social development that will impact their future. This influx of talent is rather dripping than streaming in, and the result is an aging work force. Data from 2012 for the U.S. shows that the median age of people working in oil and gas was almost 45 years (compared to 42.3 years across all industries), which means that 50 percent of the employees were over 45 years old.

Oil and gas is an industry that does not stand alone with the combination of reduced financial resources and talent resources. Decades ago, the industry was considered attractive and future proof, and although it remains out of the question that human civilization will be able to live without this industry at least for the next decades, its public image today is more that of a dirty and backwards-oriented business than one that leads into a shining future.

The degree to which this perception is justified is open to personal opinion, of course—I am not here to judge, but to talk about project management. Further declines are foreseeable, as alternative sources of energy, but also of chemistry, pharmacy, and other industries, are in ascendence, and their cost efficiency is growing at a pace that promises tough competition to drilling and transporting oil and gas. What matters for our discussion is how difficult it is in times of decline to keep team spirit alive across collaborating companies. Single interests often trump common interests, and this particularly happens in industries in decline.

The same is probably true for industries in "gold-rush" commotion. The expectation of a future supremacy in a fast-growing market as in innovator, or at least a very early adopter, is a strong motivator for players to secure expected gains for themselves against others also who want them. Players then hammer their stakes into the ground in the form of protected intellectual property, domination of markets, and take-overs of pioneer companies at often ridiculously high prices. The small teams that dominate these phases can be very productive and inventive and will be perceived as disrupting by others; and with not much business history in their backyard, they do not tend to be very oriented toward fairness and trustworthiness, and they do not have much time to contemplate such questions anyway.

Another risk for cooperation in cross-corporate project teams are fossil conflicts, mostly in the form of grudges and distrust from older projects. While the projects have ended some time ago, the resentment has not.

4.4.2 The Five Dysfunctions

As discussed above, a common consequence of protective, confrontational, and competitive attitudes in PSNs is dysfunctional teaming across the organizations involved. Patrick Lencioni described the "The Five Dysfunctions of a Team",[4] and we commonly find these dysfunctions when business relations deteriorate as both a cause of the deterioration and its effect:

- **Dysfunction #1: Absence of trust.** During the offer phase, companies had to boast about their strengths, skills, abilities, success records, reference customers, etc. to win the business. Building trust, however, requires accepting weaknesses, one's own as well as those of others. Relational contracting, open communications, and active assumption of accountability are necessary elements of well-working teaming behaviors, but they can also lead to vulnerabilities with possible legal and financial consequences for the parties involved, such as liabilities, expensive damage claims, and the readiness to fix errors that one has made. People may be afraid of these risks, and when they work with the wrong parties, these fears are often not unfounded. When teaming partners are not taking these risks, they will damage the project. It is a difficult balance that needs a lot of attention.

- **Dysfunction #2: Fear of conflict.** Teaming partners may find consensus on decisions that each of them in secret does not support; it may be due to a lack of courage to question the consensus and be then decried a nay-sayer and objectionist.[5] Progress may be an issue; the partners have already spent so much time discussing, it is time to start working, even if the contents and goals of such work may be still unclear; and it may also be unclear who does what, and also, when things go wrong, who fixes what.

  A further reason for false consensus may be business interests: In certain situations, a contractor may benefit from erroneous decisions. When a construction project gets delayed, the companies that provide the temporary fencing around the site will probably benefit, so why should these companies object to decisions that slow down the project and support their share of the business? Culture can also be an issue: Certain cultures consider the contractor not as a project partner on eye level but as a subordinate. High power distance between customer and contractor may then discourage the subordinate to talk about risks and upcoming problems in the project that are visible to the contractor but not the customer.

---

[4] (Lencioni 2002)
[5] The famous narrative in Jerry B. Harvey's *Journey to Abilene* (Harvey 1974).

- **Dysfunction #3: Lack of commitment.** There is a tendency by many people to keep agreements unclear and resolve problems not upfront but when they occur. A common reason is a lack of time to develop clear rules, negotiate them, and sign off on them. Another reason is that the parties try to avoid committing to the other parties and then turn to wishy-washy agreements that are impossible to enforce and lead to embitterment. A common example in project management is the assignment of resources to projects in a percentage value as to when the resource will be available and whose job it will then be to manage it, which gives no party any certainty.
- **Dysfunction #4: Avoidance of accountability.** Holding people accountable for their achievements and for their failures is a strong bond in teams, in that it connects their individual contributions to the team's results. People and organizations doing a good job want it recognized, as much as those doing a poor job should also know that it is not ignored. Given the cost of poor performance of companies teaming in a project, the investment in the tools that help avoid such ignorance are often easy to justify: colocation, meetings of teaming partners, and scoreboards that visualize joint project achievements.

  The rolling award fee contract presented above is also a tool to help show contractors how pleasing—or unsatisfying—their work is. As a customer, one should then not shed any tears over the fee paid to the contractor, as long as the benefit gained from the great work of the contractors was higher.
- **Dysfunction #5: Inattention to results.** The goals that the teaming partners in the project must achieve together often become invisible after a while. This is particularly true in time of excessive error fixing and crisis management, when people's minds are too consumed with an army of problems that all require their attention and working time. To avoid this dysfunction, the contribution of each of the organizations to achieving common goals should be kept visible, as well as the happiness that these achievements create for the customer and the other teaming partners. The example of the rolling award fee contract discussed in Chapter 3 also makes it clear that the fee should not be given as a kind of a present, but should be communicated as a fair share of the benefit that the customer gets when the contractor helps them stride towards the common goal.

4.4.3 How Competitiveness and Cooperation Interrelate

An interesting statement by Lencioni[6] highlights the complex interdependency of competing and completing: "Teamwork remains the one sustainable competitive advantage that has been largely untapped". Collaboration and competition going hand in hand—is this not self-contradictory? Probably not, once one digs a bit deeper.

There are actually two classes of competitive behaviors: *outpacing* and *undermining*.

- **Competitive behavior class 1: Outpacing.** The competitive party is just faster than other parties, leaving them behind on the common playing field—in essence, a peaceful kind of competition. Sometimes, the outpaced party may admire the faster one, as speed can be a kind of leadership behavior, and the faster party can be a role model for the slower one. However, being outpaced can also lead to bitterness, and the outpaced may

---

[6] (Lencioni 2002)

turn to bullying behavior to deflect from their defeat. This bullying behavior can destroy any project; it is a form of *undermining competition*.

- **Competitive behavior class 2: Undermining.** The competitive party weakens the other party. This form of competitive behavior is aggressive. An example from sport is boxing, in which a participant has two reasons to land punches: (1) To make the points, and (2) to weaken the punching power, speed, concentration, and balance of the opponent. The party applying class 2 competitiveness will use any opportunity to diminish the other party's resources to disorganize and to gain benefits to the disadvantage of the other side.

4.5 Interfaces Among Contractors

4.5.1 Customer Projects with One Customer

When I managed my first project for a prime contractor in the manufacturing industry with a still small number of only two subcontractors, the project manager of the customer advised me to focus on the interfaces between my organization and the subcontractors and also among the subcontractors. He considered me rather young and inexperienced to manage the PSN; he may have identified my inexperience as a major risk for his project, and with the blessing of more than a quarter of a century that has passed since, he was probably right.

Due to some major change requests from the customer side, the project grew further, and after some months, it had turned into a project with five subcontractors. The customer-side project manager's advice was soon proven right. Complaints and conflicts occurred between the organizations involved, mainly circling around the workload delineations (who must do what?) and the accountability for deliverables (who must fix what?). A further area of concern was security—no company likes to see its own staff exposed to hazards caused by work done by another one.

Figure 4.3 shows the number of interfaces among three contractors—one of them the prime contractor—and the customer. This was the original setup of the PSN. The number of interfaces is six. During the project, when the change request brought additional work that required employment of three more subcontractors, the number of interfaces increased to 21.

The example shows how the number of technical, social, and interpersonal interfaces among customers and contractors grows exponentially with the number of parties involved.

In the example, I was still lucky to have a small number of contractors and only one customer. Some readers will say that the project was still a simple one, compared with the projects that they need to manage, with scores of customers, contractors, and subcontractors, and they are right. Looking at the complexity that these project managers must manage, finishing megaprojects successfully is a great achievement, even when this is associated with major delays, cost overruns, and functional deficits.

Figure 4.3 shows at the top the formula used to calculate the number of interfaces from the number of contractors, assuming that there is only one final customer organization. Using this number, it is interesting to see how the interfaces develop for even larger PSNs. This is shown in Figure 4.4. It is interesting to see how even small changes in the number of contractors translate into much larger changes in the number of interfaces.

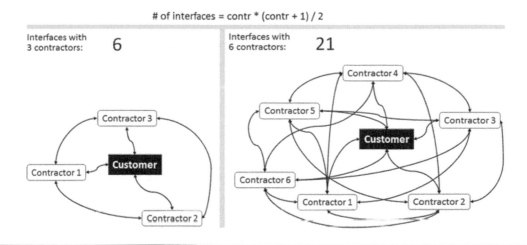

Figure 4.3   The number of interfaces in a simple PSN (left) with only three contractors is six, including the interfaces with the customer. The PSN on the right-hand side has twice the number of contractors, but the number of interfaces has grown by a factor of 3.5 to 21.

These interfaces are an aspect of PSN complexity, and their growth indicates that this complexity changes non-proportionally with changes of the PSNs. This aspect is often overlooked.

4.5.2 Customer Projects with Several Customers

One can grow complexity even further by increasing the number of customers. An example is the Joint Strike Fighter Lockheed Martin F-35 Lightning II, a fifth-generation fighter aircraft, which was developed in different variations for nine participating states—Australia,

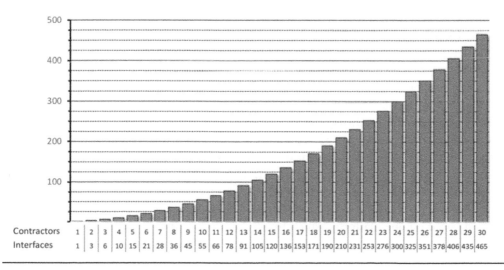

Figure 4.4   For a project with one customer and a larger number of contractors, managing the interfaces among the contractors can be among the most challenging tasks.

Canada, Denmark, Italy, the Netherlands, Norway, Turkey, the United Kingdom, and the United States—and three non-participating customer states—Israel, Japan, and the Republic of Korea.[7] In three of these countries, the aircraft were ordered by different military branches, raising the number of customers to a staggering 18—customers who each had different requirements against which the aircraft needed to be designed:

Australian Air Force	Canadian Air Force
Danish Air Force	Israel Air Force
Italian Air Force	Italian Navy
Japanese Self-Defense Air Force	Korean Air Force (South Korea)
Netherlands Air Force	Norwegian Air Force
Turkish Air Force	UK Air Force
UK Navy	US Air Force
US Marine Corps	US Navy

Complexity arose from the different desires of each of these customers. There was also an expectation that the participating countries would get access to new technologies—a desire that was not in the business interests of Lockheed Martin.[8] Additional complexity was added by the concurrent design of highly different types:

- Conventional take-off and landing (CTOL, named F-35A) for normal take-off and landing
- Short take-off and vertical landing (STOVL, named F-35B) for short take-off and vertical landing
- Carrier variant (CV) for carrier-based operation (F-35C)

Traditionally, such versioning would be developed in sequence, thus simplifying the process by designing one version after the other, basing each version on the previous, but decisions made for the F-35 were to develop them side by side.

The number of project customers and aircraft versions added to complexity in financing, particularly when the program exceeded its budget and discussions arose as to which country would have to bear which percentage of these overruns. Further discussions arose over the distribution of development and production work, which brings jobs to the participating country and influence over the program, but conflicts with other countries that want to have the work as well.

The program was originally considered a cost saver because of the large number of 2,953 aircraft that would be made,[9] allowing the distribution of development costs over a high number of items to reduce the cost per single aircraft. This benefit seems more than consumed by the cost of complexity that occurred during development—complexity to some degree caused by the large number of customers and also by the number of different versions that needed to be developed based on one platform, which finally became a salmagundi of compromises. The

---

[7] (Lockheed-Martin 2017)
[8] (PM Network 2005)
[9] (supplychainbrain.com 2010)

difficulties of accomplishing the business case calculations due to the delays that arose from the complexity massively added to the cost problems.

What can we learn from this example? When I meet project managers in projects with two or more customers, I often hear complaints about the organizational complexity that such a structure develops. It is hard for them to solicit decisions by these customers, who in turn often come with change requests that are not coordinated with the other customer(s) and have the potential to be damaging to the project. It then becomes very difficult to follow a sound change request management process, which is necessary to protect the project and at the same time promote those changes that are necessary, or at least beneficial, to it.

Similar complaints come from customers in such projects. The customer should be king, but in such projects, too many "kings" want to make decisions, often in a competing manner. They develop antagonisms and alliances, and finally the noisiest wins, not the one who is actually right. Big egos are a problem in many projects, but in projects with multiple customers, the clashing of such egos is often unavoidable.

Complexities in projects do not add up, they multiply. Adding organizational complexity to technical and interpersonal complexity may make the project no longer manageable. It is a clear recommendation to avoid such situations and develop structures that allow the project to be performed for just one customer or to serve the customers in a sequence that allows the focus on just one "king" at a time.

Projects with multiple customers bear another risk for each of them, particularly in projects that are found on the tabloids' front pages: When the project is about to be troubled or even fail, there will be a lot of finger pointing, assigning blame to specific stakeholders, and often, these fingers point to one or more of the customers. However, this is not necessarily the one who should assume responsibility; too much money is involved to let the buck stop easily at the party to whom it belongs.

I observed that having multiple customers to a project is more often found to be a cause of troubles than a blessing for the project and recommend avoiding it. The larger the project, the more complex it is, the higher its financial burden, the larger is the temptation to go for multiple customership, and the stronger the resistance against that should be.

4.6 Risks from Outsourcing Projects Under Contract

4.6.1 Risks from Differences in Software Tools Used

Project management terminology is often used inconsistently. A common example of such a term is *milestone*. There are three interpretations commonly used of what a milestone is:

- A synonym for *phase gate,* located between the end of one phase and the beginning of the next. This interpretation comes with a problem: gates have durations, sometimes months, used for different kinds of reviews and approvals; a milestone should have the duration "zero".
- A synonym for *deadline,* interpreted as a milestone with a fixed date. The problem is that a milestone can occur earlier or later, but a deadline is hard, often impossible to move. This interpretation leads to a loss of information.

- A point in a schedule that marks a desired, planned, forecast, or actual achievement, such as the finishing of a deliverable or the commencement of a work item. The achieving of the milestone triggers some reactions, such as increased management attention, test runs, or payments. This last interpretation is definitively the best.

Scheduling functions in modern project management software all have deadlines, but some use the second definition, others the third. This can bring problems when the solutions or the people working with them have to cooperate.

Modern organizations that perform projects commonly work with software solutions that are made to support project, program, and portfolio management. Most of these solutions have been around for a while and have grown over time from simple solutions for a limited number of project management tasks to highly complex systems with broad functionality. Such software can be helpful when it is mastered professionally, but it can also cause its own problems when the user does not understand and control the program. Many functions are placed invisibly "under the hood", which means that a lack of understanding of the software can lead to errors in the planning, implementing, and tracking processes. These errors will be hard to find and to understand, and this in turn can make it tedious to fix them.

These risks from insufficiently understood and mastered software are made worse when companies decide to work together, but use different software solutions. In the past, simple data exchange protocols where used, such as Odette in the automotive industry, which communicated a limited amount of highly standardized data, so that different systems could easily communicate with each other as long as they supported the protocol. But today, the expectation is that software programs must interact in a much tighter fashion—ideally, as if several programs were just one.

The combination of different project management software programs can have unexpected effects that can impact the cooperation as much as the interpersonal and social causes of conflicts described later can.

- **Data silos.** Not only are the companies in the PSN often silos, the same is true for their data. Each of the companies collect the data, and as they use different systems with different data models and processing algorithms, data will also differ. Cooperation needs a joint view on data and on the information that is built on it.
- **Software as an element of standardization.** This is often overlooked—software standardizes terminology, processes, and more. A company may have a project management glossary, but people rarely look at it. However, they use the software frequently and are familiar with its interpretations. I described above how software uses terms differently. When users with different interpretations come together, they may be unaware of these differences, which leads to misunderstandings. They may identify the difference, but then come into conflict about whose interpretation must be used. Both are problems that can be hard to overcome.
- **Tampering with information during data export/import, synchronization, and consolidation.** Software vendors generally promise that software products work well together, but differences in their data models and processes are hard to ignore. When one software imports data developed in another software, when one program must synchronize data with another program, or when a backend solution consolidates plans that were developed

with front-end programs from other vendors, some data is often changed. The more complex the project, the more likely it is that this will happen, and the harder it will be to find the changes and reconstruct the original plan in the new environment. Complexity of such plans grows over time. The import and consolidation functions may work well early on, and the problems turn up much later, unexpectedly and therefore particularly damaging.

- **Different mechanisms for versioning and change tracking.** A tool to keep a history of a project for documentation and to support lessons learned are *versioning mechanisms,* which can also be used to develop plan scenarios to improve the change request management process. Different software products have different mechanisms for that, and whereas communicating current data from one program to another may be more difficult than expected, this is particularly true for structured historical information.

- **Out-of-the-box functions versus customization.** Many of the problems described here can be addressed with customization. The problem then is, whose software will be the "leading system", and whose software will need adaptation and customization. This customization can become quite extensive, leading to difficulties when software updates are needed that may require rebuilding the customization in the new version. Having the leading system is generally the easiest.

- **Different focus areas of software.** Project management software is often strong in certain areas but weak in others. Project managers used to software with strengths in some areas may expect that software used by other project managers has a similar focus, which sets false expectations. Examples of such areas include:
  - Scheduling
  - Resource management
  - Cost management
  - Team functions and communications
  - Support for rolling wave approaches and agile methods
  - Document management
  - Visualization of data
  - Approval workflows

- **Access rules and push and pull approaches.** Documents may be distributed (pushed) or held available for download (pulled), and software vendors answer the question of push or pull of data differently. Someone who is used to getting information pushed may not expect the need for downloading and may therefore not have important information at hand. Someone who has had access to all data may be surprised to have access removed. Someone unfamiliar with distribution mechanisms may find that data has by error been sent to the wrong people, possibly with very negative consequences. Again, the problem is that one software sets expectations that the other does not meet.

- **Physical location of data.** Another fundamental difference among software programs is where the data is stored—on the user's PC, on an enterprise server, or in the cloud, at an unknown location on a server somewhere in the world? Any versioning mechanism has its own storage strategy, and not being aware of the strategy may cause errors.

All these risks can be managed, once they are identified and understood. One response to them could be to provide one software platform for the entire *project supply network* (PSN), but this would then have the negative effect that project managers must learn different software, and they are rarely given time for a seminar and then to get familiar with the program. Another problem is the shortage of software that addresses the specifics of supply networks by allowing a specific company view and a cross-company view for the entire project.

4.6.2 Risks from Interpersonal and Social Conflicts

I discussed this point above in the section that is dedicated to sellers, asking: Do you present yourself and your company as the one that the customer will desire to work with?

The question actually goes much deeper than that. The contractual relationship in a project binds the customer and the contractor organizations and creates a mutual relationship that may last from only days to possibly years. This duration may even be hard to predict, as small projects tend to grow larger over time, and business encounters that were planned to be short term may then become lasting for much longer than originally expected. There is a corollary in the binding nature of a contract for the entire organization. Your project schedule is just the schedule of your project. The same applies to your human resources (HR) plan of the project, your risk register, your communications plan, your diverse management plans, and so on.

These documents are more or less binding for the project. A contract with a vendor, however, is binding for the entire organization. The same is of course true for all other project-related contracts, such as those with customers, business partners, insurance companies, and so on. The binding nature of the documents[10] rises to a level far beyond the project, and so do obligations and entitlements that the contract delineates.

The intended partnership with sellers brings many risks to the project—and the entire organization. Procurement means, in essence, tapping external assets and turning them into resources for the project—assets that the customer does not have, or that are at least not available at the time the project needs them. Sometimes, assets turn into liabilities, and the mutual dependency of a customer–contractor relationship brings the risk that the problem of one partner in the business becomes the problem of the other one as well. Among the risks for the customer, the following are some truly serious ones:

a. **Bringing a diversity of business interests into the project.** Each contractor has its own business interests that the company cannot simply neglect for the project. The first one is, of course, that the project must bring money home. Another one is the desire of the contractor to keep its employees working, ideally with billable work. A third is the desire to avoid overworking employees and burning them out, particularly when the contractor performs a number of customer projects in parallel. Averting liability and damage claims is another natural business interest. All these business interests may come into conflict with those of the project, the customer, and of other contractors.

b. **Bringing fossil conflicts into the project.** In certain fields experts are rare, so it can happen that they meet repeatedly in project after project. Unresolved resentment and bile among contractors' employees from earlier projects can hamper the current project, and because the project manager is unaware of the invisibly simmering conflict, no

---

[10] I described above that from a legal perspective, a contract may be verbal or written. In a commercial understanding, a contract is generally a document. I am using here the commercial definition.

action is taken to get over the old grudge. It is like an underground unrest that one day may break out, surprising all others involved.

c. **Bringing a diversity of cultures, legal systems, and moral compasses into the project.** This is particularly true for projects with international contracting, but even within a country, different understandings by people involved of what is right and what is wrong, what are acceptable attitudes and behaviors and what are not, may bring conflicts into the project.

d. **Disruptions by incompatible egos.** In a PSN driven by a "Mission Success First" culture, managers from different contractor companies mutually respect the professionalism, responsibility, and independence of their colleagues from the other contractors. Solutions are found by common sense and consensus, not by forcing. One troublesome contractor, however, can disrupt this culture and finally jeopardize the entire project.

e. **Becoming dependent on psychopaths and sociopaths.** One would assume that these "snakes in suits", driven by antisocial personality disorder, are rather to be found as patients of psychiatric services and in jails. Instead, their ability to switch empathy on and off as it best serves their desires, their skills in successfully manipulating people, and their highly competitive and predatory nature, unrestrained by implicit or explicit codes of conduct and by compassion, can make them successful in key positions and allow them to attain the highest positions.[11] There may be rare situations in which they can be beneficial to a project; but in most cases, they are more likely to damage it, because they will never subordinate themselves to a "Mission Success First" approach or confuse their personal desire to look successful with the interests of the project and its stakeholders.

f. **Breakdown in communications.** When corporate counsels sense the risk of legal action, they tell their employees to reduce correspondence and verbal communications outside four-eyes settings to the unavoidable minimum, because all communicated messages can be used against the company.[12] From a legal perspective, this may appear appropriate, but from a project perspective, the partial or full breakdown of communications can be disastrous.

g. **Dependency on dubious organizations.** I have previously quoted the South-African attorney Guido Penzhorn, who wrote in reference to a corruption case in a dam project in Lesotho, Africa: "Clearly, once you involve yourself in the murky world of bribery, it is not open to you to simply opt out whenever you like".[13] Assuming that project managers generally want to avoid any associations with corrupt practices in their projects in order to maintain their independence and professional integrity and to protect their project, their first step is to vet vendors before entering into an agreement with them.

---

[11] There is an interesting elaboration on the topic by Will Black, analyzing their success secrets and giving recommendations on how to manage them (Black 2015). Project business management is a perfect playing field for what he calls "Psychopathic economics", owing to the temporary nature of the business, which makes it difficult to develop a long-term record of a person's conduct.

[12] This again is different in legal systems: In the USA and other common law countries, courts can subpoena the submission of all documentation relating to a specific case. They would normally not have this power in civil law countries.

[13] (Penzhorn 2004)

A simple but diligent web search can sometimes be an eye opener. When considering whether to give work to a consortium that has just been founded for the project and has no business history, one can still do some research on the venturing organizations. It is not uncommon that murky organizations join consortia to hide behind the good reputation of other venturers, often without the knowledge of the others.[14]

h. **Breach of confidence.** There is always a degree of uncertainty as to whether contractor employees will maintain the same level of confidentiality that one would expect from one's own employees. There is also uncertainty about the degree of general trust that they deserve on top of that, in particular when the vendors are new or are incumbent but have just had a change in ownership. The mental and emotional distance to the interests of the customer and the project is generally higher for contractors' employees than for one's own, who are expected to give their full loyalty to their employer.

i. **Default of a contractor.** A defaulting contractor is among the worst experiences that a project manager can have during procurement. The problems do not start with the moment of insolvency but months earlier, when the contractor makes strange decisions that its manager can easily explain; the interests of the customer, however, no longer turn up in these explanations. The intention to avoid insolvency tops everything else in at such a time—particularly the desire to have a happy customer. Another uneasy experience I had years ago was observing a contractor finishing a project and having its product ready for final handover, then going into insolvency. At this point in time, the product became part of the insolvency estate, with the effect that handing the urgently awaited results over would have been illegal and also technically impossible, because there was no organization left to transfer it. The customer company had to buy the results out of the insolvency estate, which meant paying a second time for the same work.

j. **Paradise syndrome.** Sometimes, people feel dissatisfied with having achieved all of their goals. Their successes in projects come as repetitious and routine, not as new challenges, and it is such a challenge that they need to function well. Protected by specific assets such as patents or specialist knowledge, they are not easy to replace in the project, and the feeling of strength and indispensability they have makes it difficult to work with them. Then, customer orientation gets replaced with pomposity, and letting others wait is considered a sign of strength, not of failure. Such a degree of saturation in people's minds is hard to identify and even harder to fight.

k. **Extended vulnerability to digital attackers.** Information technology (IT) systems are vulnerable against malicious attackers, causing unprecedented digital and also financial damage. Every network interface is a weak spot in the system, but even more vulnerability comes from the people involved brought into the project by the contractors. It is difficult to develop the discipline in one's own people to resist clicking e-mail attachments and other possibly malicious pieces of software and to make them take care of the digital trails that they leave on the internet, but with external staff, this gets even more difficult—in particular, when the customer has no disciplinary power over them, possibly does not

---

[14] Another indicator may be the position of the seller's home country in the Corruption Perceptions Index of Transparency International, the global association against corruption (TI 2016b). It gives at least an indication as to the degree of integrity culture in the homeland of the seller as an early warning signal.

even know them. Modern malware needs access to only one vulnerable networked computer, printer, or other item connected to the corporate network. It has the capability to then self-propagate across the computer network and infect all other vulnerable computers. At the time the malware is identified, the harm can already be enormous.

l. **Customer is not king.** For a small contractor who serves only one or a small number of customers, the customer is king. The customer is the indispensable source of income, and the contractor will avoid as far as possible any activity that risks draining the source of this stream. For a large contractor with a multitude of projects for a vast number of customers, the individual customer is no longer the king but is rather similar to a supermarket customer in a waiting line. In the worst case, the customer is more like the airline passenger who was dragged along the aisle out of the aircraft to free his seat for airline personnel who needed to be taken to another airport.[15]

m. **Unknown subcontractors.** It is difficult to manage subcontractors the customer knows. It is generally impossible to manage those who are unknown to the customer. A direct contractor has hired the subcontractor without making the customer aware. The subcontractor brings new risks into the project, and the customer has not many options to intervene. Even if the contract with the direct contractor prohibits subcontracting without the knowledge and acceptance by the customer (which I generally recommend), it will be hard to identify infringement of such a clause and then enforce it when the customer has become dependent on the contractor and so also on the subcontractor.

The last risk can become a problem that is difficult to manage: Many project managers are indeed not fully aware of the size and complexity that their PSN has generated, because their focus is mostly limited to the direct contract partners. The legal principle is called the *doctrine of privity of contracts,* which says that a prime contractor has a legal relationship with the customer, or several customers, and also with one or more subcontractors, but that there is no legal relationship between the customer(s) and the subcontractor(s). Figure 4.5 illustrates how the prime contractor acts as a proxy for all contractual connections between the customer side and the subcontractor side.

Does this mean that there should be no relationship between a customer and a subcontractor at all? The prime contractor may prefer such a non-relationship, because of the risk that direct agreements between the two may bring them disadvantages. One can compare this with a dealer of goods, who wants to keep the source of the goods unknown to the customer, and vice versa, in order to avoid them doing business directly, circumventing the dealer. However, if the prime contractor wants to act as an intermediate for all communications and decisions, a lot of time will need to be spent to transfer all the information back and forth, and energy and management attention will be consumed satisfying the desire to ensure that no decisions are made that would bring them a disadvantage. A customer of mine sometimes refers to this uncomfortable situation as "ham in the sandwich".

Dealers can alternatively name the source of a product to a customer to benefit from the brand and the marketing power of the original supplier. They argue towards the customer that they are the entry point to these goods and provide additional services. An example are car dealers, who are mostly independent companies, but proudly show off the logo and the colors

---

[15] As actually happened in April 2017 (*Business Insider* 2017).

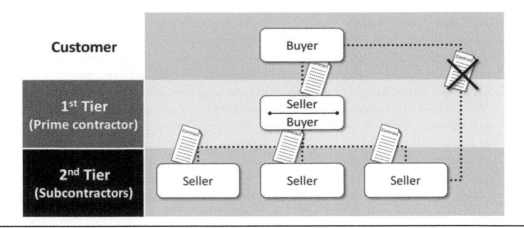

Figure 4.5   Privity of contracts: The customer has no contractual relationship with the sub-contractor.

of the car company they represent. They have contracts with the car company (or an importer) that protect their business, giving their contract partner the benefit of proximity to the sales market and the customers. This behavior can also be found with prime contractors in Project Business Management, who promote themselves to a customer as the gate to the goods and services of other sellers. In such cases as well, the customer has a contract with the prime contractor, not the subcontractor, but the organizational relationship will not be limited by that.

From the customer's perspective, limiting the relationship with a subcontractor to an indirect channel via the prime contractor leads to indirect communication, something often referred to as "telephone" or "broken telephone" in children's parties, when one child whispers a message into another child's ear, who does the same with the next, etc. In the children's party, it is then fun to compare the original message with the final one. In a project, this is commonly much less funny—and possibly much more expensive—and the next question will be, who needs to take the blame for the misunderstanding and the follow-up problems.

Mosquito Corp. was a fashion corporation with a large number of outlets distributed over 25 countries. They had planned to replace their outdated inventory management system with a new solution ordered from Silverfish, Inc., who had a standard software solution that seemed to fit the needs of Mosquito to a major degree—but not fully, so some tailoring was needed, including the adaptation to the national business rules in the various countries in which Mosquito operated, and in which the software would therefore be implemented. Silverfish did not have the capacity free in-house to do the tailoring, and its business was anyway focused on solution development and marketing. So Silverfish hired Mealworm Ltd., a small company with a lot of competency that was just completing another project, promising free capacities for the tailoring. Mosquito, the customer, was not aware that the subcontractor, Mealworm, had become part of their project, and they were not interested in the details and risks that came with this subcontracting. They considered their software update in good hands and did not ask any further questions.

Silverfish and Mealworm worked well together right from the start, and the project proceeded quickly, very much to the confidence of the customer. About a third of the way into the project, Mosquito requested a major change to the software. Their business made some unexpected shifts, from almost 100 percent in-house production of their goods to greater use of goods that they could buy cheaply from overseas suppliers. The purchasing part of the business needed some alterations in the software so as to not lose track of goods during deep-sea container transportation—goods that were already owned by Mosquito but not yet distributed to countries and not yet available for placement in shops. As Silverfish had originally promised this kind of additional functionality, it was clear to Mosquito that the additional functionality would be delivered without price increases or delays.

Because of the amount of code modifications and expansions needed in their software, Silverfish decided to respond to the request by tailoring, not by altering, the standard software, which would have taken too long for the running project. The change request was therefore passed on to Mealworm, who made some estimates of how it would interact with their adaptation work and the many country-related versions they were about to create. The last point appeared to be the most difficult, and Mealworm would have to strengthen its human resources quickly to implement the change without delays. In practice, the task proved even more difficult than expected, because significantly different versions of this additional code became necessary for the various countries.

It was clear to Mealworm, the subcontractor, that the additional work needed to be paid by the prime contractor, Silverfish, who could not bill the change to its customer, Mosquito; so Silverfish would have to accept that the additional work would eat into their margin from the project. Considering that Silverfish also earned money from the software licenses sold to Mosquito, the loss seemed to be something they would teeth-gnashingly accept.

But there was another issue: delays. It was impossible to do the additional work with the staff assigned by Mealworm to the project, and the search for additional staff was very time consuming. They needed employees who would only have a brief ramp-up phase to become productive, but all they could find were beginners and developers from other subject areas, people who would have to go through a lengthy learning curve to understand both the software and the customer requirements. The project schedule had no allowance for that kind of low productivity phase. One should add, it would have been the current development team's job to tutor the new people, which would slow them down as well before the additional productivity could become effective. Voices at Mealworm insisted that further developers would not speed up the project, but rather slow it down. Mealworm was heading toward a major project upheaval, particularly when some of the key developers felt overworked, deprived of sleep, and finally left the company. Silverfish had a profit problem from the project, but Mealworm got broken by the work pressure.

In a PSN, a seemingly simple change request can trickle down the network, multiply there, and then develop its own dynamics when a technical change translates into an organizational, interpersonal, and even legal challenge. The example PSN from the case story above is just a simple supply chain, with only three organizations and two contracts. In real life, many PSNs are much larger, much more complex, and much more dynamic. Their opaque nature, as in the example, can make managing them difficult, particularly when changes and variations need to be managed, as was necessary here.

In the case of the project for Mosquito, it took a while for the customer to even notice the problems in their project. A set of deliverables came in late, but as Silverfish had been timely with all other deliverables during business development and early in the project, this was just considered a singular faux pas, nothing to worry. Unaware of the troubles at Mealworm, the subcontractor who actually did the work, the customer was still feeling comfortable with Silverfish and its competency in implementing its own software. During visits at Silverfish, everything looked professional and well-controlled. The upheaval at Mealworm was not visible to the customer.

Further delays raised the awareness at Mosquito that something was going fundamentally wrong. The moment came when the roll-out should have been implemented in the first country, but no working software could be supplied by Silverfish. Two more implementation dates passed. When the customer asked the contractor about the problem, they were told of minor flaws in the solution that needed to be fixed, and that Silverfish would not have a problem to make up the time losses later in the project, so there was no reason for panic. After some investigation by Mosquito, and especially after a slip of the tongue by an employee of Silverfish, they became aware that another company was on board, and that the picture of an easy and simple project was only a facade. The opposite was true—like a vicious circle, the project was running deeper and deeper into troubles, and the customer was kept utterly unaware of the growing crisis for a long time.

An open talk among all companies involved in the project at an early date would have helped to find a joint solution to prevent the project from running into crisis and would have avoided the mutilation of the business relationship.

Mosquito's software implementation was in the end six months late. Silverfish could re-negotiate a final price that avoided a financial loss from the project but gave them no significant margin from the combined product and project business. Mealworm had to replace some of their best developers, who left in anger and frustration and found new jobs easily at other companies. The opaqueness of the business setup led to losses for all parties involved.

One may say that the customer should have invested more management attention into the project right from the start. Mosquito was happy to have its management attention free for other tasks, operational as much as other projects. Management attention is often the scarcest and most valuable resource. They considered it sufficient to rely on Silverfish's management, who in its "ham-in-the-sandwich" position was unable to reconcile the needs of the three companies regarding schedule, costs, and people involved and who did not allow the participating parties to fix issues together until these had grown to project disaster. One may also say that they paid the price for their proposal manager's promising the availability of functions off the shelf that the product did not have, making the change requests look simple and easy to the customer, when they were actually difficult to implement.

Mealworm, the subcontractor, tried for too long to fix the issues of their customer, Silverfish. The case story is also an example of how in projects under contract, the problem of one company may turn into a problem for both its customers and its contractors, if one does not intervene early before the misfortune grows beyond what is manageable. A joint approach of the three companies to prevent an issue from growing into a major crisis would have helped protect the business benefits that all three parties had expected, each of them participating with their specific assets. Agreeing on "Mission Success First" and actually implementing it would have been beneficial for all three companies and for the joint project.

When I talk with project managers from Project Business Management, most know similar stories, and among the risks that come with outsourcing project work under contract, lack of transparency and insufficient communications among the companies involved seem to be the most common root causes of project crises.

4.7 Avoiding Crises in Project Business Management

All parties would have to contribute to the avoidance of crises in order to have an effective prevention system in place. A crisis-free project is faster and cheaper, much less rework is needed, and it is easier to assess project status. Open communications simply mean that a project manager has the chance to know what is going on in the project and where it stands. As soon as communications get restricted by political behavior, a project manager will no longer have such information.

This section is mostly directed to the sellers' side, so let us address the risks described above one by one. With all the following advice, I recommend using them in a situational style. They will often be helpful, but not all rules can be applied in all situations, and specific moments can require attention beyond them. A project manager needs an open-skill form of situational awareness, being able to read the dynamics of a situation and to respond accordingly.

   a. **Managing a diversity of business interests in the project.** In general, openness may be the best approach. Talking about diversity at the right moment in a respectful, empathetic, and humorous tone can take the sting out of it and help develop mutual empathy. The rolling award fee contract described above, or an individually developed contract that gives contractors a share of the benefits from good cooperation, may unify the business interests of the companies involved, strengthening the "Mission Success First" culture and underpinning its practical significance for the project in financial terms.
   b. **Identifying fossil conflicts in the project.** Projects are temporary by nature, but grudges can be enduring. When one is tapping assets of other organizations, one may also tap legacies with them. Fossil conflicts from earlier projects are often hard to identify; people may not be prepared to talk about them. Observing them when they first meet in the project may give some clues, and again, talking may help. The problem occurs most likely in small industries, where the same few specialists come to together in various projects, and where these experts compete for acceptance and for business. Asking contractor staff if they have met the employees of the other contractors before may be a good starting point for an investigation into bitterness inherited from earlier projects. Once identified, it may be possible to resolve the conflict and reconcile the parties involved. In the worst case, it may be necessary to remove someone from the project before it becomes impaired by the conflict. A project manager implementing a "Mission Success First" culture cannot allow the project to be impacted by inherited conflicts that cannot be resolved.
   c. **Overcoming a diversity of cultures, legal systems, and moral compasses in the project.** The best way to avoid cultural differences and different moral compasses from

impacting the project is to address them in a tone that is driven by understanding, tolerance, and again humor. The ice is thin—what was a valid cultural description yesterday may have meanwhile become a cliché. Regarding legal differences, it is important to obtain professional advice; there are too many caveats. Many project managers, educated in rather technical or organizational disciplines, have very limited understanding of their own legal system, so how should they understand a foreign one? One should also take care that a legal expert familiar with the laws of their home country may consider them just as valid in other countries—and fail.

d. **Handling incompatible egos.** If your PSN consists of ten companies, and each is managed by one big ego at the top, this makes ten big egos in the project. Not all organizations will be driven by such a kind of person, and not in all of them will the person be in contact with the project. However, add to them the egos that are found at lower levels as supervisors and project team members, and incompatibilities can become a major problem. Inside an organization, structures have been developed that more or less successfully cope with such problems, but in the PSN, which temporarily spans different organizations, no such structures are in place. It is recommended to communicate, right from the start, what the customer expects from the member companies of the PSN, to observe whether the suppliers meet these requirements, and to address and correct inappropriate behavior swiftly. The rolling award fee contract may also be helpful to give organizations a business case to make themselves compatible, because it gives the contractor representatives a monetary incentive for making themselves compatible in mutuality. The award fee is then in essence their share of the financial benefits that come from fewer misunderstandings, errors, and rework. If big egos fully support the mission goals, they may be among the most effective allies a project manager can have.

e. **Avoiding dependency on psychopaths and sociopaths.** This is a truly difficult task. People with antisocial personality disorder (ASPD) are generally hard to identify. During business development, they can be charming, interested, and empathetic, and their argumentation can be highly plausible; but when it comes to meeting obligations, they show their true selves. There are no early warning signals.[16] The most important protection is a contract with easy termination clauses for a case in which a contract partner turns into a nightmare, such as "termination for convenience" or "extraordinary termination for good cause", listing inappropriate and damaging behavior among such causes.

f. **Keeping communications alive.** When contractors are afraid that everything they say and write may be used against them one day, in the project or even at court, they may stop saying and writing down all the things that the project manager should know. Another reason for communication breakdown may be understaffing on the contractor side—everyone is busy with the project work—or on the customer side, so that contractors feel their communications are not welcomed by the distressed representatives of the client. Recommendations: Develop a communications infrastructure that invites and supports frequent and intensive exchanges; reduce communications to what is actually necessary while also keeping the "Mission Success First" spirit alive; and if you use

---

[16] Thanks again to Will Black, author of *Psychopathic Cultures,* who gave me some great insights to the problem field.

award fee contracts or similar agreements, make sure that communication responsiveness and proactiveness are among the criteria for the fee.

g. **Avoid dependency on dubious organizations.** Before you enter into an agreement that makes you potentially dependent on the seller as a contractor, research the history of the company: Have there been cases of unacceptable behavior, including bribery and blackmailing? If it is a consortium or another group of companies, research the individual venturers. During presentations and negotiations, do they openly or secretly offer bribes to your purchasing staff? In international business, a solid resource for understanding from which national integrity culture the company comes can be found in the already mentioned Corruption Perceptions Index (CPI) of Transparency International, the world association against corruption, which can be accessed online.[17] Two more interesting resources are the Press Freedom Indices of Reporters sans Frontieres[18] and Freedomhouse,[19] two global professional associations of journalists. Media in countries ranking low on these ratings are likely not to communicate the full story on the company with which one wishes to enter an agreement.

h. **Prepare for the possibility of breach of confidence.** The basic recommendation should be not to deal with people and organizations that are not trustworthy. This is easier said than done, especially when the prospective business partner is new, or when owners or management have changed and new people run the business. Trustworthiness is slow to build and quick to destroy. It is therefore not enough to know the contractor, but one must know the people, and they must know that they do not work in isolation but have the attention of the team. Breach of confidence is mostly done by people who feel left alone in their job.

i. **Take the risk of default of a contractor into account.** A defaulting contractor may be a rare event, but when it happens, the damage can be enormous, and as discussed before, the problems generally begin long before the actual insolvency is formally stated. A good solution is to ask the vendor for the authorization to obtain liquidity information from the company's bank before contract closure—at least for sellers who are to be assigned with mission-critical work. When this has been given, one may decide whether one will actually use it; the peace of mind from the authorization may be sufficient. The bank information is not a 100-percent guarantee. A performance bond by an insurance company will give more certainty, but it is costly. I remember one case in which the insolvent contractor was bought by the customer, who could then go on with the project, but this was a decision beyond the project manager's authority.

j. **Watch out for paradise syndrome.** The success of today may be the cause for tomorrow's failure. The seller may have shown an impressive success record during business development, but sometimes it is this success that leads to complacency, smugness, and a degree of arrogance among management and employees. Contractual terms that allow easy separation from the contractor seem to be the best protection, even when the dependency on the company may be very high at a given stage of the project, making it technically difficult to change to another contractor.

---

[17] (TI 2016b)
[18] (RSF 2017)
[19] (Freedomhouse 2017)

k. **Protect your project from the extended vulnerability to digital attackers.** Connecting networks of a customer with a supply chain network, which is essentially a number of temporary project business partners, brings risks for all parties involved. It is difficult to manage malware risks in one's own organization, but almost impossible to manage them in other companies. Data protection experts have the tools and the processes to protect the organization, but not all organizations have these experts, and when they have them, using them in projects may not be intended. It should be. Problems seem to happen rarely, but when they turn up, their impact can be massive.

l. **Make sure the customer remains king.** With a large contractor, the customer may no longer be the center of all attention and care, but just one among many. One finds one's own company sharing the contractor's assets as project resources with other customers, and being able to use them only when they are not blocked by other customers. When mandating tasks to other companies, particularly to large ones, this expectation should be communicated. Key people assigned to the project must be protected from cost-cutting measures, such as firing or giving concurrent assignments to several customer projects.

m. **Ensure you know all subcontractors.** Over the various tiers of the *project supply network,* companies may become involved with the project that one would not want to work with, and one may not even be aware of their presence. They could be impostors, elements of organized crime, direct competitors, or companies whose competitive behaviors negatively affect the supply network.

The points discussed above are contractual matters but relate also to project management and to the way one does business. Many project managers rely for procurement activities on the organization's purchasers and counsels. This is often not enough. The project has specific interests that these groups my not be sufficiently aware of. A further point of concern should be that contracts regulate the distinction between compliance with and breach of the legally agreed-upon scope of cooperation. For a successful project however, contract partners have to be proactive and responsive to the formal agreements and jointly strive for "Mission Success First". For a project manager in a complex procurement situation, seeking the advice of these corporate functions is advisable, but the project manager should ensure that, on top of legal and commercial aspects, the needs of the project will be regarded.

One may summarize the points above in three basic rules:

- Know the participants in your PSN, know them all, and know them well.
- Maintain common sense when the complexity is growing in the PSN beyond the expected.
- Keep up the "Mission Success First" attitude over the PSN.

The risk is high that the project was started with high hopes and objectives and with the intention to care for everyone who is prepared to contribute to the success of the project. Without a lot of watchfulness on the PSN, this noble goal will fall apart before one's eyes, and achieving the project goals timely and on budget and not causing more operational disruptions than absolutely necessary (and accepted by line managers) becomes finally impossible. The project manager's reputation will suffer, and if this is a project presented on newspaper's front pages, so will the reputation of the organizations involved.

4.8 Supportive Action, Provisions, and Enabling Services

There is a common assumption that it is the contractor's sole responsibility to ensure timeliness and to meet other requirements, such as adherence to budgets (in cost plus and T&M contracts), operational disruptions on the customer side, and delivering what is agreed upon and needed. The assumption is often not right—the customer may be a central influential factor for meeting the contractors' objectives.

Contractors depend on customers in various ways. When I am talking with contractors, they cite many reasons for how customers have caused or at least contributed to delays, budget overruns, problems on a technical or personal level, and other missed objectives, among them:

- Confusion about what mission success actually is
- Confusing objectives and constraints
- Missing institutionalized conflict resolution mechanisms
- Missing a network-wide project management information system
- Poor delivery of enabling services and provisions

Supportive actions by the customer are an essential prerequisite for project success in many *project supply networks*. These actions include discretionary or contractually mandated activities under the customer's responsibility that enable contractors to do their job. They may be required to support an individual contractor or a group of them inside a PSN. Sometimes, supportive inaction may be required, which means to avoid actions that would disrupt contractors' work for the project. In this context, I will also address *enabling services and provisions,* which are mostly contractually stated obligations rather than discretionary. They may also be implied in the contract without being explicitly stated, or they may be construed into the contract from hindsight, in a worst-case scenario by a judge who has to decide in a lawsuit.

4.8.1 Understanding "Mission Success First"

The non-governmental organization (NGO) Greenpeace, dedicated to environmental activism, is an interesting example of successful project management in a complex environment.[20] For all its operations and projects together, it has its mission defined by just one photograph. It shows a beach with the sea in front and palms in the background. It stands for islands that elevate only a few meters above sea level, and that will disappear when the sea rises, which is an inevitable effect of climate change.

For every decision made at Greenpeace, the question is asked whether it will impact the these islands, and how. The photo represents the societies most vulnerable to rapid climate change and defines a mission statement based on their fate.

A project manager who wants to institutionalize a "Mission Success First" attitude in the PSN should begin with a simple clarification: What constitutes mission success? Or to simplify further: What is this mission?

---

[20] Here, as in other parts of the book, it is not my intention to discuss political matters, but to use examples to highlight core aspects of Project Business Management.

The definition may be quite simple: It may be a new piece of software, machinery, infrastructure, or any other kind of asset that the customer receives and that helps run the business. In such a case, mission success is meeting specifications in time and at the budgeted project cost.

The definition of mission success may go much further: The project has been commenced to meet a business case, a legal or contractual requirement, or another goal defined by management and other stakeholders. In such a case, it is more the operational cost and benefit of the project result, the lifecycle cost and benefit, or even the *total cost of ownership* (TCO) and *total benefit of ownership* (TBO)[21] that define mission success.

The definition of mission success may reach even further and include subjective aspects that are hard to measure: Do users "enjoy" working with the new software? Does it give management and workers confidence that the new safety system is in place? Does the new product increase the perception by customers that its maker is a successful and trustworthy company?

Whatever the mission success (and failure) criteria are, they should be communicated to the PSN early in order to ensure that contractors can understand what "Mission Success First" actually means and what entitlements and obligations can be drawn from the statement.

4.8.2 Objectives and Constraints

Constraints are hard limitations to the decision and working domains of project teams. A project to run an election campaign must be finished by election day. A family that has $500,000 available for a new house is on the direct path into private insolvency when costs are about to exceed this number. A team member in Project A, who will leave the team by the end of the year to work for Project B, must have all work for Project A done by that time so as not to bequeath unfinished results to the current project that its team may not be able to complete. Most constraints have their origin in factors that are external to the project and that are hard—often impossible—to influence. Often, they are mandated by physics or by law and are commonly reflected in the contract or another type of agreement. Constraints may originate in national calendars that disallow certain team members from working on festive days. The most influential type of constraints in many projects are deadlines and funding limitations, whose overdoing spells debacle for the project. Constraints are often the red line between success and failure.

Objectives, in contrast, are arrangements that have been defined or agreed upon based on desires or voluntary promises and therefore describe what a project team aspires to achieve, or what achievements are expected from it by key stakeholders. Mostly, constraints are just there. They must be identified by the team and should be documented in a way that they will not be ignored or completely forgotten. Objectives are defined by the team, alone or in a decision process with mutual input from those who have a vested interest in them. Separating constraints from objectives is identical to making a distinction between *want* and *must* (or *must not*). A seemingly simple thing.

It is strange, but *want* and *must* are often confused. Some examples of when *must* means *want*:

- I recently heard a manager telling a subordinate: "I <u>cannot</u> accept your behavior".
- A British bus company states: "[. . .] we are <u>unable</u> to accept £50 notes".[22]

---

[21] See the definitions of these terms in the second chapter, beginning on page 136.
[22] (Yellow Buses 2017)

- In the famous science fiction movie "2001: A Space Odyssey" from 1968, HAL, the computer with artificial intelligence, manages the spaceship on its way to Jupiter but is turning into a killer. When one of the astronauts returns from a brief trip with a pod, he commands "Open the pod bay doors, HAL", to allow him to return to the safe and life-supporting interior of the vessel, but the computer responds: "I'm sorry, Dave. I'm afraid I <u>can't</u> do that".

In these examples, individuals with the power to impose their will hide behind a seemingly factual constraint outside their domain of influence. An example of the reverse—couching a hard command within "want" language—are managers who communicate an objective, saying "I wish you to . . ." or "I prefer that you . . .", when they actually mean "you must . . .".

This even has a cultural aspect: Edward T. Hall described the difference between what he calls *low-context cultures* and *high-context cultures*.[23] In low-context cultures, the expectation is that a said statement simply means what it says. A "yes" means "yes", a "no" means "no". Constraints in these cultures are communicated as hard borderlines, objectives as desired results. In high-context cultures, however, such direct and unambiguous communications are rather avoided. Speakers in such cultures will rather fear the tendency of such messages to hurt people and to direct a spotlight on the speaker: Saying "I want something" places accountability on me. A statement "I must" or "I cannot" takes accountability off the speaker. In high-context cultures, indirect statements are rather expected. This does not mean that they may not lead to misunderstandings there.

Inside one's own organization, most people have time to learn what their managers mean when they make statements that conceal their true meaning in such ways. Externals do not have this experience. When managers talk with externals, it is important to have precise communications, in which the said statement is identical to the meant message. The question of whether the communication of a date, an amount of money, a time span during which certain work can be done and resources utilized for the project, and other similar requirements should be considered a hard *must-be* constraint or a much softer *want-to-be* objective is important so that the contractor understands how much flexibility the requirement has and to what degree it needs to be prioritized. As a customer, one does not want the contractor to misinterpret requirements and to focus on soft goals and specifications while missing hard ones. Clear and direct communication is at the heart of contractor management.

How are constraints and objectives linked? Table 4.2 shows how, along different dimensions or knowledge areas, objectives and constraints are separated by reserves. An example is the deadline stated in the time dimension in the table, which is a strict constraint. In industries such as automotive, the start of production (SoP) deadline is actually one of the most pressing deadlines that one will have. The objective to be ready for the SOP three months earlier creates a time reserve that one can use to fix problems or cover delays. Communicating the deadline to the contractor should ideally be done in a different way than communicating the objective. If the contractor has an unforeseen problem, the objective allows some time to fix it. When the contractor schedules against the deadline, there is no such buffer time available.

A clear communication of what constitutes a constraint and what is an objective is supportive action to the contractors, as it helps them make the right decisions when different

---

[23] (Hall 1989)

Table 4.2 Objectives and Constraints are Separated by Reserves[a]

Dimension	Objective statement	Constraint identified	Reserve
Time	Start of production (SoP) has been targeted for January 01, 2019.	SoP deadline has been imposed for April 01, 2019.	⇨ Three months schedule reserve
Budget	A cost estimate at project start has been approved: $10,000,000.	There is a funding limitation of $12,500,000.	⇨ $2,500,000 monetary reserve
Scope	15 functions are planned as 'wanted'.	12 functions are specified as 'critical'.	⇨ Three 'nice-to-haves' (reserve for de-scoping)
Quality	Control limits: Dimension x ± 1 mm	Specification limits: Dimension x ± 1.25 mm	⇨ Quality control reserve: 0.25 mm[b]
Resources	10 team members have been planned.	12 team members are available if needed.	⇨ Two bench resources
Resource availability	Bill is expected to work four weeks on the task.	We have booked Bill for five weeks	⇨ One week resource reserve
Operational disruption	We plan to stop production for four weeks.	Management has agreed to a production stop of a maximum five weeks.	⇨ One reserve week for the project
Procurement	The contract has been awarded to a contractor.	A second contractor is on standby for a fee.	⇨ One backup contractor
	The delivery has been ordered for 1 April.	The delivery will be needed for the project on 8 April.	⇨ Eight days feeding buffer
Risk	The work package is planned for $10,000,000.	An insurance has been bought to hedge the project from losses from certain risks.	⇨ The benefit of the insurance[c]
Safety	The power plant must be protected from a 3.7 m tsunami.	The power plant has been built on a location 10 m above sea level.	⇨ 6.3 m safety reserve

[a] (Lehmann 2016b, p. 200)
[b] This means that a results corridor of the measured dimension of ±1.25 mm around Dimension x would be tolerated, but when the results deviate by more than ±1 mm, correcting the process would be considered.
[c] This reserve is generated externally to the project by the insurer, but it would serve as a monetary reserve when the insured event occurs and the insurer pays.

requirements make a trade-off necessary. It helps contractors prioritize and eases communications inside the PSN, because all parties have the same understanding of what constitutes a *want* or a *wish*, and what is actually a *must*.

4.8.3 Enabling Services and Provisions

Mayfly Ltd. is a provider of optical and digital services to hospitals, including cameras, endoscopes, screens, and processing systems as supportive infrastructure, but also as communication and documentation equipment for operation rooms (ORs). These components are installed on the customer's premises in complex system integration projects and connected with the customer's other digital systems, which may be legacy systems in older hospitals or new systems when the hospital is newly built. One of their largest customers was the Titan Beetle Clinic, a large-scale new building to replace three outdated legacy hospitals with modern infrastructure and several special disciplines concentrated under one roof—a greenfield project that promised substantial improvement of the health service for the people living in the area.

The size of the new installation required a lot of preparation by Mayfly, including ordering equipment and materials and booking their own staff and subcontractors timely. Mayfly did all these tasks for the customer. Two weeks before the appointed day of installation, when two truckloads with materials were already stored in by the contractor, its project manager, on the way home from another project, drove by the construction site of the new hospital, which should have been almost finished—the integration of the digital systems normally follow some weeks after the end of the construction project.

He was surprised and alarmed to find there nothing but an open construction pit and some first concrete walls inside that would later become the basement. He immediately phoned his contacts on the client side and was told that the construction was more than a year behind schedule, that the delay was not the responsibility of the construction company, and that this had been communicated in local newspapers.

The project manager asked why he was not informed of the delay, but was in return asked why he had not made himself familiar with the progress of the project. He explained that he had several other customer projects to take care of and these had consumed his time and his energy. But if the delay would have been communicated earlier, he would have been able to rebook subcontractors and reschedule the delivery of the materials and equipment.

The further course of the project became difficult. The manufacturer of the digital equipment launched a new product generation for the core items, which the customer desired to have instead of the originally ordered equipment. A change request by the customer was submitted but rejected by the contractor, as this would have devalued the outdated warehoused equipment. Later, when the building was finished, the installation of the electronics by the contractor was done late, adding further delay to the project. The contractor booked the installation personnel this time rather late and, given that the subcontractors meanwhile had other orders, seller and buyer had to accept that these were no longer easily available. The customer could not claim damage payments from the contractor for this delay, as the contractually agreed date for the installation was over a year ago, and the contractor had been prepared to deliver by that time.

It is a simplification to just say that a customer project is one in which the contractor provides deliverables to the customer—products, services, etc.—and the customer pays the contractor for that. The business relationship is much more complex. Figure 4.6 shows how the customer has greater obligations to a prime contractor, who in turn can have the same obligations to subcontractors, and so on.

As shown in Figure 4.6:

- **Enabling services** are services that the buyer has to provide to the seller as a prerequisite for the seller to deliver what was contractually agreed timely and in the expected quantity and quality. Simple examples are badges with the necessary access rights to customer facilities (and the company restaurant), transportation services that the customer provides

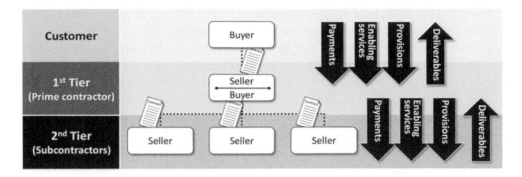

Figure 4.6   Commonly found obligations in a simple two-tier PSN.

to the contractor, regular updates on project information by the customer, among many more. Freeing up space in an office or on a construction site for the resources of the contractor is another common example. An enabling service that often causes difficulties is access to knowledgeable customer employees, combined with the direction given to these employees to actively support the contractor. A project management information system with access for contractors that enables them to monitor project progress, as far as relates to their work, would also be an enabling service, in that it allows early decision making by the contractor, ideally to the benefit of the project. In rare cases, the amount of enabling services by the customer may exceed those that the contractor has to do for the project.

- **Provisions** are items that the customer must deliver to the contractor as a precondition for the contractor's work. It could be raw materials, tools and equipment, work permits, and many more items that the customer is obliged to supply.

As a customer, one should always assume that the contractor monitors and documents the completeness and timeliness of provisions and enabling services. In a conflict with the potential to need remedy in court, such documentation can make the difference between winning and losing, or when it comes to a settlement, how beneficial this is for the party. This means in return that a customer should also document the timeliness, correctness, and completeness of services and provisions arranged for, just to be prepared for the worst case, which might be rare, but can then be very damaging.

4.8.4 Institutionalized Conflict Resolution Mechanisms

"Nursing contractors is not part of my job description". I heard this statement some time ago from an attendee in a class when we discussed what institutionalized conflict resolution mechanisms could look like. In his business life, acting as project customer's representative to contractors, he often experienced conflicts, but as a well-educated engineer, he preferred not having to deal with them. When these conflicts turned into problems for the project, he was very capable of identifying the culprit and make everyone point the finger at that company, but this did not lead to a more efficient or effective project.

Of course, the statement about nursing is fundamentally correct, but it is not very helpful, when contractors' problems one day translate into project troubles on the customer side and

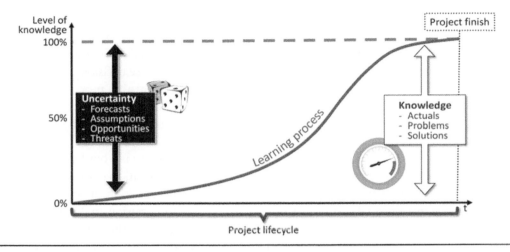

Figure 4.7   A project begins with high uncertainty. Then, stakeholders follow a learning curve.

when it is too late to find simple and inexpensive solutions. It is not untypical that these problems grow over the course of the project. Replacing "nursing" with "conflict resolution" helps find a solution for the project, protecting its productivity and the integrity of the systems it puts into place. But one should do this early.

It may be a good idea to remember Figures 4.7 and 4.8 when the task is to deal with sellers. They describe some general developments over the lifecycle of a project, the learning curve that all stakeholders are passing through, the loss of options, and the increasing costs when chosen options need to be implemented.

These developments can be particularly troublesome when, in project networks, legally binding decisions must be made early, when the knowledge to make such decisions is not yet available but will be developed over the course of the project.

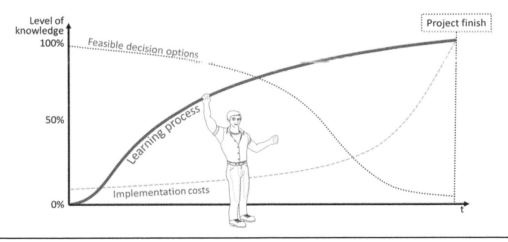

Figure 4.8   The development of feasible decision options and the costs to implement them follow an opposite development to the learning curve. The response should be to accelerate the learning curve.

It begins with the learning curve that the team will go through. The beginning of a project is characterized by a deep a lack of knowledge. Early decisions must be made at a time characterized by uncertainty, and many of these decisions will not be based on hardy knowledge but on assumptions and in consideration of the risks that come with these decisions. This is true on both sides, customer and contractor, and all over the PSN. The learning processes during the project will come with lessons that are valid only for the specific project; others will become part of the experience base that project managers collect over time, thus adding to the proficiency and professionalism of the persons and the organizations involved.

Some of these risks will have been assigned to a party in the contract, but learning is essentially unpredictable, and many risks will turn up long after the ink has dried on the contract signatures. During this learning curve, and with it the reduction of uncertainty, two more developments occur: The number of feasible options for decision making goes down, and the remaining ones get more expensive.[24] Figure 4.8 describes the approach to address this dilemma, obtaining knowledge about the project, its environment, and the stakeholders involved as early as possible.

For Project Business Management, this quickening of the learning curve includes gathering knowledge about the member companies of the PSN, including answers to some basic questions:

- What companies are members in the network?
- What are their respective business interests?
- How well are they organized and financed?
- Who are the key people involved on the contractor side?
- What motivates these people as individuals and also as a team?
- What frustrates them?
- How does the contractor assess and respond to their successes and failures?

In essence, these questions can be reduced to two: How much *should* we know about contractors, and how much *can* we know? Figure 4.8 mandates gaining this knowledge early, and one must expect high resistance—a contract party may consider such questions an infringement of their business autonomy. Why is it nevertheless important?

There is an interesting analogy: Think of the PSN as a major surface aquifer, in which the customer is similar to people, communities, or entire countries living near the estuary, and the contractors to those midstream and at the tributaries. Living downstream, one can simply enjoy life near the water and ignore what is happening upstream, but the risk of being impacted by activities there is high. Rivers and their tributaries can be polluted and diverted. Their water can be distributed over fields where it evaporates, so that it will no longer get to the river mouth. There may be a dam upstream, whose damage can cause a disastrous flooding downstream.[25] Water management measurements upstream, such as flood control, can also impact the availability of water downstream, resulting in phases of water scarcity alternating with times of unmanageable overabundance.

---

[24] This is described in detail in my book *Situational Project Management: The Dynamics of Success and Failure* (Lehmann 2016b, 50–58).

[25] Example: The Johnston Flood of 1889 in the USA, which killed 2,200 people (Clarke 2007).

Those living downstream must take an interest in what happens upstream, and ideally vice versa. One could expect wars to happen over water, in particular as large rivers cross country borders, and the different interests of the riparian countries may fuel conflicts. Research shows, however, that while complaints are common—some justified, some not—even vicious war enemies generally come together and resolve issues about it.[26] A major reason is that activities downstream can impact life upstream too. For example, land-locked upstream countries depend on the downstream riparians to allow them access to the sea, and with this to international maritime logistics on which they depend for export and import. Dams built downstream can furthermore block fish from swimming upstream, heavily impacting ecosystems that many humans rely on, such as fishermen and recreational businesses.[27] The dependency is mutual and existential, and it does not matter whether the countries and communities involved are direct neighbors or separated by others in between.

The analogy to PSNs is interesting: The contract parties are also both organizationally independent and factually interdependent at the same time, and decisions made in one part of the PSN may influence organizations in other parts, both positively and negatively. Whereas for contracts, the legal doctrine of *privity of contracts* applies, as described in Section 4.5.2 (on page 217)—the mutual influences that members of the PSN can have on each other, positive and negative, do not respect this privity. Solutions to conflicts inside these networks can therefore not be solely contractual but must also be managerial.

How do riparians resolve their conflicts?

As an irritant, water can make good relations bad and bad relations worse. Despite the complexity, however, international waters can act as a unifier in basins with *relatively strong institutions*.[28] The historical record proves that international water disputes do get resolved, even among enemies and even as conflicts erupt over other issues. Some of the world's most vociferous enemies have negotiated water agreements, or are in the process of doing so, and the institutions they have created often prove to be resilient, even when relations are strained.

Strong institutions to resolve conflicts are obviously the key to success. These could include:

- Cross-company focus groups that help resolve technical and organizational issues. This conflict resolution is generally the fastest.
- Project internal review boards that have the power to make decisions that are accepted as binding by the members of the PSN.
- Project external mediation and arbitration institutions established to perform alternative dispute resolution (ADR). Involving an independent party takes longer but helps find a mutually acceptable solution when strong emotions are involved.

For a singular customer–contractor relationship, selecting one of these resolution levels may be sufficient. For large and complex PSNs, it may be more appropriate to have a staged system in place, starting at the focus-group level and including the possibility of proceeding to external help when necessary. Such a system would then try to resolve conflicts at the most appropriate level.

The river example gives an indication of what one should further consider when one implements such an institutionalized system:

---

[26] (Kramer, et al. 2013, 4–12)
[27] (Holmlund and Hammer 1999)
[28] *Ibid*, emphasis added by me.

- The conflict resolution institution(s) must be strong enough to make swift and acceptable decisions, and the value of the institution must be accepted by all members of the PSN, ensuring a fair distribution of obligations, benefits, and risks.
- All parties must generally be addressed by the institutionalized conflict management system. Parties ignored or excluded will feel additionally frustrated and may then act against settlements found.
- The parties in a PSN are not monolithic business entities but societies made of people with different opinions, attitudes, and personal objectives. Salespeople and project contributors in a contractor company, for example, may have different views on the project, and their conflicts can impact a project as much as conflicts between companies. The understanding on the corporate level alone does not guarantee the full functioning of the PSN.
- Timing is an issue: Between the emergence of an issue and the resolution by the institution, a lot of time can pass, and a party may be tempted to act unilaterally during such a time to establish *faits accomplis* that trigger or amplify tensions and make a mutual solution impossible. When two parties do that, new conflicts may arise that the institutionalized conflict resolution may not be able to settle.
- Corruption is another threat for the functioning of the system. Corruption is sometimes considered a system "greasing" the wheels, but in reality, it is rather the sand in the gear box that relocates assets dedicated to the project into private pockets.

4.8.5 Project Management Information Systems in PSNs

Another core element of managing a PSN is communications among many parties. A project management information system with easily manageable access rights and functions, such as a centralized document repository, team collaboration services, a cross-company team calendar, and quick and confidential person-to-person messaging, would be helpful.[29]

In order to install such a system, a first question would be, who should be responsible for its installation, operations, and management. Then, it would be necessary to identify the requirements that the system must meet and what solution (or combination of solutions) is available that best meets the requirements.

Particularly in international projects, the question of where the data will be physically located can become another point of controversy. Data on the system will to some degree relate to the companies and persons involved, and objections by these stakeholders will have to be taken seriously. One would in addition need to consider possible legal constraints which have been imposed for the protection of privacy and of the companies' confidential information. They often reflect the differences in the understanding of people's and corporations' ownership of private data, and as PSNs process such data in networks that span corporate limitations and often country borders as well, such questions should be addressed when the network is developed. Complaints and accusations about inappropriate protection of this data will in most

---

[29] While I am writing these lines in the summer of 2017, there seems to be no single system dedicated to cross-corporate PSNs that integrates all these functions. The more software vendors will identify that there is a market for such a software solution, the more likely it is that one may exist in the future.

cases be directed to the PSN's customer, whose efforts to protect such data will be put down as inattentive and neglectful.

Information management systems often suffer from not being used at all, or not being used by all those who should. One reason for that can be lack of familiarity with the systems. A formal introduction to all parties using it for collaboration will probably be necessary. Another reason may be the transparency that such a system creates. I will address this later in more detail.

One more interesting aspect of a project management information system that spans the PSN is the support for the previously described institutional conflict management system. With a sound database, combined with correct interpretation of this data, such a system can also act as an online forum for discussions in preparation for effective conflict resolution activities, reducing the need for traveling and lengthy face-to-face meetings until they become necessary.

4.8.6 Specific Business Interests of In-Between Contractors

Prime contractors can do an incredible service to customers and subcontractors. They can mediate conflicts between people who develop project results and those who will use them. They can further be experts in both offer management and vendor selection, bringing together the best buyers and sellers. They can also channel the flow of information between complex communication networks on both sides—customer(s) and subcontractors—thus helping align this flow with the strategic goals of the project. They can bring order into chaos.

Prime contractors can also be the most massive obstacle to a project business manager to know all contractors involved in the project, to develop organizational and technical systems to manage these networks, and to ensure that conflicts among parties in the PSN will be resolved before they impact project success.

As mentioned before, prime contractors are in a business situation that reminds one of the "ham in the sandwich". As shown in Figure 4.6, they pass through deliverables in one direction and payments in the opposite. They also have to ensure that provisions and enabling services are arranged by the customer and bestowed on the subcontractor. It is possible that their job is considered as organizationally beneficial—valuable for the project and worth the costs. It may also be that the prime contractor is rather regarded as dispensable, and the costs for the organization are regarded as wasted. A prime contractor who is also productive for the project with its own resources and adds contractors' assets for those pieces of work that the company cannot do on its own is at much less risk of being seen as superfluous and squeezed out of the project, but the risk is high if the prime contractor is only an in-between dealer, passing all work to subcontractors. Such a prime contractor, who does not tangibly contribute to the actual completion of project deliverables, is always at risk of losing out when the customer and the subcontractor decide to do business directly and reduce both project costs and communication channels.

A common strategy for such a prime contractor to protect its business is to conceal the subcontractor from the customer. To achieve this goal, subcontractors are then contractually mandated to hide their true relationship from the customer. They will appear as business units and employees of the prime contractor, whereas they are actually independent. If the work of the subcontractor can be done without direct customer contact, the customer may not even know that another company has been made a part of the project.

4.8.7 Work Flow Management Across a PSN

Scarabaeus Ltd. is a manufacturer of extruded aluminum tracks used for a variety of applications from electric light rails to construction and engineering. They hired Woodlouse S.A., a specialist for production facility development, as a prime contractor to build a new production line for high-precision tracks of lengths between 5 m and 10 m, which were mainly used in application areas in which trueness to measurements was considered more important than a cheap price. The construction would take place at their existing premises as a greenfield project next to existing production buildings. The project included the construction of the building and the development and installation of the manufacturing line inside that building.

In order to meet a challenging deadline, a decision was made to fast-track the production line development and erection by overlapping the two processes. The production line consisted of four sections, named Sections A through D, and it had been originally planned to first fully construct the building, then develop the four sections of the production line, and then to finally fabricate and assemble the tracks inside the building. Now the decision was made to start the development of the production line when the building design was finished, and to make and install each section when its design was finished. Figure 4.9 shows a comparison of

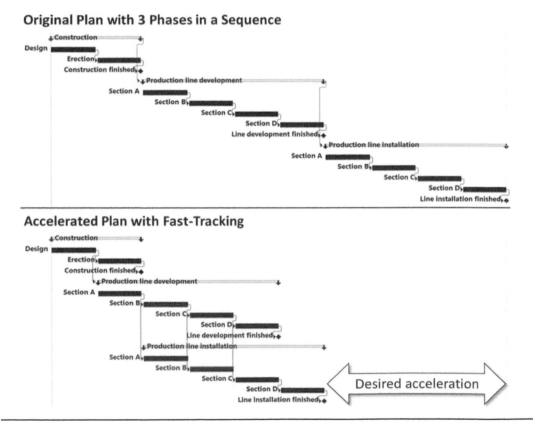

Figure 4.9   The workflow plan of the project in the case story before and after fast-tracking. The bar lengths are not proportional with the durations of the work items.

the two workflows for the project. The expectation was to save roughly 40 percent of the time by overlapping phases that were originally planned to be performed in a sequence.

Fast-tracking is a common approach to accelerate a project by performing project phases and activities in parallel instead of one after the other. In projects performed by PSNs, the allure of this approach is even stronger, because the limitation of internal resources no longer restricts it. If more resources are needed, they can be hired, and most will cost the same whether they work earlier for the project or later.

Fast-tracking comes with many risks, and the case story of Scarabaeus and its prime contractor, Woodlouse, became a good example for that. These risks can be identified and managed, but if this is not done, it will be very surprising when they occur.

The problem was that Section A was too long, not in time but in physical length. There were options to make it 27 m or 15 m long, but as the shorter options were more expensive, the longer were chosen. This dictated the position of Section B, which took the raw production output from Section A and processed it further, and so on. The production was planned as a linear sequence starting at one end of the building and ending at the other.

Relatively late during the project, it turned out that the building was 10 meters too short for this production line. There would have been solutions to shorten the footprint of the production plant, but the need to implement them was identified too late. An alternative would have been to make the building longer, but this was impossible—it backed directly up to the neighbor's ground. Major changes in the design of the production line became inevitable, which delayed the project far beyond the originally planned handover date. Both the project and the production became costlier, and the production output was reduced by roughly 15 percent.

A root cause of the problem was that the subcontractors that had the different work packages assigned, such as designing the building, erecting it, designing the production line, etc., did not talk enough with each other. Partially unaware of the workflow in the project, they just did the job for which they were paid. The prime contractor was also not too interested in promoting too much communication; that would have meant additional work for them, and the customer considered the project in good hands.

There is a modern tendency to ignore workflows in projects. Practices such as earned value technique (EVT), agile methods, and others ignore how work items depend on each other in a project. It also happens that project managers plan for such dependencies, but the teams, their own staff as much as the contractors', do not adhere to these plans, running a kind of cookie-dish project management[30]: "Can I have the chocolate cookie today?"

Cookie-dish project management has some disadvantages:

- Work is done out of sequence, so that work that relied on information from previous work, which was not done, has to be redone.
- Idle times occur when a team member wants to start work as scheduled, but cannot, as predecessor work has not been done.
- Difficulties arise when trying to assess project progress, which can only be measured when activities are done in a planned order.

---

[30] A term I learnt some time ago from a student in a seminar, referring to a project in which team members select for themselves the tasks that they like instead of those that are due in the schedule.

- Leadership issues become murky when work is not done in the order that the project manager has planned, or when there is no such plan at all.

The difficulty with work-flow planning involving different contractors is that they may not be available for the project when they have other project customers to serve. Agreeing on availability slots can be difficult, and the work schedules that are then developed can be volatile—for example, when one contractor is late because work on another project was not finished on time. One may argue that this is the same problem that a project manager has in an internal project that must share resources with other projects and with operations. Such an objection is generally correct, but the challenge grows in the PSN, with its multitude of contractual agreements, conflicting business interests, and often insufficient communications between the companies. These factors make dealing with variances against a networked schedule even more difficult and can lead to damage claims and other negative results. It nevertheless needs to be done, and project managers managing PSNs must ensure that work flow definitions are being adhered to so as to avoid crises and unmanageable projects.

4.9 More Control on the Project Supply Network

4.9.1 Naming, Nominating, or Approving Subcontractors?

This topic was discussed in Section 2.12.2 (on page 125) in the context of contracting as seen from the seller side. What effect will these techniques have on management of the PSN?

When the PSN has some weak spots, their failures can trickle through the PSN and finally damage the entire project. In order to know contractors, subcontractors, and all parties involved in the PSN—at least all critical ones—some organizations use contractual mechanisms that give them a degree of influence on who is part of the project and who is not.

- **Naming.** Naming means that the prime contractor is directed by the customer which company is to be used as a subcontractor. The difficulty in this model is that the prime contractor must typically guarantee the performance of the subcontractor, which may be a company that is more or less unknown to the prime contractor before the business relationship and with which no rapport has yet been built. Not a good precondition for cooperation.
- **Nominating.** The customer gives the contractor a list with approved subcontractors, from which the contractor can chose one for the project. The problem remains that these are commonly unknown to the prime contractor, which makes it hard to select one and to take responsibility for that company's performance.
- **Approval.** The prime contractor selects the subcontractor of choice and presents the company or person to the customer, who can approve the subcontractor or not.

There is also a *reverse nominating* that occurs from time to time. As a trainer, I have had situations in which major customers, mainly software companies, have asked me to offer my services as a trainer. I was then given a list of training providers, vendors whom the customer had approved. I was to select one among them as a prime contractor, send my offer to that contractor, who would then reoffer me, of course with an additional charge for their work. It was good business for these companies; they had little work to do, seeing that I organized the entire seminar, but they received attractive price surcharges for that. The benefit for the customer was that no new contractor needed to be listed.

Naming and nominating can be effective means to ensure the inclusion of long-time partners of the customer in the PSN. Approval helps keep unwanted companies out of the PSN.

4.9.2 Coaching, Consulting, Mentoring

Project procurement brings people and companies into the project that have not been in a business relationship with the customer before. Some of them may have to work temporarily at the premises of the customer; others will work over a distance. Often, they must get an understanding of the rules that apply when they work with this customer. Some rules will be formal, such as a Code of Conduct, meeting policies, or other kinds of ground rules that apply for the project. There are also informal rules, such as, "The person who took the last cup of coffee from the coffee pot brews the next one". Rules may also follow local or regional standards, such as how to greet people and how to respond to greetings.

Rules may deal with matters of workspace security, data protection, restrictions on walking around customers' premises without attendance, and many more. Such rules may be specific to the project or may be valid for the entire company and its suppliers.

Lack of awareness of such rules can cause tensions inside the PSN. Expectations on behavior are not met, and discussions that should focus on technical or organizational matters start focusing around interpersonal problems and misunderstandings.

In such cases, it can be a valuable proactive measure to coach or consult people on the expected behaviors of the organization sponsoring the project. A customer may do that to protect just the project or the entire organization from disruptions. I have experienced such events as highly effective. They were done in a physical classroom setting or over a meeting platform and lasted between two and four hours, depending on the amount of knowledge that the temporary staff needed to learn. In projects with a strong fluctuation of contractors and, with them, of people, these meetings were repeated, becoming more effective with each iteration, as the customer company learned over time where more emphasis was needed and what issues were less problematic than originally expected.

4.9.3 The Other Stakeholders on the Customer Side

Project managers are rarely alone when they manage complex PSNs. To this job, which some have compared to herding cats, internal stakeholders will add the dogs that frighten the cats further and make the herding job even more difficult. There is good reason for these stakeholders to ensure influence. As mentioned above, a schedule is binding for the project. So are scope statement, WBS, HR plan, and many other documents that the project team generates to organize the future of the project. Contracts, however, are binding for the entire organization. If a major conflict arises, it will not be a conflict solely of the project but will escalate immediately. Some typical parties affected are listed below.

- There may be a **purchasing department** in the customer organization, whose most common focus is to achieve low prices. Most of them will not look at operational costs of the project results, at lifecycle costs of project and operations, or even at total cost of ownership, but solely at the amounts on the contractor's invoices.

- Often a **legal department** is involved, particularly to sort out contractual details with contractors.
- **Upper management** will want to be involved as well, given the strategic nature of contracts for many organizations and of the selection of PSN members. One should also remain aware that, although a project schedule binds the project, as well as a human resource plan or most other elements of a project management plan, obligations stated in contracts are binding for the entire organization and, in the very worst cases, conflicts over them may need to be remedied at court.
- **Politicians** may further influence projects—not only government projects—by enforcing the use of contractors located inside their constituency. From a politician's perspective, this may be understandable; from a project management perspective, these contractors may be the project's weakest spots.
- The various **regulatory compliance departments**, particularly in large organizations, can massively impact the management of PSNs. The management of safety, professional integrity, equal opportunity, and others do not restrict their work to their own organization. They are aware that when the finger pointing begins, the buck will finally stop at the customer.

The ability for sound stakeholder management is an asset in every situation in which PSNs need to be managed. The multitude of stakeholders can make the task truly difficult.

4.9.4 Professional Integrity in the Project Supply Network

There is another reason to look at the last of these internal stakeholder groups on the customer side. Project management is unfortunately not free from corruption. This seems to very dependent on the industry—some industries have a tradition of rejecting bribery and other forms of corruption; in others, it seems to be the norm.

Research done by Gallup International in 2010 with 770 participants from business, law, accounting, banking, and chambers of commerce in 14 countries were asked to what degree the respondents consider players in their industry sectors willing to pay or extort bribes.[31] The "cleanest" sector was agriculture; on the other end of the scale was public works and construction—the industry with the highest perceived willingness. Figure 4.10 shows the results of the research.

From my experience in project management and as a trainer, I can confirm that there are differences among industries. I was dealing with project people from electronics, IT and software, automotive, aerospace, and others, in which bribes would never be considered or discussed. But I also remember an advertising agency whom I had contacted asking for a proposal; they asked me openly if they should calculate a surcharge into their offer, which I would receive. I also had personal experience with the pressure that dubious corporations and people can apply by suing whistleblowers and writers for defamation.

There are important reasons to take corruption seriously in project management, as was discussed above referring to corruption inside the PSN. Here, corruption is addressed that relates

---

[31] (TI 2000)

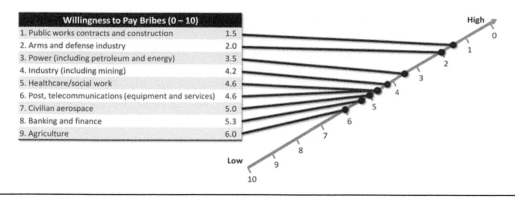

Willingness to Pay Bribes (0 – 10)	
1. Public works contracts and construction	1.5
2. Arms and defense industry	2.0
3. Power (including petroleum and energy)	3.5
4. Industry (including mining)	4.2
5. Healthcare/social work	4.6
6. Post, telecommunications (equipment and services)	4.6
7. Civilian aerospace	5.0
8. Banking and finance	5.3
9. Agriculture	6.0

Figure 4.10   Preparation to pay or extort bribes in industry sectors from a Gallup survey in 2010.

to procuring staff on the customer side. Such employees, asking for a bribe or accepting it when it is offered to them, render ineffective all attempts to create a "Mission Success First" attitude across the PSN—the attitude that builds upon, and enhances, open communications and trustfulness. Corruption creates an environment of distrust and paralyzes communications. The virtuous circle that creates an environment in which the PSN acts as a team turns into a vicious circle in which the project manager has no idea of what is actually going on in the project. While the project manager wants to create a "Mission Success First" culture, corruption places an "un-culture" against that, which secretly sets personal enrichment first.

What can the customer-side project manager do to protect the project from corruption by his or her own staff? A look into history is helpful, such as when the rich city of Venice in the Renaissance had similar problems: public agents who required palm greasing by citizens to do their job, do it timely, and do it correctly. The city distributed a network of *bocche dei leone*[32] letterboxes that gave whistleblowers a safeguarded opportunity to send messages to the city magistrate on corruption inside its ranks. Accusations had to be supported with evidence— one does not want to invite defamation and slander, and anonymous accusations were only followed up in cases of very serious accusations to the disadvantage of the entire municipality.

Whistleblowing has proven a strong protective mechanism when any form of corruption occurs, but the risks for whistleblowers are also high, and they are open for abuse. A contemporary form of a *bocca di leone* could be an encrypted communication channel that protects the whistleblower's anonymity and ensures that allegations are followed up with enquiry—and potential litigation, if the allegations are found to be true. Such a channel may be internal or may use an external party, such as a lawyer who ensures a trustful but effective process. It is also advisable to verify that measures taken to support whistleblowing are developed in accordance with applicable laws, and also to make sure that all parties in the PSN are aware that such virtual form of a *bocca di leone* exists, as a signal that the project manager is dedicated to keep the project clean of corruption.[33]

---

[32] Singular: *bocca di leone,* Lion's mouth.

[33] Associations such as Transparency International have measures to protect whistleblowers, and their local chapters may be good contacts for advice on how to make a project corruption-proof (TI 2016c).

4.10 Chicken Races and How to Avoid Them

"Chicken race" is a term from game theory, the discipline that deals with conflicts between individual and common interests. There are various forms of chicken races. A common sample is a race in which two "players" drive cars at high speed toward the edge of a cliff, and the "first to jump [out of the car] is the chicken"[34]—the coward. If no cliff is at hand, the drivers can also run the cars toward each other, and the first one who swerves has lost. Sometimes the players stand on a railway track awaiting a train, and the first who jumps off the track is the chicken.

In *project supply networks,* chicken races are also found, and they are not rare. Damselfly Ltd. is an example of a subcontractor who has a contract with prime contractor, Grasshopper, Inc., as part of a project to build a new production line for an automotive company in an already existing building. The program to erect the production line consists of 100 subordinate projects performed by different subcontractors, dealing with special machinery, infrastructure, control software, and other items. All contractors have the same deadline, which must be met in order to have all results handed over to enable timely start of production (SoP). To give readers a full understanding of this kind of business situation: Ninety-nine timely handovers by subcontractors but one being late does not constitute 99-percent success, but is 100 percent failure to meet the deadline. Figure 4.11 visualizes how one late project contractor delays the SoP beyond the deadline. In environments that apply this kind of scheduling, pressure on suppliers is extreme, as the consequences of a single project's delay are significant.

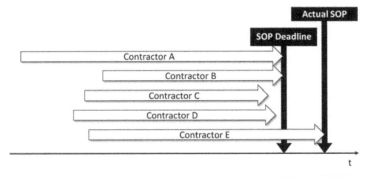

Figure 4.11   In a project or a program consisting of (sub)projects with a common deadline, such as a project to enable timely start of production (SoP), one late project delays the SOP, the deadline of the program.

A month before the deadline, the project manager of Damselfly's project did some assessments and forecasts on the project and found out that it would no longer be possible to meet the deadline. Delays from late deliveries of standard components that were out of stock at the company's supplier added to absenteeism of employees during a flu epidemic, and it became obvious that timely delivery would no longer be possible. She should have told that immediately to her contacts at Grasshopper, the prime contractor, to allow them to talk with the

---

[34] A famous example is shown in *Rebel Without a Cause,* a 1955 film with James Dean and Corey Allen as Jim and Buzz, the young men driving the cars.

customer and find a solution earlier, when this would have possible at low costs and still with a greater number of opportunities.[35]

The manager of Damselfly had a different opinion on that. He did not allow her to send a notice on the delay to the customer. He speculated that Damselfly would not be the only contractor to be late. In a *project supply network* with a hundred contractors, there must be more with similar problems. Damselfly could inform the customer and be the culprit for the late start of production. But, if someone else was the culprit, Damselfly would be given more time to finish work and would not have to worry. The strategy was successful; Damselfly got "off the hook" when another contractor could no longer hide its delay, just some days before the scheduled SoP.

No one knows how many further contractors were in a similar situation and how many champagne bottles were opened when these contractors received the message from the prime contractor, Grasshopper, asking them to delay the delivery by a month. The managers at Damselfly took this experience as a confirmation that not informing the customer in such a situation may be safer, protecting the company from troubles and from damage claims.

The project managers from the prime contractor and the customer had planned to start production immediately the day after the deadline, when all elements of the line were in place, tested, and ready for operations. It was planned to begin with five days' pilot production. After that, actual production would start at a low initial production rate, which would be slowly ramped it up over three months to allow for a capacity reserve, in case that was necessary to fix initial flaws that are normal for new production lines.

The production plan had been developed based on the reports by the contractors that they would be able to finish their sub-projects on time. Based on the production plan, commitments had been entered into by the customer with clients for the first deliveries. To meet these commitments in spite of the delayed start of production, the ramp-up phase was shortened to only two months. This then led to reduced operational performance of the production for more than half a year, because not enough time was left to find flaws early in the system and to fix them.

The all-over delay for production caused by such flaws led to overall production delays of several months, until the intended full productivity could be achieved. In this way, the partial communication breakdown caused delays in the production that added up to months. Some flaws caused errors in the production output that were only identified when the products from the line had already been shipped to the market, causing additional costs and a loss of client confidence.

Early communications would have limited the delay to just one month and would have allowed Grasshopper to talk with the customer, who in turn could have talked with the clients of the product, telling them early that their deliveries would have a short delay. The late communication allowed for late response, which added further delays and costs.

A solution to protect the customer from a chicken race, in which the contractor who responds early loses, are regular maturity checks,[36] in which the progress and the ripeness of the development on the contractor side are assessed, and in case of delays, measures are taken early to either accelerate work at the contractor or to prepare for the late delivery on the customer side. The difficulty in the example was that the customer and the subcontractor had no contract. The legal doctrine of *privity of contracts* (introduced on page 217) was then expanded to all

---

[35] (Lehmann 2016b, p. 52–53)

[36] In Germany, falsely referred to as "quality gates" or "Q-gates"; they are actually not gates, and they examine maturity rather than quality.

communications in the project, which suited the business interests of the prime contractor but disintegrated the entire project to the disadvantage of the customer. In a managed *project supply network,* the customer talks with the prime contractor's subcontractor(s), and vice versa, based on the trust that the business relationship is built in a way that the prime contractor's business interests will not be damaged by this openness.

4.11 Closing Contracts and Projects

A good friend of mine was hired by the Backswimmer AG insurance company some years ago as their new CIO. On the first day in the new company, she occupied her office and was introduced to the employees, her new subordinates. She had to become familiar with the corporation, its infrastructure, its processes, culture, and attitudes, and with the people she would have to work with from now on, including internal staff but also business partners and long-term contractors. One of the first issues that she was confronted with was an invoice from Stonefly GmbH, a contractor from a recently closed project, over a significant amount of money for post-project services provided to Backswimmer. The new CIO was assured by the employees that no billable service was provided by Stonefly and that the invoice should be rejected. Backswimmer then sent a letter to Stonefly, rejecting the invoice, but the company insisted on its validity and on payment. Backswimmer's CIO then accepted the invoice without further proof and submitted the invoice to the finance department for payment.

At Backswimmer, everyone was convinced that the invoice was baseless, but the new CIO could not devote management attention to the case. She needed to become cognizant of the new job and of the corporation, and because the running operations of the insurer's IT as well as a major portfolio of running projects needed to be governed, dealing with the case would have been too distractive for her. In other words, Backswimmer found itself unable to defend against the claim.

The case story shows the risks from having project contracts that are not formally closed, but rather die over some period of time. When all work has been finished and all payments have been made, either party—seller or buyer—could use, or rather abuse, them to make claims that are unfounded but may be hard to fend off. Contractual conflicts consume time, energy, and management attention, and closing contracts formally with a note that all contractual obligations are met (with the possible exception of warranties and operational post-project services) can make both parties' minds free for new ventures. Figure 4.12 describes how formal close-out ends the contractual relationship between seller and client, except for those obligations that survive the end of the contract.

4.11.1 Handovers and Acceptances

The term *handover* describes the physical or virtual process that takes place when the deliverables of the project are handed over to the requestor, which in a contract project is the customer or customers, if there is more than one. A project may have one handover or many, often referred to as *staged deliveries* (see Chapter 1, Section 1.3.3, page 15).

Acceptance, in contrast, is a formal managerial decision to approve the deliverables as being in compliance with specifications and other requirements, possibly linked with the obligation on the contractor for some minor rework. There may also be a multitude of acceptances—for

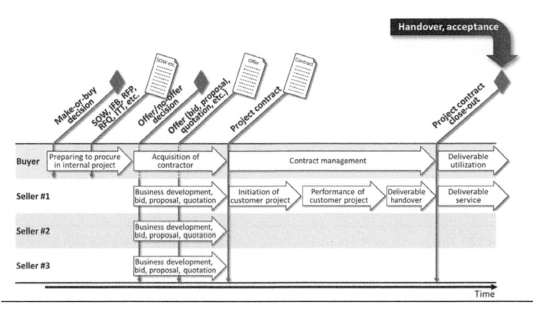

Figure 4.12   The common process flow in project procurement ends with formal contract close-out.

instance, in system engineering, when drawings, system components, sub-systems, and in the end the final system are accepted.

Handover(s) and acceptance(s) generally precede formal contract close-out; they may be done either in a single process of deliverable reception against a signature or independently. It is not uncommon that formal acceptance precedes the handover by weeks. In other cases, however, handover may have happened long before the project deliverables are finally accepted.

Formal project close-out, furthermore, depends on the final settlement of payments. Other obligations should also be ended by that time—for example, the return of temporary provisions, such as facilities and equipment that the contractor borrowed from the customer to work for the project. Enabling services, such as the access of contractor employees to the corporate restaurant, should also have been ended when the contract is closed out.

This is rarely done in projects, but I strongly recommend having a specific close-out document, signed by both parties, that the contract has been formally ended and that all obligations have been met, except long-term obligations that are listed in the document, including post-project services and warranties. In the case story above, the insurance company would have been protected from the late claim by a document in which the parties declared the contract closed out, or it would have given a reference as to how new claims could arise in the contexts of warranties and services.

4.11.2 Contract/Procurement Revisions in Project Supply Networks

In very large projects, it is not uncommon to announce a procurement revision early and perform it after the end of the procurement lifecycle. Figure 4.13 describes the relationship of the procurement and the contract lifecycles from the perspective of the customer. It also shows how each of the lifecycles can be terminated by use of a post-audit—a revision.

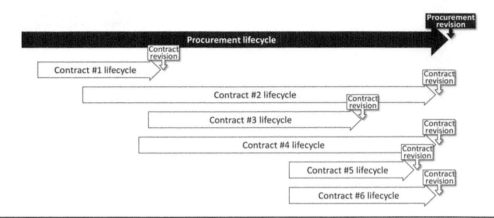

Figure 4.13 The procurement lifecycle includes the various contract lifecycles plus some time before and after the actual contracting period for preparatory work and final organization of documentation and other deliverables. Revision at the end of each lifecycle can help communicate a culture of "cleanliness".

The basic intention of these revisions is to ensure truthful behavior by both customer and contractor personnel. Project management under contract is a major challenge to the professional integrity of all parties involved, particularly for employees who make direct contract decisions and who have the capacity to make them alone. A lot of money is flowing, there are business interests involved, and errors made can lead to costs for a party when they are uncovered, which may tempt it to sweep them "under the rug" and bribe or blackmail people who know of them. Questionable payments, kickbacks (partial paybacks of invoiced sums that are not listed in all documents), and the involvement of dubious persons as proxies in the business are strong signs that more scrutiny is appropriate. Auditing documentation is an important task, but more important is early communication that such reviews will be performed, as a clear message that the PSN will look into shady behavior and will not allow underhanded activities.

4.12 Do We Need New Approaches to Contracting in Project Supply Networks?

History can teach interesting lessons for Project Business Management. Around the year 1900, countries, particularly in Europe, had a complex system of mostly bilateral contracts developed, in which they ensured mutual economic cooperation, non-aggression, and support in cases of conflicts, including wars. Many contracts were secret, and at least to my knowledge, no one had an overview of all contracts and how they would interact.

On 28 June 1914, Bosnian Serb Gavrilo Princip assassinated Archduke Franz Ferdinand of Austria and his wife Sophie, Duchess of Hohenberg in Sarajewo, which is today the capital of the Republic of Bosnia and Herzegovina, but was at that time a province hub of the empire of Austria–Hungary. The assassination was originally an internal event in the Austria–Hungarian empire, but became soon an international affair, when Austria–Hungary delivered the so-called "July Ultimatum" to the country of Serbia, which had actively supported the assassination.

The terms of the ultimatum included an authorization for Austria to investigate the crime in Serbia. They were written to be unacceptable by Serbia, which then rejected its full implementation. Austria–Hungary declared war on Serbia, Serbia called its ally Russia for help, Austria–Hungary its ally Prussia, and so on. The alliances snapped in, and the war soon spread all over the world. In the four years until its end in 1918, 16 million people were killed, and the political landscape of the globe was changed.

Unresolved problems left over from this war led directly to the Second World War, 1939–1945, which was even more disastrous. And again, World War II began with a secret bilateral contract—the Molotov–Ribbentrop Pact—in which Nazi Germany and Communist Russia buried their hostility temporarily and agreed to invade Poland and distribute its land between them.

After World War II, most countries had learned the lesson of the risks that come with bilateral contracts and developed multilateral treaties, which have effectively secured peace over decades. The most prominent of them is the United Nations, but many other treaties regulate regional or industry-specific cooperation. Most wars that occurred in the last decades involved countries that were not members of such multilateral treaties. There are forces today who want to turn history back to the time of bilateral contracts, ignoring the fact that they dismantle the most important peace mechanism that humanity has developed, while the risk of a devastating war for all countries involved has massively increased.

Before the First World War, it was considered normal that a state contract involved two, possibly three countries, and large treaties were a much rarer thing. One can compare this to the situation in Project Business Management, in which we commonly see PSNs built from a complex system of bilateral contracts, each with one organization acting as a buyer, the other as a seller. Sometimes, sellers join forces and create a consortium, a temporary joint venture, but then again, the customer has a bilateral contract with the consortium, which over times develops the characteristics that are typical for a firm.

Figure 4.1 (on page 200) showed how a consortium is used for indirect contracting. Some or all subcontractors are also investors of the consortium, which acts as a separate entity and contract partner to the customer. It is a prime contractor and a temporary joint venture at the same time. In such a consortium, the customer is normally not expected to also be a member and venturer, but there is no rule which disallows that.

There are some examples of PSNs that are developed as customer-led consortia (CLC, sometimes also called a *project alliance*). In a CLC, the customer does not have a seller–buyer contract with the consortium but is a core member. A prerequisite for a contractor to work for the project is to join the consortium, which may include a financial investment or be limited to an approval to join, which is then accepted by the consortium members.

The contract between the members of the consortium is a multilateral CLC treaty, and the share of workloads and payments is agreed among the members in a cooperative fashion outside this treaty. Such a CLC is more similar to a club or a political union of nations than to a classical project procurement contract. It is more complicated to develop, because it needs the acceptance of a multitude of corporations, not just two. Its benefit is (1) that it gives the openness that allows for easy re-arrangement and change of work assignments without the need to adjust a large number of individual contracts, and (2) that it enables the situational application of different practices, such as agile methods when work is exploratory and "the way is made by walking", or predictive when decisions need to be made with sufficient lead time to book scarce resources early or place orders for work today that takes the contractor months to finish.

Another benefit of a customer-led consortium is the acceptance of a "Mission Success First" approach across the PSN; this approach is the glue that keeps the consortium integrated. It is this mission for which the customer provides the financial and organizational resources to the CLC, and access to these resources for a contractor includes the acceptance of the consortium goal as their own goal. In other words—if a contractor wishes the membership to be successful, it must support achieving the consortium goal.

This leads to another benefit: the increased maturity of decision processes in a PSN. Decision maturity follows a four-stage model from data to action, and one may use the acronym DIKA for it. The model is shown in Figure 4.14.

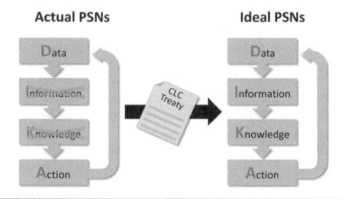

Figure 4.14   The DIKA maturing process of decision making, beginning with (raw) data and ending with action, which in turn provides new data.

- **Data.** The project returns raw data to the project management team, such as delivery dates, costs, efforts, and other numbers. The data of a project typically refers to its past time, when the data has been taken, and possibly to its present.
- **Information.** The data is interpreted. Causes of variances from baselines (expectations and plans) are identified, facts are separated from opinions, and positive news is separated from uneasy ones. The project is assessed for impediments, risks, and other issues. Forecasts are developed and analyzed based on the data at hand, which add predictions for the future to the comprehension of the past that the data gave.
- **Knowledge.** The team prepares its behavior: Where are people allowed to go on with their work without interference, where is corrective action needed? Is it necessary to update parts of the plan, or the entire plan? Is it necessary to implement risk responses, including contingency plans and fall-back plans? Knowledge leads from comprehending to acting.
- **Action.** The decisions for actions made in the previous step are implemented. The actions deliver new data that restarts the process.

In a PSN with its common opacity and dynamics, it can be difficult to put data into the context that is needed to understand it and to develop information. Without such information, it is difficult to develop the knowledge that is necessary to make well-founded decisions on actions to be taken. Instead, the PSN will be influenced by misunderstandings. In

Section 2.4.3 (page 68), I discussed Conway's law, which in essence says that the functioning of a system built by a number of teams will depend on the communications among these teams. Misunderstandings between the teams will lead to malfunctions in the interplay of the system components that the teams develop.

PSNs deliver data and require actions but can make it hard to process the data in a way that the actions can be considered mature—that is, informed and knowledgeable. A well-designed CLC may help overcome this maturity gap by improving the maturity steps of information and knowledge.

A customer-led consortium based on a multilateral CLC treaty is not easy to set up and manage, but it is a great tool to place completing over competing. It is a tool to take a *project supply network* not as an inconvenient consequence of tapping resources of sellers external to one's own company, but instead as an opportunity to design it, replace opacity with transparency, manage the dynamics of supply networks, and jointly with the contractors seed and grow the "Mission Success First" culture that helps the project in such a complex setting to finally be successful.

Chapter 5

Project Business Management and Crisis Management

5.1 The Power of Documentation

Cash flow considerations can be critical to project success. Companies can default and run into insolvency due to a lack of profitability, but also due to insufficient liquidity. One may say that it does not matter if a company goes bankrupt, as long as it is not their own, but in a complex *project support network* (PSN), this may be too shortsighted; one contractor in dire straits may financially damage other contractors, subcontractors, and the customer, as well as jeopardizing the entire project.

Springtail, Inc., is a training provider in the field of management seminars, including project management. They had an ongoing classroom business with open seminars, for which customers could book seats, and had a second business model doing qualification projects for companies to support their strategic objectives by training professionals, mostly from project management and sales.

In the year 2014, Springtail ran into a financial crisis due to massive costs of online advertisement. The online ads were sold by web advertisement providers in the form of auctions on keywords: When a keyword is used by a user, the providers' system selects those advertisers that pay highest and then shows these companies' ads. Springtail, in competition with other providers, had to increase their bid to the advertisement providers; their competitors reacted and did the same, and the costs for advertisement exploded within weeks.

Springtail then tried to reduce costs elsewhere: They put pressure on trainers—actually freelancers as subcontractors; Springtail had no employed trainers—to reduce their daily rates, which resulted in their best trainers ending cooperation. Springtail's other approach to make the classes cheaper looked at locations. Most open classes were performed in hotel rooms, and

Springtail developed an obsession to place as many students in one small room as possible. Small rooms are simply less costly than larger rooms. In many hotels, seminar rooms facing noisy streets are cheaper than those on the side of the silent backyard. As Springtail booked these cheaper rooms to further cut costs, the crowded classes could not even open a window to let fresh air in. All these measures were, in sum, insufficient to cover the cost explosion on the marketing side, but Springtail was so dependent on its online marketing that it could not develop other sales channels, and with an office staff stripped down to the minimum, it would no longer have the resources for that. Springtail, Inc., was on its way to bankruptcy.

At the same time, the company performed a major qualification project in project management and proposal management as a prime contractor for Sandfly Corp., an international manufacturing and trading company that was building a new service branch. The objective of the qualification project was to introduce an internal business development process and a project management methodology to all project managers of the new service branch and bring all project managers to the same level of proficiency and mastery of the methodology.

The business with Sandfly was commercially robust and profitable; the customer, however, noticed the growing impact on the project from Springtail's other business with open seminars. Accepted and performing trainers left the project, because they had refused to work further with Springtail, and new trainers were brought in—some of them quickly selected when imminent seminar dates needed instructors—who had to be available short term and had to accept the low daily rates. Some trainers were not booked a second time as a result of bad reviews from participants; others were not prepared to work for Springtail again after they had experienced the working conditions in brimful seminar rooms on noisy streets. These new and often temporary trainers also had no introduction to the strategic goals of the customer; they just performed their seminars unaligned to the client's overall business intentions. The customer, Sandfly, had expected a professionally managed qualification project, but at one point in time, the impression was that the only professional aspect of the project were the invoices of the contractor.

Sandfly then asked Springtail for a meeting to discuss the problems and find solutions. Appointments were made, and the chief executive officer (CEO) of Springtail was invited to them but did not attend. The customer's managers were waiting alone in their meeting room, frustrated and uncertain about what to do next.

They gave Springtail a last chance for a meeting before cancelling the contract. This time, Springtail's CEO turned up and immediately began a long monologue, explaining the causes for the company's often strange behavior. His explanations sounded reasonable, but the interests of the customer no longer appeared in them. Sandfly cancelled the contract. Springtail took Sandfly to court on damage claims for the loss of profit but lost the court case, because Sandfly had documented the decaying business, including the two missed meetings. The judge who presided over the case called Springtail's claim for payments "inconcludent", which is a judge's form of saying "utter nonsense". The qualification project nevertheless failed. Roughly one third of Sandfly's project managers had passed the training, the others had not, and the mission goal of unifying proficiency and mastership of the methodology was therefore not successfully achieved.

5.2 Introductory Questions

The following questions are written in the style of a certification test. They are intended to give you an understanding of the contents of the following text section and the questions that will be discussed in it. It may be interesting for you to answer these questions before you read the section, and then again once you finished it.

---

1. What effect can even moderate budget overruns and delays in a small number of projects have for the performing contractor organization?
   a) They can make it impossible to implement agile methods across the organization's complete portfolio.
   b) They can incapacitate the organization technically and organizationally.
   c) They can turn a planned profit from an organization's customer projects into a loss.
   d) No effect, the other projects can effectively cover the monetary disadvantages for the organization.

---

2. What is the spotlight area of benefit engineering by a contractor?
   a) Generating benefits for subcontractors in the project to increase their commitment to the project.
   b) Generating benefits for project team members to increase their commitment to the project.
   c) Reducing project costs to avoid or mitigate losses in a customer project.
   d) Increasing monetary or intangible benefits for the customer as an element of active issue/crisis management.

---

3. There are several reasons why organizations that perform customer projects as well as those that manage complex PSNs should have a person dedicated as a project business manager. Which of the following is NOT a reason for that?
   a) Projects with more than one party cooperating under contract need managers who can oversee the entire business process from the beginning to the end, removing fences between contributors to this process.
   b) Project business engineers are particularly competent in increasing pressure on business partners, whereas project managers rather focus on the technical details of the project.
   c) Business relationships in supply networks often extend over national borders, and on top of the technical, social, intercultural, and interpersonal skills, managing these projects needs legal understanding.
   d) Contributing partners in PSNs tend to turn to competing behavior. The project business manager's job is instead to promote the behavior of completing to achieve mission success.

4. In an internationally distributed PSN, contract partners are tied together to generate common success. What can stand in the way of achieving this success?
   a) Ignoring the business interests of other contributing members of the PSN.
   b) Hesitation in going to court against a poorly performing PSN member.
   c) Balanced trust between the extremes of naïvety and permanent suspicion.
   d) Having project business managers to manage the business relations.

5. Benefit engineering is an approach that can be utilized by all contributing partners in a PSN. How can a customer use benefit engineering with a contractor to the advantage of the project?
   a) By using the understanding of the contractor's needs to offer desired benefits in return for improved performance or additional project scope.
   b) By threatening scope and price reductions in return for dissatisfactory performance and results from the contractor.
   c) By approaching the contractor's project manager directly and offering the person benefits in return for additional benefits for the customer.
   d) By squeezing out prime contractors from PSNs that bring no benefit to the project but act as proxies between customers and subcontractors.

6. Benefit engineering applied by a contractor-side project business manager is based on a proposal to the customer to increase benefits from the project for both parties. It can fail for many reasons. Which of the following is NOT among them?
   a) The customer may not have the freedom to accept the proposal due to external constraints.
   b) The contractor does not have the resources to implement the proposed change and deliver the benefit.
   c) The customer representatives like the proposal but are angry that it is not linked to a payment made directly to them.
   d) The contractor has insufficient knowledge of the customer, who may respond negatively to the proposal.

5.3 The Dynamics of Success and Failure in Project Business Management

Cicada LLC is an engineering company that performs a small portfolio of six projects for its customers. Their business year follows the calendar year, and in November, 2012, they made some projections and calculations for the business year 2013 in order to forecast the costs that the company would have to bear for doing the projects and how much it could expect to earn from them. Table 5.1 shows the calculation for the six projects, including the forecast for general and administration costs (G&A), sometimes referred to as *indirect* or *overhead* costs. These

Table 5.1 The Cost–Revenue Plan of Cicada for the 2013 Business Year, Projected at the End of the Previous Year

	Price to customer	Cost for contractor	Margin for contractor	
Project 1	12,500,000	8,150,000	4,350,000	
Project 2	153,000,000	119,000,000	34,000,000	
Project 3	9,800,000	8,200,000	1,600,000	
Project 4	231,000,000	165,000,000	66,000,000	
Project 5	16,000,000	10,900,000	5,100,000	
Project 6	93,000,000	54,000,000	39,000,000	
Total	515,300,000	365,250,000	150,050,000	
	General & administration		98,000,000	
	Profit from projects		52,050,000	10.1%

include all the costs that cannot be assigned specifically to any particular project but occur in order to enable the organization to be in business at all.

The prediction was a profit of $52 million from the projects. At forecasted total revenues of $515 million, this equals 10.1 percent. The business promised to be profitable, and some reserves were in the profits that could cover risks—lessening the company's profit, of course, but with 10 percent profit left, these monetary contingency reserves seemed well applied.

After six months, in the middle of the business year, Cicada revised the forecasts on the costs and revenues based on assessments of the first six months and on modified projections for the remaining months of the business year. Two projects, Project 2 and Project 6, were cost-wise over plan, and these cost overruns could not be balanced out by saving costs elsewhere. Cicada management eased its emotions by noting that these were only two out of six projects, and as revenues were expected to remain unchanged, the additional costs seemed manageable.

Table 5.2 shows the updated numbers. The profit from the projects dropped to roughly a quarter of the original prediction, but in order to end the project with a happy customer, this low profit was considered to include an investment into the organization's future.

Table 5.2 The Cost–Revenue Plan of Cicada, Revised in July 2013 for the Entire Year

	Price to customer	Cost for contractor	Margin for contractor	
Project 1	12,500,000	8,150,000	4,350,000	
Project 2	153,000,000	*139,000,000*	14,000,000	
Project 3	9,800,000	8,200,000	1,600,000	
Project 4	231,000,000	165,000,000	66,000,000	
Project 5	16,000,000	10,900,000	5,100,000	
Project 6	93,000,000	*73,000,000*	20,000,000	
Total	515,300,000	404,250,000	111,050,000	
	General & administration		98,000,000	
	Profit from projects		13,050,000	2.5%

Table 5.3 The Cost–Revenue Plan of Cicada for the Given Business Year

	Price to customer	Cost for contractor	Margin for contractor
Project 1	12,500,000	8,150,000	4,350,000
Project 2	153,000,000	139,000,000	14,000,000
Project 3	9,800,000	8,200,000	1,600,000
Project 4	*214,000,000*	165,000,000	49,000,000
Project 5	*13,300,000*	10,900,000	2,400,000
Project 6	*58,000,000*	73,000,000	-15,000,000
Total	460,600,000	404,250,000	56,350,000
	General & administration		98,000,000
	Loss from projects		**-41,650,000 -9.0%**

At the end of the business year, some more issues turned up that affected reductions in the revenues from some of the projects. Caused by delays, portions of the project work moved into the next business year, and so did the payments that were linked with them, while the team was busy with rework and sometimes idle times[1] that could not be billed to the customer. Another factor impacting revenues were damage claims by the customer and price reductions for late and incomplete deliveries. Table 5.3 shows the updated values by the end of the business year, which still showed positive margins for five out of the six projects.

The expectation of the organization to bring money home with the projects had turned into an overall loss. Despite the fact that five out of six projects brought a positive margin home, the total margin was not able to cover the organization's G&A costs. Cicada incurred a massive loss from its projects. This is a business situation that the company would not be able to survive for long. They were expecting to secure the company's future, its impact on the market, and the jobs of its workers and employees. Instead of bringing money home with their projects, they ravaged the entire organization, jeopardizing its jobs and depriving it of the financial assets the organization would need to develop further and to keep pace with fast-changing technologies.

Project business management is high-risk business for all parties involved. The complex interplay of scope, time, costs, as well as people and business interests, makes the success of one party the success of others, but also the failure of one the failure of all. In projects performed under contract in both simple customer–contractor settings as much as in complex PSNs, projects may be able to develop virtuous circles, in which success at one moment leads to successes in the next and in which the role model of one party's being communicative, open, and trustworthy is understood and followed by the other parties, so that a system of cooperating partners based on mutual good faith is created, focusing more on completing the mission than on competing with each other. The motto from Dumas's *Three Musketeers* may come to mind: "All for one, one for all".[2] It was also the motto of the Protestant party in the early days of the Thirty Years' War (1618–1648), and it became the motto of the Swiss cantons (states) to act

---

[1] I discuss the widely overlooked effect of non-billable idle times in my book *Situational Project Management: The Dynamics of Success and Failure* (Lehmann 2016b, 41–43).

[2] (Dumas 1844)

together to rebuild Switzerland after major flood disasters in the Alps in 1868. Today it is written in the cupola of the Swiss Bundeshaus in Bern, the home of the parliament of the Alpine republic, which is one of the oldest existing democracies in the world.

An interesting example of an organization that changed its fate from a long series of failed projects to a stunning run of successes is the US National Aeronautics and Space Administration (NASA). By the end of the 20th century, NASA had a streak of bad luck, when several missions failed, among them the Mars missions described in the first chapter and two space shuttle disasters (Challenger in 1986, Columbia in 2003). In the years since, NASA still had some failures, among them the Orbiting Carbon Observatory (OCO) in 2009, which was dedicated to bringing valuable data about the carbon load in the atmosphere, the root cause for climate change. But the vast majority of missions was successful, and it is easy to see results from these missions by visiting NASA's galleries[3] and having a close look at distant planets, into the universe or back to Earth from a distance.

The motto of NASA in the late 20th century had been faster-cheaper-better, and the ambiguity in this motto about what was priority left teams in PSNs helpless as a basis for decision making. NASA developed its new motto—"Mission Success First"—in the year 2000.[4] It may be hard to believe that changing a motto (and with it the attitudes of individuals and organizations involved) can increase the success rate of projects, but one should not underestimate the effect of a clear guideline for prioritization in decision making. Attitudes lead to behavior, and behavior leads to performance and results, so effectively changing attitudes can finally lead to better projects. The "Mission Success First" motto prioritizes, but it also gives people a sense of hope and optimism—achieving mission goals is obviously possible, but it also raises confidence in their ability to meet these goals and resilience against tendencies that threaten to undermine them. Some readers may consider this discussion illusionary and more a bed of roses than factual reality, but research has shown the validity of the concept, which is today often subsumed under the title *psychological capital* (PsyCap).[5]

Instead of such positive PsyCap, based on a "Mission Success First" attitude, as NASA has shown to be successful, one can often observe a vicious cycle of mutual distrust and miscommunication that destroys the cross-company team spirit, incapacitates the project and its participating organizations, and covers the mission goals behind a fog of opaqueness, misgivings, and suspicion. Then, competing replaces completing. The venom of distrust sometimes attacks the project openly; in others, it creeps in secretly, poisoning the relations among the organizations and finally becoming visible when it is almost too late to respond appropriately.

Some readers may still believe that this vicious cycle is a product of the phantasy of a book author, but we can see troubled projects around us that were heavily burdened with personal and cultural incompatibilities of the parties involved, and when a project can suffer from the difficulties that a party suffers and that disable it from meeting its obligations, it is not rare that these difficulties are not only of technical or organizational nature but lie in the contractual and relational interfaces between the organizations, the interplay of business interests and attitudes towards one's own company and the entire project.

---

[3] (NASA 2016)

[4] (NASA 2000)

[5] (Avey et al. 2011)

Troubled projects abound, many burdened with personal and cultural incompatibilities among the parties involved. When one party's difficulties in meeting its obligations threaten the health of the project for the other parties, these difficulties can often be traced not only to technical or organizational deficiencies, but also to the relational interfaces—the interplay of business interests and attitudes—among all the parties.

The previous chapters of this book focused on avoiding such major conflicts, which can finally translate into project crisis. The last discussions in this book will concentrate on techniques to bring the project back on track when the crisis could not be avoided.

5.4 Causes for Conflicts in Project Supply Networks—A Survey

Between 14 June and 4 July 2017, I made another microsurvey, asking project managers in supply networks what sources of troubles they found to be more or less frequent in their projects.

I received 302 responses from project managers globally, who classified themselves into three groups:

- Employed: 208   68.9%
- Self-employed freelancer: 77   25.5%
- Others: 17   5.6%

Participants who answered "others" were asked to specify. They were mostly managers and consultants/coaches. Figure 5.1 describes the distribution of the roles that the participants had.

I asked participants to state on Likert scales from 0 (= never) to 5 (= frequent) how often they recalled causes of disruptions of work in *project supply networks*. Figure 5.2 lists the causes of conflicts according to their frequency, as reported by the participants.

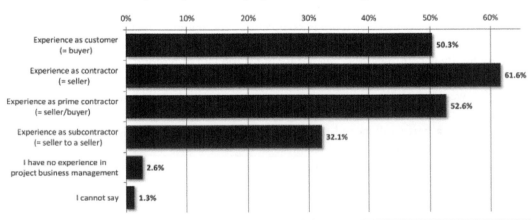

Figure 5.1   Many participants of the survey have collected experience in multiple roles, so that the numbers do not add up to 100 percent.

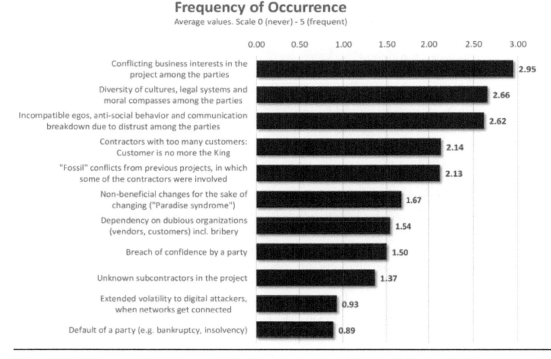

Figure 5.2   The most frequent cause of disruptions among the participants of the survey were conflicting business interests.

Conflicting business interests were the most common cause, followed by diversity of cultures, legal systems, etc. and by incompatible egos and antisocial behavior. Other causes were much rarer but were also reported, and some of them can have extremely detrimental effects on a project.

I gave participants the opportunity in a free text field to report experiences outside the strict structure of the numbered Likert scales and received some more interesting statements. Here is a small selection:

- "Our dependency on a large number of subcontractors for every project poses large risks for our business success and our customer's business success. Global internal vendor management organization and tools are needed to vet and monitor vendor quality and control the risks".
- "It seems that the ego of my long-term customer is the main drawback to the entire account. She is unpredictable, nonsensical, and narcissistic. She does not keep up with her record keeping (on her side) no matter the updates given and frequency. Thus, when she gets called out for not having her part together, she attacks me".
- "Serious lack of common understanding due to 'lack of time'".
- "The subcontractor is not fully exposed to the contract/agreement and the agreed scope of work of the main contractor".
- "Arguing about material specs after signing contracts".
- "When a contractor sells too much projects and then cannot give the correct service".

- "I'd add that it's often convenient for project managers to blame contractors".
- "Roles and responsibilities are often misaligned or crossing, causing communication breakdown".
- "Expectations are different".
- "In all projects, vendor (contractor) analysis is very important. The PM must in all cases look at the vendors and look at suitability to tasks as well as to the other vendors and/or contractors. The overall project is always a group effort, so one bad link in this group can sometimes cause the group to disintegrate".
- "Very often poorly structured communication channels".
- "Subcontractors do not have the same understanding of project goals as the [prime] contractor".
- "Poor understanding of risk and lack of appreciation of the benefits of risk sharing, risk responsibility. Lack of preparation by buyer's organisation for proper governance and project support. Subcontractors making changes in agreed or contracted baselines without reference to prime contractor. Subcontractors doing 'end runs' to get around the prime and making unacceptable agreements with buyer. Buyers taking the view that contract compliance is more important than a collaborative approach to mutually rewarding outcomes, on the basis that a contract cannot write in all project eventualities or challenges, which will take flexibility and collaboration, not rigid thinking".
- "The recognized issue in any form of partnering or alliancing is the challenge of aligning diverse interests and expectations to create a common set of objectives for everyone".
- "Common cause is the unreasonable expectation of a fixed bid on unknown requirements".
- "Having pre-selected and well investigated supply networks mitigates many of these issues. The tight management of a supply chain is critical to success, especially for larger and more complex initiatives".
- "A common game is schedule low-balling,[6] based on the experience that other vendors will also not be able to meet their deadlines".
- "Complex contractual process up to sixth level is a big issue while the rest all are culture specific".

5.5 Benefit Engineering

Benefit engineering rivals with cost engineering. Cost engineers focus on the costs of performing the project, benefit engineers on the benefits that the project creates for their own organization, for a customer, or for another organization involved in the project. When I talk with managers of project managers in small engineering companies, project service units, project management offices (PMOs), and other business entities that perform projects for paying customers, how they are handling problems with projects, particularly when these projects are about to make a loss, the answer is generally that the project managers must find ways to cut

---

[6] Offering timely delivery at dates that are understood as not feasible, expecting that other vendors will also not be able to deliver on time; see discussion on "Chicken races" above (starting on page 242).

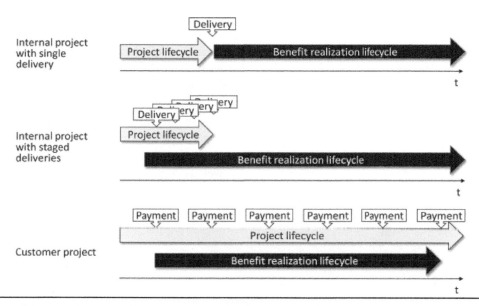

Figure 5.3 Repetition—for an internal project, benefits from the project are typically expected for the future. In a customer project, the contractor expects the benefits during the project lifecycle, beginning with the first payment and ending with the last.

costs. Unfortunately, there are limitations to this approach. Requirements from the contract, from law and regulations, are often hard limitations, and there is also the desire to make the customer happy, which sometimes strangely evaporates in such situations.

Another limitation is that cost engineering cannot help with problems such as overly pressing deadlines or unfeasible technical requirements. Just the opposite—meeting such requirements mostly increases project costs. Benefit engineering goes beyond these limitations, but comes with new caveats.

By increasing the benefits for the customer and its internal stakeholders in a deliberate, measured, and calculated fashion, benefit engineering can be a strategy to take a project out of crisis. Figure 5.3 is a repetition from the first chapter. It shows three different benefit lifecycles:

- The first assumes a single delivery at the end of the project, which finishes the project and at the same time allows the recipient of the deliverables to start gaining benefits.
- The second assumes staged deliveries, which lead to an overlapping of project lifecycle and benefit generation.
- The third situation describes a customer project whose benefit generation begins with the first payment by the customer and finishes with the last.

Benefit engineering can be among the most powerful tools that a project manager has at hand to drag a project from upheaval into a better controlled state. It builds on a give and take with these stakeholders, on good faith on mutuality and on an in-~~deep~~depth knowledge of the parties involved including their intentions, desires, fears, constraints, and all the other factors that influence their decisions.

5.5.1 Benefit Engineering—The Process

Figure 5.4 shows how cost engineering and benefit engineering address different time phases in a typical project business lifecycle, which begins with the make-or-buy decision on the buyer side, moves over business development and offer phase to contract closure—at which point the buyer becomes the customer and the winning seller the contractor—then moves further over the phase in which their contractual work is being done, to the handover of the deliverables and final closeout of the customer project. At this point, the buyer and the seller may finish their business relationship; it may also be that a period of operational business will follow, which may include management of warranties as well as ongoing services and deliveries.

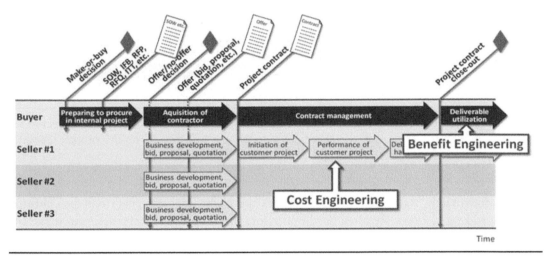

Figure 5.4 Cost engineering addresses project costs, mostly on the side of the contractor. Benefit engineering addresses and increases the benefit.

Traditional cost engineering addresses the performance phase in this typical business lifecycle. There, it focuses on the costs of the seller, who is now a contractor, to reduce costs and free additional profits. It may also be linked with delaying the moment at which certain costs occur, in order to protect the contractor's credit line. In contract types such as time and materials (T&M) and cost reimbursable (cost plus) contracts, it may be the customer who does cost engineering, because this is the side in which cost risks are located and which has to bear most cost overruns. For our discussion herein, that topic of interest will be contractor-side cost engineering.

The limitations and risks of cost engineering have been discussed above, and they can restrain the options for cost engineering decisions to a degree that it is no longer effective. Its benefit is that it is much simpler than benefit engineering (the topic of the following discussion), which necessitates a very good understanding of the needs, wishes, and constraints on the customer side and also of their own organization. Benefit engineering includes an intellectual challenge to propose the right changes—those that meet the needs of both customer and contractor, that bring benefit to the project, and to which the customer will respond positively. Without deep investigation into both businesses, benefit engineering can backfire and, instead of resolving the problems, create new conflicts or increase existing ones.

The basic question of benefit engineering is quite simple:

"How can we propose a change to the project that is beneficial for the customer and allows the adjustment of price, fees, deadlines, and other terms of the contract that make it impossible for the contractor to perform a successful project".

Benefit engineering includes a number of activities that must be performed to identify a beneficial solution, assess its feasibility and favorability, sell it to the customer, and, once it has been agreed upon, implement it. Figure 5.5 describes the process steps that should be taken when benefit engineering is done for a customer project.

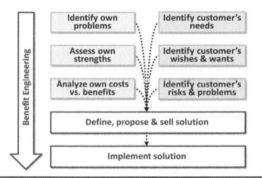

Figure 5.5   Benefit engineering builds on a deep understanding of the customer as well as one's own organization.

Step 1: Identify Own Problems

It sounds obvious that benefit engineering begins with a clear understanding on the side of the contractor as to what their own problems are. This first step may already be a very difficult one, because one of the hardest tasks of a project manager in any project is to know where the project actually stands and where its weak spots are. This knowledge requires trust between the contributing stakeholders in the project by getting to know which people and organizations are trustworthy. As team members and subcontractors may be new to the project manager, and not much time is available to know to what degree these people are trustworthy, knowing whom to trust and whom not to trust can be hard. The same is true in the other direction—team members and subcontractors need to develop trust in the project manager, and the time they have to develop this trust is also very short. Many companies therefore replace a network of mutual trust with a complex system of contractual agreements, internal processes, and formal reporting systems. While there is of course merit in good documentation, one should not overlook that contracts and similar documentation become valuable when problems between the parties have occurred and when these parties come into conflict. They do not prevent the problems and conflicts.

One's own problems may also come as a surprise. The author remembers a case when half of the project team left the contract organization to work with a competitor, and the remaining team would not be able to meet contractual requirements. The leaving team members did not give notice at all, so the project manager could not respond by bringing new employees into the team timely enough to fill the sudden capacity gap.

While it may be difficult for the project manager to understand timely what their company's own problems actually are, it is important for the person to develop a clear understanding before any proposals are developed and presented to the customer. Otherwise, the benefit engineering will not help the contractor to resolve the problems as they are, but will lead to a WOMBAT investment—a Waste Of Money, Brain, And Time—while the actual problem remains unaddressed and is likely to grow even further.

Step 2: Identify the Customer's Needs

As a project manager on a customer project, one is really in a good position to identify the areas of concern that benefit engineering can address. Managers in a modern enterprise have a far more complex task than those in former decades and centuries. Management attention is the scarcest resource in most organizations, not only for projects they are performing, but also for the multitude of other tasks that these organizations must accomplish. The ability of humans to direct attention to multiple tasks at the same time is limited, and when these tasks are becoming more tedious and challenging, the number of tasks a person can manage concurrently goes down further. Table 5.4 gives an impression of how the requirements on managers have changed over time. It is also observable that many scandals have happened in which organizations are involved whose management was simply overwhelmed with the number and complexity of issues they needed to take care of. These scandals also show that there is not much understanding to be expected when managers who are usually able to cope with this complexity fail to do so.

A modern organization is not a fine-tuned organism, in which all functions work together in a collaborated fashion, creating effectiveness and efficiency and meeting all requirements explicitly or implicitly imposed by stakeholders. Instead, it is a hodgepodge of compromises, workarounds, makeshifts, and temporary solutions that were created to meet immediate needs a long time ago but, although they should have meanwhile been replaced with solid solutions, as they were working sufficiently well, their due replacement became a sacrifice to other tasks that seemed more urgent.

As the urgent has always been the greatest enemy of the important, important tasks too often remain disregarded. We should add that these insufficiencies may not be based on facts, but rather on perceptions. An example: Due communications with shareholders of public companies on issues that will impact the value of shares has always been an important task for these companies. In the wake of treacherous activities by a small number of companies at the end of the 20th century, the USA enacted legislation to protect shareholders called the Sarbanes–Oxley Act (SOX). Corporations invested billions to become SOX compliant, and time pressure was high, because the law included a deadline for this compliance, threatening top managers (CEOs, CFOs) with jail if their companies did not meet these dates. The dimension of these investments in SOX compliance shows how the important task of shareholder protection and information has been repressed by other management tasks that were perceived as more urgent.

Benefit engineering has become a more promising task with the management environment changed and with the dominance of the urgent over the important in the perception of managers. Many important things are left unaddressed in organizations that add up to inefficiencies, lack of effectiveness, and risks to the organization and its environment and that finally make it hard for management to understand what is going on inside their own organization. Someone in the firm, or the agency, association, etc., may benefit personally from these neglected issues,

Table 5.4 Requirements Commonly Placed on Managers Today, Compared with Those in the Past

Typical Focus of Management Attention

In the past	Today
- Poorly informed customers - Employees doing mostly well-defined manual work - Small number of long-term suppliers - Few sources of raw materials - Simple, static markets - Controllable competition - Small number of laws - Integrity considered a secondary topic - Safety and security issues mostly ignored - Static processes - Well-predictable future - Decisions driven by perceived importance	- Heterogeneous, global customers with easy access to information - Talent gap for talented employees that do mostly intellectual work and are able to adapt quickly to changing requirements - Complex and dynamic supply networks, often developed ad-hoc - Thorny competition for many raw materials - Fast-changing global markets with disruptive innovations, often surprising incumbent players - Dynamic competition - Unmanageable "jungle" of national and international laws and regulations - Professional integrity scrutinized by various stakeholders - Safety and security issues have become mission critical - Ever-changing processes with a high degree of adaptiveness & agility - Future driven by disturbances & uncertainty - Decisions driven by perceived urgency

and they will do their best to ensure that management attention is not dedicated to them. It is a kind of creeping sabotage against the vested interests of the organization.

Step 3: Identify the Customer's Wishes and Wants

In the next step, the project manager who performs benefit engineering needs to develop an understanding of the driving forces of the customer's decision making: the needs and wants that have the potential to give the customer's managers sleepless nights and lengthy discussions during the day.

Management of an organization that burns more money than what is available for it will listen to proposals that can help reduce costs or gain additional income. A corporation that makes a lot of money may not be open for such proposals and may be more interested in suggestions with long-term strategic impact. Managers may respond much more quickly and more decisively to proposals that protect them from errors, that are punishable, and for which they will be held accountable in person, than to risks with low impact to the detriment of someone else.

Managers have strategic goals that they consider worthy to follow up, and if the proposal supports meeting them, it may be attractive to these managers; if it makes it difficult instead to achieve these goals, the proposal may be rejected. Managers are often measured along abstract metrics, so-called KPIs,[7] and proposals that make it difficult or impossible to meet KPI goals are also likely to be rejected. This will be even stronger when KPIs are linked with monetary incentives and promotion.

Benefit engineering is susceptible to corporate politics, and in order to do it successfully, the project manager needs to understand these politics and to navigate the project inside them.

Project managers are in a very good position to do that. They touch things that have not been touched for long time. They open cabinets, look into books, analyze processes and deliverables, and do many more such activities that no one has done for quite a while. They develop a good understanding of the deltas between the necessities and intentions of their customers' managers on one side, and the organizational reality on the other. They are not only in a good position to do that, it is an essential part of their profession to identify technical, organizational, legal, and interpersonal issues in the customer's organization and to actively transform them. This is the skill for which their employer has hired them, and for which the customer has awarded the contract to this employer.

Step 4: Identify the Customer's Risks and Problems

Not all proposals that a project manager would make and that are desirable for the customer are feasible or favorable. A project manager, for example, may try to redefine a deadline with the customer for the project. Some deadlines are easy to move. For others, this may be impossible. The project manager should therefore be aware of the constraints that make it impossible for the customer to accept the proposal, and also of the risks and problems that implementing the proposal would bring to the customer.

An important aspect of benefit engineering inside a "Mission Success First" culture is that the project manager on the contractor side desires both to bring a benefit to the customer and

---

[7]   KPIs: Key performance indicators.

to make this clearly communicated. This behavior is grounded in the deep knowledge of the customer that the project manager gains during the project, and the same is true for the desistence from proposals whose implementation would be detrimental for the customer. Benefit engineering is not a quick and easy task, but is one that requires a lot of consideration and a deep understanding of the consequences, both intended and unintended. This is true for any technique that is used in project management, and the ability to develop this understanding and act accordingly together with stakeholders is one of the distinctive factors of good project managers in any case.

Step 5: Assess Own Strengths

The best benefit engineering proposal will lead to failure if the contractor organization lacks the skills needed to successfully implement it. While this may sound self-evident at first glance, it can become a major problem in a customer project. The contractor may have the resources in house to carry out the proposal, but these resources may be booked by another project and be unavailable at the specific moment they are needed. The same is true when subcontractors are needed to implement the benefit engineering proposal. The resources to carry out the proposal must have the necessary skills, they must be prepared to do the job, and they must be available at the right time and also at the right location. They may need special infrastructure to do the job, which must also be available, and someone may need time to introduce them to the project, the customer, the team members who are already working on the project, subcontractors, and so on. If new personnel have to be brought into the project to perform the additional work, some time may be necessary to allow team building with the already assigned team members.

It is quite possible that these new team members, their tools, and their access to infrastructure must be taken away from other activities, such as operations or other projects. The managers responsible for these other activities may perceive this as disruption of their work, and the new team members may prefer to go on with their existing work instead of being assigned to a project in trouble, possibly in crisis.

Benefit engineering builds on understanding and engaging stakeholders not only on the customer side, but also inside the contractor organization.

Step 6: Analyze Own Costs and Benefits

The costs of benefit engineering may be significant. Developing the proposal, presenting it to the customer in a convincing and persuasive manner, implementing it as a change—assuming that the contractor has standing procedures for managing changes and protecting the project—and communicating such change to stakeholders involved is a costly set of activities. The proposal will probably include additional work, which adds further costs.

The benefits from this change for the contractor must of course exceed these costs. They may just be monetary benefits. Nonmonetary benefits can include aspects such as

- Avoiding breach of contract situations
- Redefining deadlines
- New agreements on operational disruptions on the customer side
- Replacing uncomfortable and unknown technologies with familiar ones
- Replacing subcontractors named by the customer with trusted ones

In essence, analyzing the costs and benefits for the contractor are similar to the considerations made at the very beginning of the business with a customer as to whether to send an offer or not. A major difference is that, during benefit engineering, no competitor drives down the price to the customer. In most situations, the incumbent contractor is the only organization who knows how to create additional benefits for the customer. Although it is not advisable to "rip off" the customer in such a situation, pricing will be more comfortable and profitable for the contractor. The only competitor to the proposal is the option for the customer to say "No".

Wrap It Up: Define, Propose, and Sell the Solution

This process step in benefit engineering may have to be done twice: Inside one's own organization and with the customer. Sometimes, convincing management at home may be the more difficult part of this job, but the focus of the process step should generally be with the customer.

Not many project contractors have a well-developed process of business development and writing of offers, which can be bids, proposals, pitches, or quotations. The various steps of business development, including compelling and convincing presentations, can be a critical success factor for a contracting organization; but as many of them are driven by engineering cultures, this is regarded as unfamiliar and out of focus. On the other hand, the better this process has been developed, which is then returning higher hit-rates[8] and more lucrative customer business, the easier it will be to implement benefit engineering.

Many project managers do not consider themselves sales people, and while some of them see this rather is a personal weakness, others insist with pride that they do not sell, they perform. It is often overlooked that to perform successful customer projects, some sales attitude is a necessary skill of a project manager. If a project manager does not have the skills of a salesperson or dislikes developing such an attitude, a solution may be to ask the company's sales staff for support.

Benefit engineering should then be done using the same tools, processes, and behaviors that have helped when the parties' contractual and non-contractual obligations were defined, the project's original scope was delineated, and all the plans were developed that have guided project management until this moment. Benefit engineering is both a project management process and a sales activity. For the latter, the fundamental difference to the original sales activity that won the contract is the proximity to and interconnectedness with the customer that has been developed over time and that excludes competition from stepping into the business.

5.5.2 Negative Benefit Engineering

Sometimes the best benefit engineering proposal may be to exclude scope from the project that was originally contractually agreed upon. The contract may include work that is not necessary and not beneficial at all. I remember a project to replace a custom-made software solution for a complex logistics system with a standard software. The core requirement was that the standard software must provide at least the same functionality that the older individual solution also had. Many of these functions were necessary when the old software was developed, but were meanwhile no longer needed. From a contractor perspective, these functions expanded the scope beyond what was actually needed, and because the customer was prepared to rebuild them in the standard solution, they meant extra business for the contractor.

---

[8] Hit-rate: Percentage of offers that lead to business.

However, as the project ran into difficulties with meeting deadlines, proposing to the customer to drop the unnecessary development work and focus on functionality that was needed gave the customer a financial benefit and the contractor the opportunity to meet the deadline.

This de-scoping may save time for the project, it may save costs for the customer, and it may help the contractor avoid unfamiliar or uncomfortable project work. It may also free critical resources for another project inside the contractor's portfolio that is in trouble or crisis. In this understanding, benefit engineering is a tool not only for project management, but also for portfolio management.

5.5.3 From Cost Engineering to Benefit Engineering: A Deep Cultural Change

At first glance, considering benefit engineering as an alternative to cost engineering to bring a troubled customer project back on track seems like a marginal technicality. As soon as one tries to implement it, one will notice how deep the cultural change is. This cultural transformation is a prerequisite for the success of benefit engineering but is also a consequence of its implementation. Instead of claim management—which means having dedicated staff combing through the project for opportunities to invoice additional amounts of money for deviations from the contract that may have happened, commonly leading to disappointments and frustrations on the customer side—a joint effort is initiated to gain improvements for the customer and the contractor.

Project managers need to look beyond the limitations of the technical and functional aspects of the project and dig deep into their own organization and even deeper into that of the customer. There, they will discover new opportunities and threats as well as new leeway for decision making and thus-far unknown constraints. They will, in addition, identify human aspects in these projects, including desires, fears, sympathy, rant, and many more. They will have to navigate in the complex dynamics of power and politics; they will improve their understanding of organizations working together under contract, the interfaces between these organizations, and how they are able or unable to build working systems together.

Project management is a learning process as long as we are part of it. Each project has new lessons for us; some of them are valuable for the specific project only, others will become part of the person's personal assets for the rest of their life. Benefit engineering is a driving force for this learning process. In contrast to cost engineering, which tries to cut costs where it hurts least, benefit engineering requires us to discover new opportunities, determine limitations and risks that come with them, uncover people who are involved in the project and would otherwise remain hidden, and develop solutions to the benefit of both customer and contractor.

Should a project manager be transparent with the customer when the proposal is made to apply changes to the project? One should probably be very careful. If the rapport between contractor staff and management of the customer is poor, this can backfire: The customer may consider the proposal a rip-off, and the poor relationship can further deteriorate. If the relationship is strong and a robust "Mission Success First" culture has been developed, in which completing trumps competing, it may be a good idea to put one's cards on the table. In such a relationship, proposing a solution for mutual benefit can increase trust and strengthen the bond of business partnership, which is the most important prerequisite for successful project business management.

Benefit engineering can be applied by different players, as the following discussion shows.

5.5.4 Contractor-Side Benefit Engineering #1: Application by the Direct Contractor

The basic principle is quite simple: The contractor-side project manager has to meet deadlines, work against a tight budget,[9] or has other challenging objectives or constraints that appear as unrealistic during the course of the project. Proposing a value-adding change may be a resolution, offering the customer additional benefits that the customer finds attractive enough to rethink deadlines, price tag, and all the other things that bring troubles to the contractor.

Benefits for the final customer are mostly expected for the future. The new facilities, software, machines, organizational structure, service enablement, or whatever the project's deliverable is, is a cost factor today but is expected to bring benefits in the future. The benefits may be monetary, operational improvements, strategic advances, risk reductions, or any other step-ups for the organization that its management considers worth the upfront investment. Some projects have only one benefit: aligning the organization with applicable laws and regulations. These projects are mandatory, not discretionary, but it may still be possible to develop a business benefit on top of the necessity for meeting compulsory requirements.

5.5.5 Contractor-Side Benefit Engineering #2: The Subcontractor's View

The subcontractor works for another contractor, often a prime contractor, who is both a buyer and a seller. Between the subcontractor and the final customer may be just one or a number of in-between contractors, and each of them is both a buyer and a seller with their own obligations, business objectives, and risks. Benefit engineering can relate to the entire buyer downstream; each of them may be open to listening to additional benefits that a contractor can provide, but care must be taken that the benefit for one party may be a disadvantage to another. Accelerating the project, for example, may benefit the final customer but may reduce the time-related revenues generated from the project for a prime customer.

5.5.6 Customer-Side Benefit Engineering

As a customer, there are also opportunities to apply benefit engineering by looking at the contractors inside the PSN and identifying ways to generate new benefits for one's own organization in exchange for additional benefits for vendors. The rolling award fee contract discussed earlier (Section 3.5.6, page 189) is an example for such an exchange: Contractors are asked to create benefits for the customer, to which a monetary value is assigned, and if these benefits are sufficiently achieved, a part of this monetary value is given back to the vendors.

5.5.7 The Caveats of Benefit Engineering

Benefit engineering can be a powerful method to bring a project out of crisis. Asking a customer for additional time, money, or other reductions of pressure in exchange for something

---

[9]   The budget in a customer project is commonly defined as price to the customer minus the amount or percentage that the contractor wants to gross from the project. The remaining amount is the budget. This works in all types of contracts.

of value can be a powerful win-win approach, one which shows how much the contractor cares for the customer and is prepared to invest competency and other assets. Offering a contractor payback for additional efforts can also fail, creating conflicts and distrust. Some rules should be followed to avoid backfiring when this method is applied:

- **Understanding constraints**: Asking for a delay when a deadline cannot be moved is futile. Soft deadlines can be discussed, but hard and rigid ones cannot.
- **Funding limitations**: The same is true for funding limitations. Before one asks for additional budget, one should check if such money can be made available at all.
- **The actual value of the benefit**: Benefit engineering will only work if the benefit is valuable for the buyer or seller and if this value is perceived as higher than the disadvantages that come with the recommended change.
- **The relationship with the buyer or seller**: In a business with poor relationships, the recommendation of a value-adding change may be rejected without looking at the value at all, based on the preconceptions that one party is about to rip off the other. Benefit engineering needs an atmosphere of trust to work.
- **One's own available resources**: Recommending a change for which one does not have the resources can also backfire. One sets expectations and is then not able to satisfy them.

5.5.8 Benefit Engineering as a Form of Change Request Management

In essence, the core tool of benefit engineering is the change request. The desire is to use a change that is adding value for all parties involved to resolve one's own problems. Change requests can indeed get a project out of trouble; poorly managed, however, change requests can drive a working project into crisis. One has to understand the consequences of the change—the wanted as well as the unwanted and the beneficial as well as the detrimental. I recommend having a change request management process in place that consists of several elements:

1. **Clear assignment of decision responsibilities.** Project managers often have the problem that it is unclear who can decide upon a specific change request. Then, a common problem is that no one is prepared to make the decision, or that two or more deciders require the decision responsibility for themselves, and not asking one of them will lead to frustrations and to conflict.
2. **A clear protective change process.** Protective means that change requests are not processed for approval before they have been assessed for their impact.[10] This practice is implemented to ensure that changes do not damage the project in an unpredicted way and is particularly important in PSNs, where a change in one work item can trickle through the project and cause changes in many other work items that are done by different contractors.

5.5.9 Project Managers as Benefit Engineers

Project managers are actually in the best possible situation to perform benefit engineering. Project managers see and touch things that no one has seen and touched for a long time. They

---

[10] Please see the description of a protective change request management process in my first book, *Situational Project Management: The Dynamics of Success and Failure* (Lehmann 2016b, 190–194).

see inefficiencies and room for improvements that others will not see. They open locked doors and look into closets that no one has looked into for a long time, and the skeletons that they find there are often a big surprise for the people involved.

As discussed above, a modern organization is not a highly efficient business organism, optimized to the maximum effectiveness at the minimum costs and friction, but is instead a complex system of compromises, work-arounds, and ignored ineptitudes. The urgent is the most vicious enemy of the important in these systems, and when managers focus on the pressing issues of the day, their attention is taken away from the matters that also need dedicated care, but that are not urgent enough and will one day be forgotten. I heard from a manager some time ago that the procrastinated issues of the company often cross her mind in the middle of a sleepless night and call for resolution, but are gone in the morning, when the normal necessities of the daily business consume her full attentiveness, which is then weakened by the lack of sleep.

Project managers on the contractor side are often in a perfect situation to help the customer or contractor in a win-win approach, if they can make the resources available for such help. The original business development for the project contract was done in a competitive situation, and vendors had, at least in theory, equal chances to win the award. During contract execution, an expansion of the scope is unlikely to be bought using competition. The contractor in place can resolve the identified issue quickly and easily and take a burden from the customer's shoulders without placing another burden of running a competitive procurement process.

If, in turn, the customer applies benefit engineering to get more performance and better results from the contractor, the situation is similar. As a customer, one sees the contractor working and observes strengths and weaknesses. One especially sees the weak spots that the contractor may be unaware of and that have impact on the company's work for the project. Benefit engineering is mutual help under the principles of "Mission Success First" and *good faith,* which allows two or more parties to be successful together and achieve a common result that one party alone would not be able to achieve; it is driven by empathy, common sense, and a focus on the things that bind the organizations together, not on those that separate them.

5.5.10 Does the World Need Professional Project Business Managers?

When the occupational profile named "project manager" was developed, a major goal was overcoming old "over-the-fence" approaches, in which projects were managed along sequences of strict phases, each of which was performed under the auspices of another business unit or a contractor. In these over-the-fence projects, no one was in place on the project management level to integrate the diverse activities and make a functioning project out of them. The integrative capacity of the project manager has proven to be a great help to improve project successes by removing the breaks in responsibility that the "fences" strike into the process. The responsible person or team in charge for the entire project lifecycle has therefore become indispensable.[11]

When it comes to customer projects, however, there are still two fences left that have not been addressed so far:

1. The fence between the business development period and the actual project lifecycle, which is also a fence between the teams involved.

---

[11] This development is described in Harold Kerzner's *Project Management: A Systems Approach to Planning, Scheduling, and Controlling* (Kerzner 2013, 48).

2. The fence at the moment when deliverables have been handed over to the customer for usage and also to contractor-side service departments that deal with warranty, maintenance, and repairs.

The fence between business development and project management is critical, because the executives involved in winning the business have different objectives to follow and different success metrics. Foremost, it is their job to get contracts signed by customers that will allow a temporary revenue stream for the contractor. When the business has been won, some remain in contact with the project, but the majority will consider their job done and focus on the next sales lead. Their job is not performing the project for the customer but making it exist. In order to achieve that, these people are often in a dilemma as to whether to give in to customers' demands either for lower prices or for deliverables and functionalities that the organization is actually unable to bring about. The professionals in business development commonly have high business acumen but low project delivery competency.

Project managers, in contrast, are trained for project management process knowledge. They know how to develop the deliverables according to the contract and to additional requirements. They manage schedules, human resources, and risks; they take care of documentation and prepare handovers and approvals. Only a small number of project managers have true business acumen coming from natural talent or as a result of dedicated training. The ideal project business manager combines these two competencies: business acumen to (1) identify opportunities for benefit engineering and other value adding measures, (2) present them to the customer in a convincing manner, and (3) project management to ensure that work is completed to meet agreed-upon requirements.

Figure 5.6 illustrates how a project business manager combines the competencies of business developers and project managers. They help the seller meet the two great objectives in the customer project—making the customer happy and bringing money home—and to the degree possible make sure that the profit center that the customer project constitutes for the contractor is actually profitable.

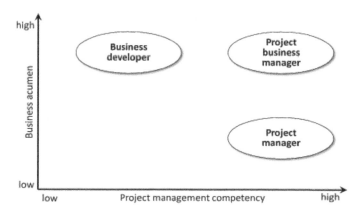

Figure 5.6 A project business manager combines business acumen, which helps win attractive customer projects and expand existing ones, with project management competency to realize the potential from the business.

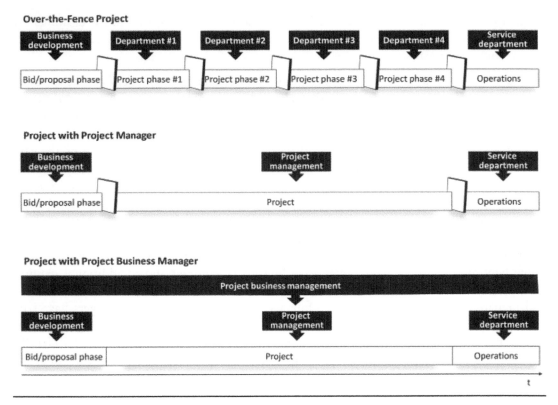

Figure 5.7 In old-style over-the-fence project management, a number of business units or independent contractors drove the project along a sequence of phases, with another unit responsible for another phase. Project managers integrated the project phases. Project business managers remove the fences and integrate the remaining phases at project beginning and end.

The ability to combine business acumen and project management competency may be a natural talent or a skill obtained through training and experience. The person may have originally been trained in project management and later added the business understanding, or vice versa. It enables the project business manager to overcome the two remaining fences in the customer project that can both be detrimental to meeting the two great goals for the contractor organization: bringing money home and making the customer happy. These fences too often incapacitate communications between stakeholders from the different task areas, and when projects run into troubles, the fences stand in the middle of mutual finger-pointing as to whose fault it was that things went wrong. The project business manager's job is also to make sure that the profit center that the project constitutes for the seller is actual profitable.

The process responsibility of the project business manager is shown in Figure 5.7. It stretches along the entire business process from the first project-related contact with the customer to the post-project service period, ensuring that the process is not interrupted by changes of responsibility and owners of knowledge on the project.

There are also requirements on the other side—the side of the buyer of the project under contract—that extend traditional project management capabilities when complex *project supply networks* need to be managed, something project managers do not learn today. Addressed

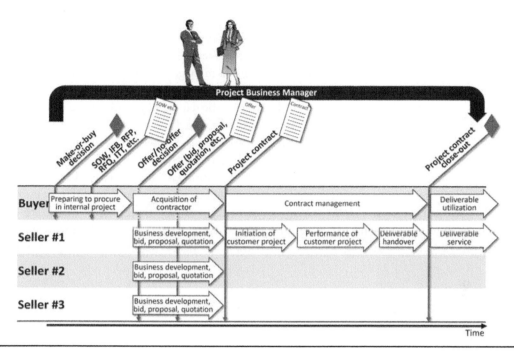

Figure 5.8   Project business managers oversee the entire process from the first contact relating to the project between buyer and seller to the final closeout and into utilization and servicing of the final deliverables.

is simple procurement management, assuming a business between one buyer and one seller, limited by the legal doctrine of privity of contracts, which in essence says that the contractor's contractor is not the customer's business. The complex, dynamic, and often opaque networks of contractors, subcontractors, etc. is rarely addressed in literature and education, and it is not well understood. The abundance of these networks, however, is increasing, and so is their complexity. Contracting work out to contractors on various tiers means tapping other organizations' assets and turning them into project resources; but with these assets come new risks that need to be identified, understood, addressed, and responded to.

As much as the buyer-side project business manager needs the ability to dig deep into the *project supply network,* the person also has the task to oversee the entire procurement process from the make-or-buy decision to the final close-out of the contract and the post-project service relationship. This is shown in Figure 5.8.

The long-term orientation of project business managers on both sides make them into natural partners in the business relationship, and it can indeed be helpful to ensure that all companies in the project have such a function. Buyer-side project business managers are interested not only in their own organization but also in helping contractors maintain effectiveness, and as this interest meets the interest on the seller side to make the customer happy, minds can meet at the point at which the project produces the most benefits to all parties.

Where does one find professional project business managers? Sometimes talented people either from business development or from project management take this wider responsibility. Business development people have the benefit that they know the project from the first contact on. There is, on the other hand, the restriction that it is their job to win new business. In small companies, the CEO may even fill this role, as this person has the interest in profitability as much as in the happy customer and the effective contractor.

What is the focal qualification of project business managers? They ensure that the project is driven by the "Mission-Success-First" attitude that better ensures a successful project, as long as all parties involved submit to it and live it in the day-to-day work of the project. They understand how to deal in business situations, when the centrifugal and disintegrating forces of contract law compete with the need of the project for intensive and open communications, reliability of the partners involved, and good faith of the parties, who rather seek to help each other to ensure a successful project that will benefit all its constituents, not just one. They understand what it takes to manage contractual relationships across country borders, in different cultures, and in diverse legal systems. Their strongest assets are probably the capacity to ask the right questions at the right moment, never assuming that their own experience and knowledge is sufficient; and the network they have built, so that they always have a person whom they can ask.

5.6 Turning a Customer Project Out of Crisis

Some time ago, I observed a particularly deep crisis in a project performed by a *project supply network* for my training customer Booklouse, Inc.,[12] a provider of large-scale hardware and software infrastructure projects for corporations from financial industries. Booklouse was the prime contractor and had 14 subcontractors as well as an unknown number of sub-subcontractors on lower tiers in a project for a multinational insurance company to install an online insurance booking system. Most of the project work was contracted on a T&M basis. The project was one among a greater number, and there was a lot of business communication going back and forth with the subcontractors to ensure that their work results got integrated to turn into the effectively working system that the customer had ordered and for which the customer was prepared to pay.

Major elements of these communications were the invoices sent by subcontractors to Booklouse, which needed to be booked, examined by a person familiar with the work invoiced, and then approved for payment before it would finally be paid. The process would then proceed with re-invoicing the work to the customer. The invoice validation process then also included monitoring incoming payments against invoices. Parts of the process were automated, but there was still an amount of manual work needed to review the invoices, compare them with the timesheets of subcontractor personnel, and ensure swift payment as well as billing to the customer. Figure 5.9 illustrates the flow of invoices and payments. It shows how any disruption in invoice validation—one of the core processes of a prime contractor—could lead to problems with Booklouse's customer, subcontractors, or the company's bank. About halfway into the project, this is what happened.

Booklouse had at that time an internal project to reduce non-billable overhead costs by offering staff early retirement with subsidized pensions. Management originally wanted to send off only specific staff members who were considered unneeded, but they were told by

---

[12] Names in this case story are of course also altered to protect my customer and my seminar attendees.

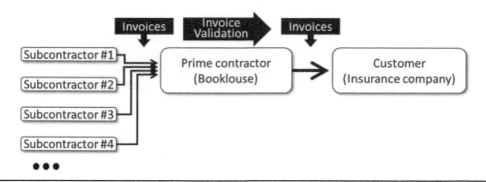

Figure 5.9   The flow of invoices and payments between subcontractors and the customer. Invoice validation is the core process at the heart of the business.

trade unions and also by their lawyers that in order to avoid discrimination lawsuits, they had to make the offer equally to all employees or none. That's what Booklouse did. Among the employees who accepted the offer were two operatives involved in the process of payments and re-invoicing. This immediately roiled the process and led to delays in the payment and billing process. Subcontractors complained about late payments, and the customer about late invoices, which impacted transparency in the cost management of their project procurement and caused lagging of information on cost overruns in the project. It also increased the work and stress level on the remaining employees in the department, who responded with higher absenteeism, which further increased the problems.

Contractors threatened to delay work if Booklouse went on paying late. As an immediate solution to the growing problem, the department reduced the manual part of the validation of the sub-contractor invoices to a minimum. Many invoices were just approved without being validated against time sheets at all, or the time sheets were accepted without verification of their correctness. Booklouse employees assumed that the subcontractors were sufficiently trustworthy to skip the process step for the majority of invoices, and given its shortage of staff, they had no alternative. The implosion of a core process can create a gap in an organization that impacts the functioning of all other processes, and without external help may lead to complete failure.

Booklouse still wanted to validate the time sheets before they sent their invoices to the customer, which led to a backlog of unchecked work documents and intensified the complaints from the customer. As a second problem, Booklouse's bank asked why the company's accounts ran deeply into the red and threatened to cut their credit line. It turned out that the savings from the two employees sent into early retirement caused costs and organizational upheaval of a much higher order of magnitude.

Problems with the customer escalated when a major invoice from Booklouse was mistakenly underpaid by the customer. Caused by the lack of validation, Booklouse found the error only late and could therefore send a note on the issue to the customer only weeks after the payment.

Things became even worse: One of the subcontractors noticed that Booklouse no longer validated incoming invoices against time sheets. This became visible when the subcontractor sent them an invoice, which due to an error was too high, but the customer immediately paid it without any delays and discussions. They tried that again, this time on purpose, and they could not believe their eyes when they saw the customer paying it in full and without debates or deferments.

Under normal circumstances, the invoice would have been sent back to correct it and resend it. The news soon spread across the company and, as subcontractor staff often talk with each other, was also shared with other companies. Some of them tested Booklouse's payment behavior too and found the rumor confirmed. Soon, several contractors began to over-bill their work by small amounts, which soon grew larger. After a while, it turned into some kind of sport to test how much could be fleeced from the prime contractor on top of justified amounts. It was like looting a burning house.

A project that was commenced with high ambitions and supposed to be completed in a joint effort of a *project supply network,* with a prime contractor and subcontractors working hand in hand to deliver what the customer needed and to the commercial benefit of all, became a plundering race, in which many suppliers were afraid that if they did not participate in the unethical behavior, others would have money that should have been theirs.

It did not take long for Booklouse to find out how much it had opened itself to supplier fraud. The tip-offs came from outside the company. Subcontractors who did not participate in the fraudulent behavior told them about the looting, and the customer, who also became aware that subcontractor management had become chaotic, signaled strong dissatisfaction.

It took Booklouse several weeks and a lot of external help to get the project, and with it the company, out of the crisis and back on track. The retired workers were offered a major payment if they would return and help rebuild the internal processes. The fraudulent subcontractors were allowed to finish their work as contracted, but got temporarily blacklisted for future projects. The customer was given a significant price reduction to not lose future business. Booklouse had been just about managing for several years, but this year, it made a substantial minus.

The way out of this crisis was a painful and strenuous effort, which took roughly two years. During this time, some people were trained in the essentials of business management, with a focus on projects. Some employees and managers needed to be replaced. Implementing professional Project Business Management finally helped the company to get back into the profitability zone and have happy customers who trust the company and are happy to enquire for new business. It also gave the organization the architectural strength to better deal with challenging business situations that would otherwise shatter the company.

Crisis management is a combination of issue management and risk management, both on steroids:

- Issue management is generally *reactive.* Problems have surfaced and need to be coped with. In a crisis situation, the problems have grown to a dimension that is almost no longer manageable, at least with the company's own resources.
- Risk management is by definition *proactive,* looking at the identifiable uncertainties in projects that may influence the future of the project. Managing the risk of crisis looks at the largest risks—those that impact not only activities but the entire project, and with it the performing organization, maybe several organizations—in the worst case, the customer together with the entire *project supply network.*

Philosopher Friedrich Nietzsche once said: *"Was mich nicht umbringt, macht mich stärker".*[13] Unfortunately, what does not make us stronger may instead kill us.

---

[13] "That which does not kill me makes me stronger" (Nietzsche 1888).

In the case story, the crisis made Bookworm stronger, but it could also have destroyed the company. This is an example of an organization that came out of a crisis strengthened and better organized. In a Project Business Management environment, this could be mostly seen from the improved profit they made and from the smiles they put on their customers' faces. Other companies terminally failed in such situations.

Professional crisis management needs people with strong backbones. A wishbone will not suffice. Crisis management deals with uncertainties—at any time a fundamental ingredient in project management, and particularly Project Business Management. Crisis management begins in a situation that needs reactive issue management, when business structures have ceased to function and have become uncontrollable. When this control has been gained again, crisis management changes its approach and its means to become proactive again. It is like dealing with a horse that has escaped a barn. It must be captured and brought back first, and then measures must be taken to avoid repetition. Defining these measures has two parts: finding the weakness in the barn that allowed the horse to escape, and then looking for other weaknesses that could cause the problem to reoccur.

Crisis management in a project does the same: identifying and removing the drivers of the crisis and bringing the project back to normal progress, and then ensuring that the crisis does not reoccur. In *project supply networks,* this task comes with a number of additional challenges and impediments:

- One of the hardest jobs for a project manager is to know where the project actually stands. This becomes even more difficult when the project is using a complex, dynamic, and often obscure *project supply network,* with member companies that the project manager does not know and can neither monitor nor control.
- Knowing where the project stands is also impacted by the different business interests, leadership approaches, communications cultures and systems, and—if the network crosses national boundaries—different legal systems, and all the other factors that impede or support the openness and trueness of communications. Data may be easy to get, but verifying and understanding such data and deriving actionable knowledge from it can become very challenging.
- Another hard task is to identify the degree of trustworthiness of players and separate the trustworthy parties from those who are not. One may be too dependent on the second group to reject them immediately from the project; but one will observe them with far more diligence and attentiveness than the others and will try to become independent of and replace them as soon as possible.
- The most difficult task is to develop leadership, rebuild a "Mission Success First" approach among the members of the PSN, and make them share ownership both for the solutions developed to bring the project out of crisis and for the results that they are intended to bring about. There is no guarantee that the companies involved and their managers will be prepared to follow this new leader, and it is the followers who will decide the leader, not the person that wants to lead.
- Once the project is taken out of the immediate crisis zone, the causes for the crisis must be identified, assessed, and transferred to risk management, where they are responded to with strategies such as avoiding, mitigating/reducing, or creating of reserves for active acceptance.

It may be helpful to bring in a third party to support crisis management. A third party has not been involved when the crisis broke out and is not entangled in emotions and quarrels among the parties. For the same reason, it is sometimes necessary to replace the project manager in such a situation, to take the person out of the line of fire and allow rebuilding of trust and rapport. The insider knowledge of the old project manager may nevertheless still be necessary to help the new person become productive in a short time.

5.7 Project Business Management—Is It an Open or Closed Skill Discipline?

In the previous chapters, I have repeatedly noted the high-risk nature of project business for all parties involved. This includes buyers, sellers, and also the "in-betweeners", such as prime contractors and other forms of intermediate contractors, who are sellers in a downstream perspective, but also buyers when looking upstream. This last group shares all the risks of buyers and sellers, and their awareness of these risks and processes developed to manage them should be particularly well developed.

Risks have their origin in uncertainties that stand at the beginning of any project, when definitive knowledge about the future is not yet sufficiently available to make decisions based on them, and when people involved must instead make decisions relying on assumptions.

Assumptions may be wrong.

Over time, uncertainties are replaced one after the other with definitive knowledge about facts. At these times, often a "wait-and-see" approach with high attention is necessary to respond when an assumption has been verified or falsified by facts, and when a risk has either occurred or can be retired because it can no longer occur. Many uncertainties originate in the environment of the project, which is undergoing change while the project is under way. Uncertainties may further derive from the people involved and their changing availability, but also their interests, opinions, and of course their health. Uncertainties lie in organizational matters. It is often observable in team sports how a day's form of the player can influence success and failure; the same is true for project teams.

Although these uncertainties occur in all projects, the projects performed by PSNs add extra layers of opacity and unpredictability, because more than one organization is involved.

In order to finish this chapter and the book, I would like to look back at the beginning of my book *Situational Project Management: The Dynamics of Success and Failure* to discuss a basic question of project management and apply it to the specifics of projects done in a contractual environment.[14]

A common approach generally follows three steps: standardization, methodology, and tailoring. The standard's focus is on professionalism. One can take cooking as an example. The standard focuses on professionalism and on the availability of kitchen equipment for the cooking tasks that are expected. Often, methodologies are derived from such a standard. Their common promise is to deliver "best practices", comparable to a collection of recipes in a cookbook. In a next step, methodologies need tailoring to adapt them to specific project situations. In the

---

[14] (Lehmann 2016, p. 9)

kitchen metaphor, a cook may vary a recipe to adapt it to the preferences of the customer or to the momentary availability of ingredients. Table 5.5 shows the three steps.

Predictive methodologies are often developed from descriptive standards and need tailoring to become appropriate for the specific project (see Table 5.5).

Table 5.5 A Common Approach Generally Follows Three Steps

	Standard ➡	Methodology ➡	Tailoring
Examples	PMBOK Guide British Standard 6079 DIN 69901 Normenreihe	Prince2 Six Sigma Corporate methodologies	Adaptations of standards & methodologies
Validity	Generic	Domain-specific	Project-specific
Approach	Descriptive	Prescriptive	Adaptive
Focus on	Professionalism	Procedures	Context & goals
Proposition	"Good practice for most projects most of the time"	"Best practice"	Best fit
Delineation of project management	Repository of processes in • Process groups • Knowledge areas	Guidelines implemented through • Phase models • Policies	Management plans • Roles • Course of action
Compare	Rulebook for chefs & kitchen planners	Cookbook with recipes	Modified recipes

As a trainer in project management for over two decades, I have worked for a variety of companies and have had the opportunity to see them address project uncertainty by implementing the three-step process. Among them were implementations that worked very well, some worked somewhat, and others did not work at all.

The last group was, unfortunately, the most common one, when processes that were put in place to avoid or address specific issues congealed and were done for the report to management—"yes, we have done that"—instead for the purpose of managing uncertainty. All approaches to the management of risks, issues, problems, and crises should be chosen and implemented with a situational focus. They should help us have a better project, not simply to follow the rulebook—not an easy task for a project manager.

The critical step was the second one, developing and implementing the methodology. Although the first and last steps are commonly situational enough to address the dynamics of success and failure in a project, the methodology, in its justified determination to standardize and align, often limits the freedom of project managers and creates a tendency towards "process blindness". In a PSN, with its additional causes of uncertainty and with its strong presence of commercial and legal aspects, this danger gets accelerated.

Can one then assume that there are "best practices" in Project Business Management? This question can be best answered by having a look over the boundaries surrounding the field of project management into sports psychology.

In sports, a distinction is sometimes made between *open-skill* and *closed-skill* disciplines. An example of a closed-skill discipline is figure skating, which is an *introspective* art. The skater learns a program to the utmost perfection, and during the performance isolates himself or herself from the audience and the environment, concentrating on their own presentation. The environment of the presentation is prepared in a way to keep it free from disruptions and perfectly static, so that the skater does not have to respond to situational changes from outside the presentation. If a disruption from outside the performance would interfere with the presentation, the presentation would be stopped, and the skater would get a new chance to do the program. Sometimes, stuffed animals are thrown by spectators into the skating arena, but only after the performance.

An example of an open-skill discipline is hockey. The players must respond to changes in a fraction of a second, including the movements of their own team, the competing team, the referees, the puck, and sometimes even the goal. Hockey is *extraspective*—the players must keep their minds open to the ever-changing environment, and the playfield is not unobstructed for them; it has instead many obstacles just in the form of the many people with whom they share the arena.

Players in hockey need situational awareness. They are under high stress, which can be sometimes seen, especially when they start brawling. One never sees figure skaters brawl.

Project management in this understanding is more similar to hockey than to figure skating. Most of the time, project management incorporates the need to swiftly adapt to changing situations, players, obstacles, and impediments, as well as to dynamically shifting organizational conditions, requirements, and resources. Resources are scarce, and whereas most methods in project management assume that they are generally available, in practice, they are not.

These difficulties become amplified in Project Business Management. The organizational conditions do not relate to one organization but to more, possibly hundreds. Changing requirements pose not only technical and often personal challenges, but also commercial and contractual ones. The scarcity of resources is even more difficult to manage, because a project manager rarely has the opportunity to look deep enough into another organization to know where resources are sufficiently available and where they are not.

5.8 Limitations of Tools for Project Business Management

In Section 3.5.6 (page 189), I described the *Rolling award fee* contract and in 4.12 (page 246) the *Customer-led consortium* (CLC) as tools to ensure that mission success is not considered in conflict by contractors with their particular business interests. When a project has developed into crisis, can these tools also be used to get the project on track again? Possibly, but the following case story will show the limitations of such means when they are being used reactively to respond to crises and not proactively to avoid them:

Papilio Ulysses is not only a beautiful butterfly from Australia and some islands in South-East Asia, it is also the alias I am using here for a company that ran into troubles as a customer in a software development project that was completely outsourced to Monarch LLC. The purpose of the software was to bring GPS-based telemetry to deep-sea containers that would allow transportation companies to track the transport and to better predict arrival times of goods not only at customers, but also at harbors and other locations, where the trucks' cargo would be unloaded for transshipping to ship or train or vice versa. It further allowed the early announcement of the arrival of trucks at ports, which reduced waiting times for the trucks and saved costs.

Papilio Ulysses contracted Monarch to develop the software for them, something the vendor seemed capable of doing. The contract was fixed price, and Papilio Ulysses considered the development in good hands. A deadline for delivery had been agreed upon, and it was only when Monarch missed the deadline that the customer learned how the vendor had run into massive problems with the project.

There were several causes for the missed deadline:

- The complexity of the project was underestimated by the vendor.
- The vendor added complexity by assigning a distributed team to the development task.
- To save costs, the team members were not given opportunities to meet face to face; instead, all communications needed to be done via video conferencing and e-mail.
- The team members were all assigned part-time to the project. They had to work for other projects as well; these projects were in similar crises, and as their project managers were crying louder, more time was dedicated to them—to the disadvantage of the project of Papilio Ulysses, of course.
- The contractor's budget derived from the fixed price was already used up. All further costs directly ate into its profit margin, and as this was also practically consumed, the vendor was about to incur a loss, which would grow larger with every piece of additional work for the project.

Four months after the contractor's missed delivery date, I had a long talk about the project with the customer-side project manager, at a time when it was clear that the software was still far from being finished and in a condition that the telemetry service could be launched. This delayed time to market had massive consequences for the business case of the customer:

- The investment decision had originally been made based on the presumption that the software would already provide income and pay back the investment.
- Papilio Ulysses was not the only company that had identified the market and tried to exploit it. Market window theory became an issue.

5.8.1 Market Window Theory

This second point may need some explanation:

Market window theory says that different market entrants over time perceive the market very differently. Figure 5.10 shows six groups:

- **Prematures.** Come to too early, the market is for various reasons not yet prepared to accept the innovation, such as missing infrastructure to support the novelty or rejection by customers and users. Their investment is failing because it is not yet time for it to succeed.
- **Innovators.** They have in common with prematures that their investment is high. They have no role model that shows them what the new product or service should look like. They cannot hire experts in the innovation or hire contractors for that. The benefit, on the other hand, will be large after market launch, as the company enters the market in a time of zero competition.
- **Early adopters.** The investment gets cheaper, as the innovator serves as a role model on what to do and what to avoid, and the first experts are available to support the development project. As a disadvantage, one can commonly observe that the earning options after market launch also go down due to growing competition.

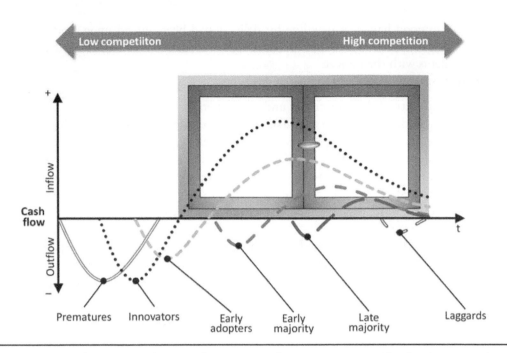

Figure 5.10   Market window theory with six groups of participants. Note the development of initial outflow and consequential inflow over time.

- **Early and late majority.** The outflow for the initial investment to launch the product is getting smaller over time, but when a degree of saturation has been reached and all potentials for optimization have been used, the investment will not go down further; but competition is growing, putting even more pressure on the price that can be achieved on the market.
- **Laggards.** They launch the product when the market window is no longer open. It may have taken them too much time to bring the product into a market, which is by that time no longer receptive to a new player. It may well be that the closing of the market window only becomes apparent to the company when the product has been launched and has been found to be a failure.

I made Papilio Ulysses aware that they were not alone in intending to move into this business, and that the delays would change the competitive situation and reduce the prices they can achieve at the time, when the market entry would be finally achieved.

5.8.2 A Method Cannot be Applied

I recommended that the project manager consider adding a *Rolling award fee* to the existing fixed price contract. The costs from expectable further delays would be massive, and it could be a solution to incentivize Monarch, the contractor, for saving this money to the customer. I also added a caveat: The disappointment with the contractor was very high, as were the emotions of Papilio Ulysses's management. While the project manager could save his project with a high degree of empathy and by developing a business case for the seller-side project manager to invest more in the customer and bring the project to a swift and successful end, his managers might not be able to develop this empathy.

It turned out that my warning was correct. Business reason would have recommended the customer to help the vendor out of the crisis by adding the award fee, thus finally also saving money for the customer and allowing market entry at a more profitable time. However, the emotional condition on the side of the customer was that one could not reward the poor performance of the contractor with additional payments.

Some months later, the customer terminated the project. It was found that there was already significant competition with similar services on the market that made it unattractive to step into the business. Papilio Ulysses did not want to be a laggard—a decision not only made for economic reasons, but, as one of its managers was quoted as saying, because they "did not want to be the laughing stock of the marketplace".

The investment in the software was written off, and, following the advice of the corporate counsel, who doubted their chance for success, no attempt was made to recover money at court. However, Monarch, the vendor, was blacklisted in the company for seven years.

The case story shows that methods that, used proactively, could prevent a project from running into crisis, employed late to steer a project out of a dilemma and to respond to a crisis may be futile or ineffective. Not because they would no longer have positive impact, but because the emotions of disappointment and rage become unsurmountable and make it impossible to apply these methods that build on empathy, good will, and good faith.

5.9 Conclusion

Looking at all these difficulties and complexities, one may ask why organizations do not do their projects alone, to simplify team structures and make the projects more predictable.

All the difficulties that performing projects under contract bring with them are obviously outweighed by a great benefit: the opportunity to tap other organizations' resources. To say it more clearly: Not many organizations are in a position to do all projects internally. They lack people, know-how, certificates, agility, and the key resource of all: management attention. Customer projects are indeed a business for contractors, who make their living from them, determined to make a profit and win the happiness of the customer. Often it is their core business.

For the customers, it is necessary to work with contractors to win agility and new capabilities.

A problem that I observe is that project business so far has not been sufficiently described. I am happy in this book to do some exploration into this omnipresent but mostly uncharted field of project management, but far more work needs to be done to truly understand it.

Successful project management is often based on the ability of project managers to do projects and manage people without formal authority. In Project Business Management, this ability is even more relevant. When managing a project for a customer, or on the customer side outsourcing work to a *project supply network*, one is rarely in a position to direct people, at least not those working in other organizations. And there can be many:

- Customer
- One's own contractors and their subcontractors
- Other contractors with their own contractual relationships
- Self-employed freelancers
- Temporary people from staffing agencies
- Government agencies
- Approving bodies

Particularly when it comes to resolving crises, it can be difficult not only to unify their particular business interests to form a joint team following a "Mission Success First" approach, but also to ensure that the behavior of these players is not driven by fear and competition but by the common desire to get the project out of the crisis and back on the route to completion.

In a workshop with a group of project managers from companies doing customer projects, the participants developed a list of sources of non-formal authority and named:

- Escalation
  - Inside their own organization
  - To management in other organizations
- Attention to people
  - Distinguishing good work from mismanaged bungle
  - Showing interest in people and their activities
  - Cross-company team spirit
  - Building interpersonal relationships
  - Creating a positive team atmosphere
  - Show and require trustworthiness
- Soft authority
  - Expertise
  - Speed in making and implementing decisions
  - Understanding of the dynamics of power and of the organizational nexus
- Prioritization
  - Creating a sense of urgency

This last point is often the critical one. Urgency is among the most powerful drivers of goal achievement. When people have to make a choice between what they consider important and what is urgent to them, most will prioritize the urgent. Developing a sense of exigency among the people and organizations we work with avoids procrastination and helps get things done. Good crisis managers are generally good communicators of urgency.

For the challenge to do business projects that are commercially and organizationally successful, we need to develop an as yet undescribed set of tools, skills, and possibly a new ethical system, which should help us define proficiency and professionalism and separate it from incompetence and ineptitude. We must reconcile the different business interests of contractors, the next payments, and the long-term interests of the customer to gain sustainable benefits. We must also see how the dependencies between customers and contractors lead to situations in which the problems of a customer can become the problem of the contractor, and vice versa, and we must also see the risk that a local issue of a company in a *project supply network* can spiral out of control and become the problem of all companies participating in the project.

The task of managing project business is a difficult one. The challenges are high, and so are the risks, and people are needed who are trained like hockey players to work as a team, to stand together against opponents, and to swiftly change their roles and be able to get an understanding of a situation in a fraction of a second and adapt their behavior appropriately.

For this book, we have reached the end of a discovery tour into a business field that for many readers has been their day-to-day job for years. We do it, but we are far from fully understanding it. I wish to close with the notion that the complex dynamics of success and failure that we have seen cannot be handled by a single person alone.

The book has described a large number of details, among them opportunities, risks, and challenges—particularly on the social, legal, organizational, and interpersonal level—that come with tapping external resources of other organizations. These details entail increased significance with the development of complex *project supply networks* (PSNs), which often span country borders, time zones, cultures, legal systems, and other differences that can make working together difficult. Raising awareness of and attention to these details can prevent many problems.

We then saw solutions that can help address the risks in PSNs, in particular benefit engineering, rolling award fee contracts, and customer-led consortia.

While these tools are helpful, we need more to fill our toolbox, and for that, we need to place the discipline that I call Project Business Management on a more professional foundation, comparable to that which project management has developed for internal projects. We need more exchange among professionals, but also research and a delineation of what we should regard as professionalism in the field and what should be considered improper and amateurish behavior. We also need to give companies help, when they are hiring people to perform customer projects or to engineer and manage project supply networks, what a qualification looks like that meets the needs of the project management aspect as well as those of the business side.

We further have to gain a better understanding of the situational aspects of Project Business Management, which are even more important than for internal projects. The changing business relationships add a further level of dynamic complexity to project management, and this layer is not a thin one.

In the end, on top of methodologies, processes, tools, and techniques, Project Business Management is built on humans: Project Business Management is teamwork. As much as organizations team to achieve results that one organization would not be able to achieve alone, is it necessary for the people involved to stand together, following the joint motto:

"Mission Success First!"

Answers to the Introductory Questions

Chapter 1:

Q1. **c)** The problems that arise from the deadline should be communicated early to appropriate stakeholders to allow for timely resolution. Early resolution still leaves room for more decision options for resolution, and the costs of these options are lower.[a]

Q2. **b)** You make yourself familiar with agile methods in order to understand if these could be helpful for your specific project situation or are detrimental. Then you report your findings to management. Agile methods are generally helpful when "the way is made by walking".[b] They can be detrimental when long-term predictions, forecasts, and estimations need to be made and the project must be based on them.

Q3. **a)** You assess whether these "best practices" are rather favorable or detrimental in the given situation. If they are not favorable or are detrimental, you reject using them. It is your responsibility to perform the project successfully. It may be helpful, when the practices are rejected, to explain the reasons why to the appropriate stakeholders.

Q4. **d)** The typical intention of creating a consortium is to build a temporary joint venture to perform a project together that a company alone would not be able or willing to do.

Q5. **c)** An internal project is a profit center; a customer project is a cost center.

Q6. **a)** Omissions and errors on the customer side can impact the contractor's success. You should recommend that the customer also build a project structure, and if this fails, find ways to protect the project

---

[a] (Lehmann 2016b, pp. 52–53)
[b] (Machado 2012)

Chapter 2:

Q1. a) Tapping the assets of another company. These assets may be persons, abilities, licenses, infrastructure, knowledge, agility, and many more.

Q2. b) The market is robustly growing.[a]

Q3. d) It is hard to build interpersonal relationships between buyers and sellers over virtual B2B marketplaces.

Q4. c) The RFP describes the objective of the items or services to be procured; the IfB specifies them in detail.

Q5. d) An MOU is a diplomatic document. The seller cannot claim any damages.[b]

Q6. b) The cost and difficulty to fix an error grows with the local, temporal, and organizational distance from its origin.

Chapter 3:

Q1. b) The contract is valid under a legal system that is probably unfamiliar to at least one party.

Q2. c) Product contract, service contract. The other contract types can be used but are not specifically described in Civil Law codes.

Q3. b) Project contracts cannot be fully complete; there will always be areas that need change and refinement later. Projects include learning processes, and some of them will need to be reflected in contract refinements and changes.

Q4. a) Contractual provisions and enabling services by the customer are common further obligations that customers have toward their contractors on top of payments.

Q5. d) A capped TCC contract is a cost reimbursable contract with cost–benefit sharing and price ceiling.

Q6. c) Improving the project and saving costs for the customer in a special moment.

Chapter 4:

Q1. d) Inattention to processes. The *five dysfunctions of a team* are:

1. Absence of trust

---

[a]   (Lehmann 2017)

[b]   The term *memorandum* is a strong signal; it means *don't forget* and was originally used in diplomatic contexts.

2. Fear of conflict
3. Lack of commitment
4. Avoidance of accountability
5. Inattention to results

Q2. a) Subcontractors will do work for the project, and the customer has no contractual relationship with them due to the doctrine of "privity of contracts".[a]

Q3. b) The customer gives the prime contractor a list with companies that are approved as subcontractors for a certain work item, and the prime contractor selects one of them.

Q4. d) Handover and acceptance can be done in one process or separately, possibly with weeks between the dates. Each in itself is important for project success.

Q5. a) Mediation and arbitration are methods of alternative dispute resolution to avoid a lawsuit when negotiation alone is not enough to resolve the conflict.

Q6. c) The contract regulates what happens when the prime contractor's subcontractor does not perform. It does not protect the project from malperformance.

Chapter 5:

Q1. c) Even moderate budget overruns in some projects can turn a planned profit from an organization's customer projects into a loss.

Q2. d) Benefit engineering on the contractor side intends to increase monetary or intangible benefits for the customer as an element of active issue/crisis management.

Q3. b) Project business engineers may be competent in increasing pressure on business partners, but their purpose is to overlook the entire procurement process and develop the appropriate degree of empathy for the partners involved.

Q4. a) PSNs can fail due to ignored business interests of other contributing members.

Q5. a) A customer can apply benefit engineering by using the understanding of the contractor's needs to offer desired benefits in return for improved performance or additional project scope.

Q6. c) One cannot fully exclude that customer representatives like the proposal but are angry that it is not linked to a payment made directly to them; in such case, their corruption should be addressed first, and the proposal be made as a second step.

---

[a]   In common law jurisdictions. Civil law has similar legal provisions.

Glossary

Agile approach	An approach to managing a project lifecycle (or certain situations during a project lifecycle) with a very short planning horizon of less than one month, using agile methods to ease frequent change. Also called "adaptive" approach. Many requirements are identified and described during the course of the project and "the way is made by walking". Contrasts with *Waterfall* and *Rolling wave* approaches.
Agile excuse	The defense of a lack of discipline and planning with the excuse "we are doing things the agile way". Contrasts with the *Waterfall excuse.*
Agile methods	Developed originally for software development but expanded and transferred into other application areas: Several disciplined approaches characterized by very short planning horizons, rather low planning depth, and leaderless organization. A typical aspect of agile methods is the rejection of having a project manager and the reliance on self-organizing teams. Contrast with *Predictive methods*.
Agilism	An ideological approach to project management that considers agile methods generally superior to other approaches, independent of the project situation. Contrasts with *Predictivism*.
Award fee	An incentive that is paid for meeting subjective criteria. An unpaid award fee is not open for appeal at court.
Balanced matrix	An overlay of one or more project team structure(s) over a functional organization structure with the project manager(s) and the functional managers on eye level. Contrasts with *Strong matrix* and *Weak matrix*.
Benefit engineering	Methods to measure and positively influence the benefits from project deliverables. May include the attempt to make a budget overrun, an increased price in a customer project, re-definition of deadlines, and other changes of fundamental parameters acceptable to stakeholders by increasing the tangible or intangible benefits for these stakeholders. Contrasts with *cost engineering*.
Bid	1. An offer in a solely price-driven competition done by a buyer to select the cheapest seller. See *Invitation for Bid (IfB)*. 2. A generic term for any kind of offer (may lead to confusion).

Brownfield project	A project that is performed in an already developed environment. Brownfield may literally refer to a piece of developed land, but the term may also be used metaphorically for similar situations in other industries. Contrasts with *Greenfield project*.
Build-Operate-Transfer (BOT) project	A business model often used in a *Public-Private Partnership* setting: A private company is licensed to develop an infrastructure deliverable (e.g., a motorway); operate this solution for an agreed-upon time (often 30 years), earning money during this time (e.g., through collection of tolls); after which the deliverable is given back to the public.
Business development	The activity by a *seller* to win a *buyer* as a *customer* under contract for a project and vice-versa.
Buyer	The organization that wants to buy from another party. Becomes the customer after contract award.
Change request	Any requested change to the project's scope, schedule, human resources, and other key data such as objectives and constraints. A change request in a customer project often leads to a contract change.
Change request management	A set of processes that describe how change requests will be managed in the project.
Chicken race	A game theoretical dilemma situation in which players wait for others to "jump first"—for instance, when they should tell a program or project manager that they are late, but hope that someone else admits to being late first.
CLC	See *Customer-led consortium*.
Closed-skill disciplne	A discipline whose participants develop proficiency mostly by introspective focus on their own performance.
Consortium	A temporary joint venture of companies founded for the purpose of managing and/or performing small, limited numbers of projects, often just one project. Some consortia have an additional purpose of running the deliverables from the project for a limited time under *build-operate-transfer* (BOT) projects and similar *public-private partnership* (PPP) schemes.
Contract change	Change that is reflected in contractual language.
Contractor	A seller, which may be a person or an organization, who has been taken under contract by a buyer for partial or complete work in a project, who then becomes the customer for subcontractors (if any).
Conway's law	"Organizations which design systems are constrained to produce designs which are copies of the communication structures of these organizations."[a] Or, in short: Systems reflect the relationships among those who make them.

[a] (Conway 1968)

Core team	Team members in a project who are expected and planned to be active in the project over most of its time, and who will perceive project success as personal success, project failure as personal failure.
Cost engineering	Methods to measure and positively influence the costs of a project. May include attempts to identify budget overruns early and avoid and mitigate them. Contrasts with *Benefit engineering*.
Cost reimbursable contract	Agreement for the delivery of goods and services with a predetermined price for the complete delivery.
Critical incident technique	An interview technique that uses moments of special relevance for the interviewed person as an entry point to dig deeper into the person's memory. Can also be used in workshops.
Customer	A buyer in a project under contract from the moment when the contract has been concluded to the moment when it is closed down. Contrasts with *Contractor*. See also *Prime contractor*.
Customer-led consortium (CLC)	A consortium which has the customer among its venturers, mostly as the lead venturer.
Customer project	A project executed by a performing organization for a (mostly) paying customer organization. Most customer projects are profit centers. Contrasts with *Internal project*.
Field change	Ad hoc change decision that becomes necessary during an implementation phase, often with an urgency that makes it necessary to circumvent a *Change request* process.
Fixed-price contract (FPC)	Agreement for the delivery of goods and services with a predetermined price for the complete delivery.
Freelancer	A contractor who works mostly alone as a self-employed individual.
Gate	A review process in a strictly sequential phase gate process, which is performed after a finished phase to review its correctness and in a second step if the project can enter the next phase. This review process can take significant time and should therefore not be confused with a *Milestone*, whose duration is zero.
Good faith	Legal principle predominantly in civil law jurisdictions that a contract must be understood not only following its words, but also by an underlying agreement not to unfairly harm the other party.
Greenfield project	Project that is performed in an undeveloped environment. Greenfield may be literally a green field, but may also be used metaphorically in other industries. Contrasts with *Brownfield project*.
Hard assets	Tangible assets that can be utilized by a project team as resources, such as money, personnel, equipment, facilities, etc. These assets and the effectiveness and efficiency of their use are commonly easy to measure. Contrast with *Soft assets*.

Hope creep	A situation in which a project is late or otherwise non-performant, because team members and contractors report that they are on schedule, hoping that they can make up the lost time with overtime work, or that someone else may delay the project. See *Chicken race*.
Incentives	Contractually agreed-upon additional payment by a contract partner, usually the customer, for meeting contractual requirements such as deadlines. Predominantly used in Common law countries. See *Penalties* and *Liquidated damages*.
Internal project	A project run by a performing organization for its own purposes. Most internal projects are cost centers. Contrasts with *Customer project*.
Invitation for bid (IfB)	A request to sellers in a competitive procurement to offer a price for a fixed set of deliverables that may include services and products. Allows selection of the cheapest seller. Often confused with other forms of competitive and non-competitive procurement.
Invitation to bargain	Non-binding offer by a seller to a buyer. Allows for withdrawal from the offer when the customer would accept it.
Invitation to pitch (ITP)	A request to sellers in a competitive procurement to offer a solution for a fixed price. Allows selection of the seller with the best offer against a given budget. Often confused with other forms of competitive and non-competitive procurement.
Key stakeholders	The subset of the project stakeholders who have direct and legitimated influence on the project.
Leadership	The authorization by followers given to an individual to lead them.
Liquidated damages	Contractually agreed-upon damage claim against a party for missing contractual requirement or constraints, most commonly deadlines. Predominantly used in common law jurisdictions. See *Penalties* and *Incentives*.
Make-or-buy decision	The decision made by an organization to use its own resources to perform a project or to procure the performance from outside the organization.
Mark 1 project	A first of its kind project for the project team, with a high degree of novelty. Contrasts with *Mark n project*.
Mark *n* project	A project with similar predecessor projects, which give the team members confidence in their capabilities and routine. Contrasts with *Mark 1 project*.
Melon project	A project run inside an organization, where traffic lights are used to indicate the status of a project: The melon project's status is "green" based on superficial perception, but, similar to a watermelon, the deeper one drills, the redder it gets.
Mission success first	The project success attitude cultivated by NASA in the early 21st century, softly ending a decade of failures.

Nash Equilibrium	A dilemma situation in game theory, wherein parties act to serve their particular interests that conflict with a common interest, and where a party must fear disadvantages when it would act to serve the common interest while others follow their own interests.
Offer/no-offer decision	The decision made by a seller to respond to a request to offer goods and services to an enquiring seller.
One-shot project	A project that gets only one chance to successfully deliver its products, services, or other kind of results.
Open-skill discipline	A discipline whose participants develop proficiency mostly by extraspective situational awareness and adaptation to a fast-changing environment.
PBM	See *Project business management, project business manager.*
PBMO	See *Project business management office.*
Penalties	Contractual deductions to "punish" a contract partner (mostly the seller) for late delivery or other forms of failure to adhere with contractual requirements or constraints. Predominantly used in civil law jurisdictions. See *Liquidated damages* and *Incentives.*
Phantom resource	A human or non-human resource that is planned to be used in a project, but (1) does not exist, (2) exists but is not available, (3) may be available but has not been formally booked, or (4) is over-allocated to more work items concurrently than the resource can handle.
Phase	A project phase is a discrete time period in a strictly sequential ("phase-gate") or overlapping ("fast-tracked") phase model. Different phases in a project may have specific teams, work contents, locations, cost centers, or other characterizing differences.
Pitch	1. An offer in a solely solution-driven competition done by a buyer to select the seller who delivers the best solution for a fixed budget. See *Invitation to Pitch* (ItP). 2. A generic term for any kind of offer (may lead to confusion).
Planning horizon	The point in the future up to which a project manager intends to plan the project. The time to the planning horizon may span anything between some days or the entire remaining duration of the project. A planning horizon may also relate to the level of detail and other aspects of a plan and is among others influenced by the time horizon, for which forecasts can be made and for which plans should be developed.
PMBOK® Guide	*The Guide to the Project Management Body of Knowledge®*, a globally accepted descriptive standard for project management published by the Project Management Institute (PMI).
PMI	The Project Management Institute, a global professional association with over 465,000 members.[b] Mostly known for standardization and certification.

[b] By August 2015. *Source:* Internal communications.

PMO	See *Project management office*.
PMP®	Project Management Professional®, a professional certification by PMI, actively held by over 750,000 individuals.[c]
Portfolio	In multi-project management, a portfolio is a collection of programs, projects, and operational work under a common management domain that share and compete for common resources. They may have different contents and may not share common goals.
Power achieving style	Behavior of leaders who prefer to deploy their own resources for tasks and bring order into chaos.
Practice	Application of specific approaches, tools, techniques, behaviors, and procedures with the intention to guide action and bring about desired results.
Predictive methods	Developed originally for engineering development but expanded and transferred into other application areas: Several disciplined approaches are characterized by long planning horizons, rather high planning depth, and organization with strong leaders. Contrast with *Agile methods*.
Predictivism	An ideological approach to project management that considers predictive methods generally superior to other approaches, independent of the project situation. Contrasts with *Agilism*.
Prime contractor	A contractor to the project customer who mandates major parts of or all of the work of a project to subcontractors. For the business with the subcontractors, the prime contractor is the customer.
Privity of contracts	The legal doctrine that a customer has no direct contractual relation with a subcontractor.
Process assets	A combination of process know-how that an organization owns and the availability of specific resources to implement this know-how. Process assets commonly include documented procedures, forms, templates, tools, databases, and documented lessons learned. Process assets in an organization (or a portfolio of projects under one management domain) are often administered by a *PMO (Project management office)*.
Program	In multi-project management, a program is a collection of projects (and other work) that are performed to achieve common goals and benefits beyond their specific goals and benefits. The projects may be run under different management domains and may take their resources from different sources.
Project	An investment that consists of a temporary and unique set of actions performed to develop required or necessary products, services, or other kinds of results.

[c] By June 2017. *Source:* Internal communications.

Project business management	The discipline of managing a customer project in a fashion to bring money home and make the customer happy. Is also used to manage complex *Project supply networks (PSNs)* on the *Customer* side.
Project Business Management Office (PBMO)	Type of *project management office (PMO)* that adds a focus on business matters in customer projects to the focus on project management methodology.
Project business manager	The person qualified for *project business management* and actually practicing it.
Project governance	A management function above the program or project manager level, mostly applied to ensure common terminology among the projects, compliance with corporate rules and processes and legal requirements. Another goal is alignment with corporate strategy and using lessons from the projects to improve this strategy.
Project management office (PMO)	An operational unit that manages standardized processes, procedures, templates, software, and other process assets in an organization, put in place to unify methodology and terminology. This office often organizes training for the implementation of methodologies across the organization.
Project management team	The team that supports the project manager in the tasks necessary to manage the project. It may share responsibility for project success and failure with the *Project manager.*
Project manager	The administrator of an investment that meets the definition of a project and has sufficient complexity to require active management under formal or informal mandate.
Project supply network (PSN)	A complex and often highly dynamic system of contracts with customers and contractors that spans over three or more tiers.
Proposal	An offer in a solution- and price-driven competition done by a buyer to select the seller. See *Request for Proposal (RfP).*
Quotation	1. An offer in a non-competitive procurement done by a buyer to select the seller who delivers an acceptable solution at an acceptable price. See *Request for quotation (RfQ).* 2. A generic term for any kind of offer (may lead to confusion).
Ramp-up phase	A period after deliverable handover from a project to an operational production or service environment, during which the operations are performed at a low rate and slowly increased to avoid being flooded by a big number of bad results and to have resources free if initial problems need to be managed. During this time, the project shares responsibility for the deliverables with operations.
Request for proposal (RfP)	A request to sellers in a competitive procurement to offer and propose a solution as set of deliverables, which may include services and products, when the buyer does not know details on that beforehand. Allows selection of the "best" seller. Often confused with other forms of competitive and non-competitive procurement.

Request for quotation (RfQ)	A request to sellers in a non-competitive procurement to offer a solution as set of deliverables, which may include services and products, and that is too small in value to justify the effort of a competition. Allows selection of an acceptable seller. Often confused with other forms of competitive and non-competitive procurement.
Rework	Laying hands again on a deliverable that was already considered finished, to repair or alter it. Hours of rework in a project are a great metric to assess efficiency and quality in a project.
Rolling award fee	An award fee that is paid in regular installments, often monthly. See *Award fee*.
Rolling wave approach	An approach to managing projects with a limited prediction and planning horizon and progressive elaboration of plans. The plans are based on early descriptions of requirements, which are expected to be refined and changed during the course of the project and which will then lead to refinement and change of the plans. Contrasts with *Agile approach* and *Waterfall approach*.
Scrum	The most popular *Agile method*.
Situational awareness	A project manager's mindfulness of changes inside the project or in its environment that necessitate swift adaptations in the approaches and tools being used.
Situational intelligence	The combination of (1) the understanding that the same practice that was successful in a given situation in the past may fail in a different situation, or vice versa; (2) the ability to adjust practices to the specific needs of the project and the current situation; and (3) the care that this adaptiveness is not perceived by others as signals of lack of authenticity or reliability.
Situational project management (SitPM)	An approach based on the understanding that the same practice that was successful in one situation may fail in another one and vice versa; applies situational intelligence to project situations.
SitPM	See *Situational project management*.
Soft assets	Intangible assets that can be utilized by a project team as resources, such as defined processes, motivation, reputation. These assets and the effectiveness and efficiency of their use are commonly difficult to measure. Contrasts with *Hard assets*.
Speed blindness	The inability to fully perceive the project environment and the dynamics, obstacles, and hazards it incorporates when a project runs at full speed.
Staged deliveries	The project does not have a single deliverable handover, which finishes the project and commences the use of the deliverables. Instead deliverables are handed over in stages, and while the team expands the scope of the product or service in steps, the team can implement feedback from the recipients (e.g., users) and incorporate it in its further development.

Stakeholder	Any person, group of people, or organization that project managers should consider during their decision processes.
Strong matrix	An overlay of one or more project team structure(s) over a functional organization structure with the project manager(s) in the more powerful position. Very common in customer projects. Contrasts with *Balanced matrix* and *Weak matrix*.
Subcontractor	A contractor who works for a client who is also a contractor to another customer, for instance a prime contractor. Subcontractors can exist on various tiers and then build complex project supply networks.
Target cost contract (TCC)	Agreement for the delivery of goods and services based on a cost-reimbursable arrangement with a target cost defined. Deviations from these costs are shared by the contract partners. Often includes a price cap.
Teaming	Cooperation of sellers to supply goods and services together that one alone would not be able to supply. Example: *Consortia*.
Time and materials (T&M) contract	Agreement for the delivery of goods and services with fixed prices for the former and fixed rates per hour, day, or similar for the latter.
TRAC	A multi-objective assessment of the commercial rationale of a customer enquiry to prepare for an offer/no-offer decision. Stands for *Time-Resources-Attractiveness-Chances*.
Vendor	Seller; a company that offers its services and products for the project. Becomes the contractor after contract award.
Waterfall approach	An approach to planning and performing a project in a predictive manner with a long-term planning horizon, ideally over the entire project lifecycle. Assumes static definitions of requirements and long-term predictability and plannability. Contrasts with *Agile* and *Rolling wave approaches*.
Waterfall excuse	The rejection of an important change because it is "not in the plan". Contrasts with the *Agile excuse*.
Weak matrix	An overlay of one or more project team structure(s) over a functional organization structure with the managers of the functional structure in the more powerful position. Very common in internal projects. Contrasts with *Strong matrix* and *Balanced matrix*.
Work breakdown structure (WBS)	(1) In a decomposed project, hierarchical decomposition of the entire project, commonly in graphical representation or in the form of indented lists. (2) In a composed project, a structuring system which captures the contributions of teams and consolidates them up to project level. (3) In traditional application, the WBS consists of planning packages, among them control accounts and work packages, which are its lowest-level elements. (4) In software, all WBS components are called *tasks*, and the lowest level may be individual activities.

| Zombie project | A project that is bound for failure right from the beginning, because no consideration was given to the match of project type and approach to the project, imbalance of obligations on the project with authorization and resources provided, or an environment in which other things were more important than project results. |

Bibliography

Admin.ch. (2017/2/12). *Bundesverfassung der Schweizerischen Eidgenossenschaft.* Retrieved 5/11/2017, from https://www.admin.ch/opc/de/classified-compilation/19995395/index.html#a9

Al Jazeera. (2013/11/9). Blatter: World Cup in Qatar Is Not Reversible. *Al Jazeera.* Retrieved 1/30/2017, from http://www.aljazeera.com/news/middleeast/2013/11/blatter-world-cup-qatar-not-reversible-2013119172121971208.html

Anon. (2016/6/27). *Panama, Canale Made in Italy, c'è anche Cimolai con i suoi 16 cancelli giganti. Il mattino di Padove.* Retrieved 1/20/2017, from http://mattinopadova.gelocal.it/focus/2016/06/27/news/panama-canale-made-in-italy-c-e-anche-cimolai-1.13731598

Apple. (2017a). *Apple Supplier Code of Conduct.* Retrieved 1/30/2017, from http://images.apple.com/supplier-responsibility/pdf/supplier_code_of_conduct.pdf

Apple. (2017b/1/1). *Apple Supplier Responsibility Standards.* Retrieved 1/30/2917, from http://images.apple.com/supplier-responsibility/pdf/supplier_responsibility_standards.pdf

Avey, J. B., Reichard, R. J., Luthans, F., and Mhatre, K. H. (2011). Meta-Analysis of the Impact of Positive Psychological Capital on Employee Attitudes, Behaviors, and Performance. *Human Resource Development Quarterly, 22*(2), 127–152.

AV-Test. (2017). *Statistics—Malware.* Retrieved 1/22/2017, from https://www.av-test.org/en/statistics/malware

Black, W. (2015). *Psychopathic Cultures and Toxic Empires.* Edinburgh, UK: Frontline Noir.

Bond, M. (2017/5). Ready for Anything: The Best Strategies to Survive a Disaster. *New Scientist* (3125). Retrieved 5/11/2017, from https://www.newscientist.com/article/mg23431250-400-ready-for-anything-the-best-strategy-to-survive-a-disaster

Bonn, S. A. (2014/1/22). *How to Tell a Sociopath from a Psychopath.* Retrieved 1/17/2017, from https://www.psychologytoday.com/blog/wicked-deeds/201401/how-tell-sociopath-psychopath

Bos, C. (2004/3/1). *Titanic—The Fatal Voyage—Ice Warnings Ignored.* Retrieved 1/16/2017, from www.awesomestories.com: https://www.awesomestories.com/asset/view/Ice-Warnings-Ignored-Fatal-Voyage-The-Titanic

BSI. (2000). BS 6079-2:2000—British Standard—Project Management—Part 2: Vocabulary. London, UK: British Standards Institution.

Business Insider. (2017/4/10). Video Shows a Passenger Forcibly Dragged off a United Airlines Plane. Retrieved 6/13/2017, from https://youtu.be/VrDWY6C1178

Casani, J., Albee, A., Battel, S., Brace, R., Burdick, G., and Burr, P. (2000). Report on the Loss of the Mars Polar Lander and Deep Space 2 Missions. NASA. Retrieved 9/6/2015, from ftp://ftp.hq.nasa.gov/pub/pao/reports/2000/2000_mpl_report_1.pdf

Chang, L.-M., and Chen, P.-H. (2001/6). BOT Financial Model: Taiwan High Speed Rail Case. *Journal of Construction Engineering and Management* 127(3), 214. Retrieved from https://www.researchgate.net/publication/245283188_BOT_Financial_Model_Taiwan_High_Speed_Rail_Case

Christopherson, S., and Lillie, N. (2005/11/1). Neither Global nor Standard: Corporate Strategies in the New Era of Labor Standards. *Environment and Planning,* 37(11), 1919–1938. Retrieved 1/30/2017, from http://journals.sagepub.com/doi/pdf/10.1068/a3789

Clarke, C. (20007/3/2). *The Johnstown Flood: the Worst Dam Failure in U.S. History.* Retrieved 6/12/2017, from ABC News: http://abcnews.go.com/2020/story?id=2918360

Conway, M. E. (1968). How Do Committees Invent? *Datamation Magazine,* 28–31. Retrieved 8/11/2015, from http://www.melconway.com/Home/pdf/committees.pdf

Covey, S. (2004). *The 7 Habits of Highly Effective People: Powerful Lessons in Personal Change* (Anniversary ed.). New York, NY: Simon & Schuster.

Cowan, P. (2014/11/18). *NSW Transport's Mammoth Sap Overhaul Hits Delays.* Retrieved 1/18/2017, from itnews.com.au: http://www.itnews.com.au/news/nsw-transports-mammoth-sap-overhaul-hits-delays-397958

Coyne, A. (2016/8/31). *NSW Transport Grilled over Growing Erp Project Costs, Delays.* Retrieved 1/18/2017, from http://www.itnews.com.au/news/nsw-transport-grilled-over-growing-erp-project-costs-delays-435586

Davis, F. (1989). Perceived Usefulness, Perceived Ease of Use, and User Acceptance of Information. *MIS Quarterly,* 13(3), 319–340.

Defoe, D. (1697). The History of Projects. In *An Essay Upon Projects.* Salt Lake City, UT, USA: Project Gutenberg. Retrieved 9/13/2014, from http://www.gutenberg.org/files/4087/4087-h/4087-h.htm

DIN. (2009). *DIN Norm 69901-5—Projektmanagement—Projektmanagementsysteme—Teil 5: Begriffe (= Project Management—Project Management Systems—Part 5: Terminology;* Own Translation). Berlin, Germany: DIN—Deutsches Institut für Normung.

DLA Piper. (2014/4). *Migrant Labour in the Construction Sector in the State of Qatar.* Retrieved 1/30/2017, from http://www.engineersagainstpoverty.org/documentdownload.axd?documentresourceid=58

Donnelly, J. (2005/June). Childe Harold Wills. *Hemmings Classic Car.* Retrieved 1/8/2017, from https://www.hemmings.com/magazine/hcc/2005/06/Childe-Harold-Wills/1281300.html

DPA/The Local. (2016/30/12). Berlin Airport Boss Admits: Chances of 2017 Opening Slim to Zero. *The Local.* Retrieved 1/16/2017, from https://www.thelocal.de/20161230/berlin-airport-boss-admits-chances-of-2017-opening-slim-to-zero

Drucker, P. F. (2013). *Managing in the Next Society* (Classic Druck Edition 2007 ed.). New York, NY: Taylor & Francis, Routledge, USA.

Dumas, A. (1844). *The Three Musketeers.* Retrieved 6/25/2017, from http://www.gutenberg.org/files/1257/1257-h/1257-h.htm#dartagnan-shows-himself

Duncan, J. (2008). *Any Color—So Long as It's Black!* Auckland, NZ: Exisle Publishing.

Eaton, B. (2014). *Success Platforms: The Breakthrough Guide to the Success Secrets of Corporations and Millionaires.* Hampton, VA, USA : Morgan James Publishing.

Fehr, E., Hart, O., and Zehnder, C. (2008). Contracts as Reference Points—Experimental Evidence. *American Law & Economics Association Papers,* 1–39.

Fortune. (2017/3/9). *The 100 Best Companies to Work For.* New York, NY: Time Inc. Retrieved 5/26/2017, from http://fortune.com/best-companies/

Franklin, O. (2014/8/12). *The Quest to Hit 1,000 MPH in an Insane Rocket-Powered Car.* Retrieved 1/16/2017, from Wired.com: https://www.wired.com/2014/12/bloodhound-ssc-1000-mph/

Freedomhouse. (2017). *About Freedom of the Press.* Retrieved 6/16/2017, from https://freedomhouse.org/report-types/freedom-press

Gatti, F. (2016/11/15). *Venezia, sta affondando pure il Mose. L'Espresso.* Retrieved 1/20/2017, from http://espresso.repubblica.it/inchieste/2016/11/11/news/venezia-sta-affondando-pure-il-mose-1.288116

George, J. P. (2007). Reimposable Discounts and Medieval Contract Penalties. *Loyola Consumer Law Review,* 20(1). Retrieved 5/16/2017, from http://scholarship.law.tamu.edu/cgi/viewcontent.cgi?article=1158&context=facscholar

GPM. (n.d.). IPMA International Project Management Association. Retrieved 10/11/2015, from http://www.gpm-ipma.de/ueber_uns/ipma.html

GUPC. (2014a/2/5). *Consortium Continues to Seek Panama Canal Solution Despite Break in Negotiations by Canal Authority.* Retrieved 1/20/2017, from http://www.salini-impregilo.com/en/press/press-releases/nota-gupc-consortium-continues-to-seek-panama-canal-solution-despite-break-in-negotiations-by-canal-authority.html

GUPC. (2014b/1/20). *Press Releases.* Retrieved 1/21/2017, from http://www.gupc.com.pa/en/press/press-releases

Hall, E. T. (1989). *Beyond Culture.* New York, NY: Anchor Books.

Hart, O., and Moore, J. H. (1998/9). *Foundations of Incomplete Contracts.* Retrieved 5/17/2015, from http://eprints.lse.ac.uk/19354/1/Foundations_of_Incomplete_Contracts.pdf

Harvey, J. (1974). The Abilene Paradox: The Management of Agreement. *Organizational Dynamics,* 3(1), 63–80. Retrieved 7/6/2014.

Hess, T., McNab, A., and Basoglu, K. (2014). Reliability Generalization of Perceived Ease of Use, Perceived Usefulness, and Behavioral Intentions. *MIS Quarterly,* 38(1), 1–28, A1–A29.

Hoffman, B. (2012). *The Ad Contrarian: Getting Beyond the Fleeting Trends, False Goals, and Dreadful Jargon of Contemporary Advertising* (Vol. Kindle book). San Francisco: Hoffman and Lewis, Kindle book. Retrieved from https://www.amazon.com/Ad-Contrarian-Bob-Hoffman-ebook/dp/B008X6XQZY

Holmlund, C. M., and Hammer, M. (1999). Ecosystem Services Generated by Fish Populations. *Ecological Economics,* 29. Retrieved 6/21/2017, from http://www3.carleton.ca/fecpl/courses/ESS.pdf

Hornby, R. (2017) *Commercial Project Management: A Guide for Selling and Delivering Professional Services,* 1st edition, London, UK: Routledge.

IKEA. (2013). *IKEA Group Sustainability Report FY12.* Retrieved 1/30/2017, from http://www.ikea.com/ms/en_JP/pdf/sustainability_report/sustainability_report_2012.pdf

IPMA. (2016). *Project Excellence Baseline—for Achieving Excellence in Projects and Programmes.* Zürich, Switzerland: International Project Management Association. Retrieved from http://products.ipma.world/wp-content/uploads/2016/02/IPMA_PEB_1_0.pdf

Juris GmbH. (2013/10/1). *German Civil Code, BGB, §242: Performance in Good Faith.* Retrieved 5/11/2017, from https://www.gesetze-im-internet.de/englisch_bgb/englisch_bgb.html#p0726

Kerzner, H. R. (2013). *Project Management: A Systems Approach to Planning, Scheduling, and Controlling* (11th ed.). Hoboken, NJ, USA: John Wiley & Sons.

Killgore, W. D., Balkin, T. J., Yarnell, A. M., and Capaldi, V. F. (2017/6). Sleep Deprivation Impairs Recognition of Specific Emotions. *Neurobiology of Sleep and Circadian Rhythms,* 3. Retrieved from http://www.sciencedirect.com/science/article/pii/S2451994416300219

Kramer, A., Wolf, A. T., Carius, A., and Dabelko, G. D. (2013/1). The Key to Managing Conflict and Cooperation over Water. *A World of Science,* 11(1). Retrieved 6/21/2017, from http://unesdoc.unesco.org/images/0021/002191/219156E.pdf#page=4

Kuhn, M., Wolf, E., Maier, J. G., Mainberger, et. al (2016/8). Sleep Recalibrates Homeostatic and Associative Synaptic Plasticity in the Human Cortex. *Nature Communications,* 7, 23. Retrieved from http://www.nature.com/articles/ncomms12455

L'Assemblée Nationale. (2017). *Proposition de loi.* Paris, France. Retrieved from http://www.assemblee-nationale.fr/14/pdf/ta/ta0924.pdf

Langley, M. (2015/11). Project Portfolio Management: A Holistic Picture. *Delivering on Strategy—The Power of Project Portfolio Management.* Retrieved from http://www.pmi.org/-/media/pmi/documents/public/pdf/learning/thought-leadership/deliver-strategy-portfolio-management.pdf

Lehmann, O. F. (2016a/12/10). Advances in Project Management: An Introduction to a Typology of Projects. *PM World Journal,* 5(XII), 1–14. Retrieved 12/11/2016, from http://pmworldjournal.net/article/introduction-typology-projects/

Lehmann, O. F. (2016b). *Situational Project Management: The Dynamics of Success and Failure.* Boca Raton, FL, USA: Auerbach Publications.

Lehmann, O. F. (2017/2). Customer Projects: What Is the Future of the Business? *PM World Journal,* 4(2). Retrieved 6/1/2017, from http://pmworldjournal.net/article/customer-projects-future-business/

Lencioni, P. M. (2002). *The Five Dysfunctions of a Team: A Leadership Fable.* San Francisco, CA, USA: Jossey-Bass.

Levav, J. (2011). *How Extraneous Factors Impact Judicial Decision Making*. Columbia Business School, New York, USA. Retrieved 4/25/2017, from https://www8.gsb.columbia.edu/newsroom/newsn/1659/how-extraneous-factors-impact-judicial-decisionmaking

Lewin, K., and Weiss, G. (1997). Defining the "Field at a Given Time". In K. Lewin, *Resolving Social Conflicts and Field Theory in Social Science* (pp. 200–211). Washington D.C., USA: American Psychological Association. Retrieved from http://dx.doi.org/10.1037/10269-015

Lockheed-Martin. (2017). *The Centerpiece of 21st Century Global Security*. Retrieved 6/7/2017, from https://www.f35.com/global

Machado, A. (2012/8/7). *Latino Poemas—Antonio Machado: Caminante no hay camino*. Retrieved 9/15/2015, from http://www.latino-poemas.net/modules/publisher2/article.php?storyid=1115

Mail Online. (2006/8/18). The Stark Reality of iPod's Chinese Factories. *The Daily Mail*. Retrieved 1/29/2017, from http://www.dailymail.co.uk/news/article-401234/The-stark-reality-iPods-Chinese-factories.html

Manera, M. (2014/7/19). Venice Is Sinking Under a Tidal Wave of Corruption. *Newsweek*. Retrieved 1/20/2018, from http://europe.newsweek.com/venice-sinking-under-tidal-wave-corruption-259878

MG Motor UK Ltd. (n.d.). *About MG*. Retrieved 7/1/2017, from http://mg.co.uk/about-mg/

Montesquieu, C. d. (1748). *De l'esprit des loix* (*On the Spirit of the Laws*). Paris, France. Retrieved from http://press-pubs.uchicago.edu/founders/documents/v1ch17s9.html

Moss, G. C. (2007). International Contracts Between Common Law and Civil Law: Is Non-State Law to be Preferred? The Difficulty of Interpreting Legal Standards Such as Good Faith. *Global Jurist Advances, 7*(1). Retrieved from http://folk.uio.no/giudittm/Non-state%20Law_Good%20Faith.pdf

Mueller, F. (2012/7/24). Apple Seeks $2.5 Billion in Damages from Samsung, Offers Half a Cent Per Standard-Essential Patent. *Foss Reports*. Retrieved 10/30/2016, from http://www.fosspatents.com/2012/07/apple-seeks-25-billion-in-damages-from.html

NASA. (2000/3/13). *Report on Project Management at NASA by the Mars Climate Orbiter Mishap Investigation Board*. Retrieved 6/25/2017, from https://science.ksc.nasa.gov/mars/msp98/misc/MCO_MIB_Report.pdf

NASA. (2016/8/25). *Image Galleries*. Retrieved 6/52/2017, from https://www.nasa.gov/multimedia/imagegallery/index.html

Newman, L. (2011). *Shipley Proposal Guide* (4th ed.). Kaysville, Utah, USA: Shipley Associates.

NTSB. (2014/11/21). *Auxiliary Power Unit Battery Fire Japan Airlines Boeing 787-8, JA829J*. Retrieved 7/25/2016, from http://www.ntsb.gov/investigations/accidentreports/pages/AIR1401.aspx

Pattisson, P. (2013/9). Revealed: Qatar's World Cup "Slaves". *The Guardian*. Retrieved from https://www.theguardian.com/world/2013/sep/25/revealed-qatars-world-cup-slaves

Penzhorn, G. (2004/3/15). *Three Strikes Against Graft*. Retrieved 1/20/2017, from https://journal. probeinternational.org/2004/03/15/three-strikes-against-graft/

PM Network. (2005). High Hopes. *PM Network,* 19(9). Retrieved 6/7/2017, from http://www. pmi.org/learning/library/launch-case-analysis-high-hopes-3259

PMI. (2004/9). Murky Waters. *PM Network*. Retrieved 1/20/2017, from http://www.pmi.org/ learning/library/proposed-construction-plan-2801

PMI. (2005/5). Case Analysis—Train Delay. *PM Network,* p. 1. Retrieved 3/8/2015, from http:// www.pmi.org/learning/train-delay-impending-failure-construction-project-3237

PMI. (2013). *A Guide to the Project Management Body of Knowledge®—PMBOK® Guide* (5th ed.). Newtown Square, PA, USA: PMI—The Project Management Institute, Inc.

PMI. (2017). *A Guide to the Project Management Body of Knowledge®—PMBOK® Guide* (6th ed.). Newtown Square, PA, USA: PMI—The Project Management Institute, Inc.

Pozzi, P. M. (2014/30). Speed Freaks: Tunnel Vision and Physiological Perception. *Psych 256: Introduction to Cognitive Psychology.* Retrieved 1/15/2017, from https://sites.psu.edu/ psych256sp14/2014/01/30/speed-freaks-tunnel-vision-and-physiological-perception/

Pratchett, T. (1990). *Eric: A Novel of Discworld.* New York, NY, USA: HarperTorch.

Reeves, J., and Murphy, P. (2014). *English Contract Law: Good Faith, a Common Law Duty.* London, UK: Clyde & Co. Retrieved from http://www.clydeco.com/site/pdf/insight/article/ english-contract-law-good-faith-a-common-law-duty/print

Rosenbaum, S. (2012/10/13). ATT iPhone Customers Hit with Massive Data "Sipping" Bug. *The Huffington Post.* Retrieved 10/31/2016, from http://www.huffingtonpost.com/steve-rosenbaum/att-iphone-customers-hit_b_1963505.html

RSF. (2017). *World Press Freedom Index.* Retrieved 6/16/2017, from https://rsf.org/en/ranking

Sacom. (2016/12/1). *Blood and Sweat Behind the Screen of iPhones—Another Investigative Report on Apple's Largest Display Screen Supplier.* Retrieved 1/30/2017, from http://sacom.hk/wp-contentuploads2016122016-biel-crystal-re-visiting-report_sacom-pdfrel2016-biel-crystal-re-visiting-report_sacom/

Savitz, E. (2012/9/28). Apple CEO Cook Posts Letter Apologizing for Buggy Maps. *Forbes.* Retrieved 10/31/2016, from http://www.forbes.com/sites/ericsavitz/2012/09/28/apple-ceo-cook-posts-letter-apologizing-for-buggy-maps

Shakespeare, W. (1599). *The Life of King Henry the Fifth,* Act 4, Scene 3. Retrieved 8/21/2015, from http://shakespeare.mit.edu/henryv/henryv.4.3.html

Shenhar, A., and Dvir, D. (2007). *Reinventing Project Management: The Diamond Approach to Successful Growth and Innovation.* Boston, MA: Harvard Business School Publishing.

Smith, R. (2008/5). Liquidated Damages and Penalty Clauses: A Civil Law versus Common Law Comparison. *Lexology,* 12. Retrieved 5/22/2017, from http://www.lexology.com/library/ detail.aspx?g=d413e9e1-6489-439e-82b9-246779648efb

Stephenson, A. G., LaPiana, L., Mulville, D. R., Rutledge, et. al (1999). *Mars Climate Orbiter Mission Failure Mishap Investigation Board, Phase I Report.* NASA. Retrieved 9/6/2015, from ftp://ftp.hq.nasa.gov/pub/pao/reports/1999/MCO_report.pdf

Supplychainbrain.com. (2010). *In an Outsourced Supply Chain, Lockheed Keeps the Raw Materials Flowing.* Retrieved 6/6/2017, from http://www.supplychainbrain.com/content/nc/research-analysis/supply-chain-innovation-awards/single-article-page/article/in-an-outsourced-supply-chain-lockheed-keeps-the-raw-materials-flowing-1/

Tetlow, G. (2016/11/22). Who are the Jams and Will the Autumn Statement Help Them? *Financial Times Online.* Retrieved from https://www.ft.com/content/4631d19e-b011-11e6-9c37-5787335499a0

TI. (2000/1/19). *Construction and Arms Industries Seen as Leading International Bribe-Payers.* Retrieved 6/23/2017, from https://www.transparency.org/news/pressrelease/construction_and_arms_industries_seen_as_leading_international_bribe_payers

TI. (2016a). *Transparency International—Our History.* Retrieved 1/20/2017, from https://www.transparency.org/whoweare/history

TI. (2016b). *Corruption Perception Index—Overview.* Retrieved 6/7/2017, from https://www.transparency.org/research/cpi/overview

TI. (2016c). *Our Work on Whistleblowing.* Retrieved 6/24/2017, from https://www.transparency.org/whatwedo/activity/our_work_on_whistleblowing

Titanic Inquiry Project. (1912/4/26). United States Senate Inquiry, Day 8: Testimony of Cyril F. Evans. Retrieved 1/16/2017, from Titanic Inquiry Project: http://www.titanicinquiry.org/USInq/AmInq08EvansCF01.php

Tuckman, B. (1965). Developmental Sequence in Small Groups. *Psychological Bulletin, 63*(6), 384–399. Retrieved 5/12/2016, from http://psycnet.apa.org/doi/10.1037/h0022100

Vance, A. (2016). *Elon Musk: Tesla, SpaceX, and the Quest for a Fantastic Future.* New York, NY: Ecco Press.

von Gerkan, M. (2013). *Black Box BER: Vom Flughafen Berlin Brandenburg und anderen Großbaustellen (= Black Box Berlin: On the Mega Airport Berlin-Brandenburg and Other Major Constructions; Own Translation).* Cologne, Germany: Bastei Lübbe (Quadriga).

VW. (2015). *Video Statement Prof. Dr. Martin Winterkorn* (English). Retrieved 7/25/2016, from https://footage.volkswagen-media-services.com/VW/videos/mpeg4/download.php?vid=20150971202titZMI_PRE

Wallbank, P. (2017/1/16). Will They Never Learn? Poor Project Management Derails Transport for NSW Project. *DIGINOMICA.* Retrieved 1/18/2017, from http://diginomica.com/2017/01/16/will-never-learn-bad-project-management-derails-nsw-project/

Yellow Buses. (2017). *Conditions of Carriage.* Retrieved 6/18/2017, from https://www.bybus.co.uk/about-us/conditions-of-carriage/

Zurek, D. B., and Gilbert, C. (2014/2/5). Static Antennae Act as Locomotory Guides that Compensate for Visual Motion Blur in a Diurnal, Keen-Eyed Predator. *Proceedings of the Royal Society B.* Retrieved 1/15/2017, from http://rspb.royalsocietypublishing.org/content/281/1779/20133072

Index

#

3PL, 33

A

absenteeism, 110, 242, 277
acceptance testing, 1
accountability, 24, 48, 81, 173, 198, 206–208, 227
ACP, *See* Autoridad del Canal de Panama
action caption, 141
activity, 4, 6, 10, 17, 22, 23, 31, 50, 66, 71, 73, 81, 99, 103, 108, 122, 135, 158, 159, 167, 180, 196, 217, 224, 225, 232–237, 246, 263–268, 272, 278, 286
ADR, *See* alternative dispute resolution
advertisement, 88, 89, 251
Affinity diagramming, 8
affirmative action, 166
Africa, 64, 77, 215
agile, 3, 6, 12, 13, 17, 51, 179–181, 191, 195, 213, 237, 247, 253
agile method, 3, 6, 12, 13, 17, 179, 180, 191, 195, 213, 237, 247, 253
aging, 3, 13, 16, 20–22, 25, 30, 43, 55, 65, 73, 98, 100, 106, 108, 120, 134, 135, 149, 150, 175, 176, 185, 197, 200–205, 209–213, 219–222, 230, 234, 238, 253, 267, 278, 281, 285, 287

AIDA, 102, 104
alchemy, 4, 5
alternative dispute resolution (ADR), 162, 177, 199, 233
Amadeus, 55
Anglo-American legal system, 164
antisocial personality disorder (ASPD), 75, 215, 222
APMP, *See* Association for Proposal Management Professionals
APMP Body of Knowledge, 95
Apple, 48–50, 81, 151
aqua alta, 78
arbitrator, 165
architecture of a project, 22, 23
Asia, 64, 70, 282
ASPD, *See* antisocial personality disorder
Association for Proposal Management Professionals (APMP), 95
assumption, 104, 110, 122, 131, 135, 137, 141, 144, 157, 164, 170, 186–189, 193, 194, 206, 225, 232, 280
assurance, 20, 150, 152
attention dilemma, 48, 49
Australia, 60, 64, 67, 72, 209, 210, 282
automotive development, 52
automotive supplier, 34
Autoridad del Canal de Panama (ACP), 79

award, 36, 37, 87, 94, 96, 100, 138, 140, 149,
    157–160, 175, 183, 184, 189–192, 207,
    221–223, 228, 266, 270, 272, 282–287

B

B2B marketplace, 86–89, 106
bankruptcy, 76, 84, 175, 252
basic trust, 131, 165, 168
behavior, 6, 8, 29, 43, 55, 66, 75, 78, 84,
    85, 88, 103, 105, 110, 112, 131, 132,
    139–145, 163–168, 171–177, 190, 193,
    206–208, 215, 218, 221–226, 239, 246,
    248, 252, 253, 257, 259, 267, 268, 278,
    286, 287
benefit, 11–15, 19, 22, 24, 38, 47, 53, 54, 57,
    58, 69, 70, 77, 83, 86, 94, 96, 102, 103,
    107, 109, 121–125, 128–131, 134–140,
    150, 154, 160, 163–165, 172–176, 185–
    192, 195, 201–203, 206–210, 217–222,
    226–230, 234, 239, 247, 248, 253, 254,
    260–278, 283–287
benefit engineering, 94, 107, 130, 135, 253,
    254, 260–273, 287
benefit generation, 58, 261
benefit realization, 58, 190
Berlin Hauptbahnhof, 7
best practices, 3–7, 42, 280, 281
bid, 29, 42–44, 82, 87–95, 98, 104–119,
    126, 130, 139, 140, 152, 161, 175, 233,
    251, 260, 268
bid bonds, 118, 119, 161
bidders' conference, 89, 90
bid manager, 113
binding offer, 115–118, 138
black box, 75, 200
blurred project, 11
BMW, 52
booked resources, 121
bottleneck, 16
breach of contract, 26, 93, 97, 163, 186, 267
breakthrough, 8
bribe, 77, 107, 150, 215, 223, 240, 241, 246
brownfield project, 9, 10
BS, *See* burnout syndrome
Bürgerforum 21, 10

burnout syndrome, 110, 166
Boeing 787 Dreamliner, 28
business acumen, 65, 273, 274
business benefit, 123, 220, 270
business case, 11, 14, 33, 47, 76, 104, 170,
    191, 198, 211, 222, 226, 283, 285
business coach, 87
business development, 11, 29, 32, 87, 104,
    105, 113, 130, 140, 153, 157, 220–223,
    252, 262, 268, 272–275
business endeavor, 33
business environment, 39, 41, 98, 119, 168
business goal, 18, 23, 57, 106, 132
business interest, 10, 29, 31, 53, 65, 68, 69,
    72, 77, 111, 123, 125, 128, 164, 169,
    171, 196, 202, 203, 206, 210, 214, 221,
    232, 235, 238, 244, 246, 254–259, 279,
    282, 286
business lifecycle, 262
business literature, 58
business model, 20, 33, 35, 86, 88, 251
business risk, 148
business-to-business, 44, 86
business transformation, 67
buyer, 4, 11, 24, 34–38, 44–47, 56, 57, 66,
    67, 85–108, 112–122, 130–159, 175,
    178, 193, 197, 229, 235, 244, 247, 260,
    262, 270, 271, 274, 275, 280
buy option, 44, 45, 54–58, 64–72, 75, 76,
    85, 94

C

cancellation terms, 152
capacity gap, 263
captive outsourcing, 19
capture ratio, 42, 43, 104, 108
case law, 162
catalogue management, 86
caveat emptor, 67, 164, 176
Central & South America, 64
CEO, 20, 48, 49, 70, 71, 150, 252, 264, 276
certification, 2, 8, 32, 44, 95, 159, 198, 253
CFO, 264
challenge, 2, 13, 17, 22, 24, 29, 38, 65, 69,
    71, 81, 99, 101, 104, 109, 162, 165, 172,

180, 192, 196, 202, 203, 216, 219, 238, 246, 257, 260, 262, 279, 282, 286, 287

change request, 17, 34, 38, 76, 87, 94, 107, 120, 124, 155, 182, 184, 193–198, 208, 211, 213, 219, 220, 229, 271

checklist, 32

CIO, 3, 75, 80, 98, 148, 167, 182, 185, 196, 198, 206, 207, 215, 216, 220, 222, 233, 241, 244, 254, 257, 272

civil code, 162–164, 172, 182

civil engineering, 30

civil law, 85, 97, 118, 160–165, 172, 181–186, 191, 215

claim management, 166, 269

claim manager, 166, 181

claims, 31, 71, 81, 107, 147, 166, 172, 193, 198, 206, 214, 238, 243–245, 252, 256

CLC, *See* customer-led consortium

client, 19, 31, 33, 38, 109, 188, 222, 229, 243, 244, 252

closed typology, 17

coffee break zone, 161

common law, 85, 97, 162–164, 172, 181, 182, 185, 186, 215

company, 3, 4, 18–21, 24–26, 32–35, 38, 41–45, 48, 50, 55, 56, 66, 69–71, 74–86, 91, 96, 101, 105, 109–111, 116, 118, 123, 127, 129, 136, 139, 144–148, 156, 160, 170, 174, 175, 178, 194, 197–203, 208, 212–220, 223–226, 229, 230, 233–235, 238, 239, 242–245, 249–257, 263, 268, 272, 276–279, 282–286

competition, 18, 19, 32, 36, 48, 76, 83, 89–94, 105, 111–114, 120, 126, 130, 134, 139, 140, 157, 164, 172, 175, 202, 205–208, 251, 268, 272, 283–286

competitive behavior, 29, 112, 139, 140, 171, 174, 175, 207, 208, 224

competitor, 20, 39, 49, 53, 83, 85, 89, 95, 100, 103–105, 119, 129, 131, 134, 137, 143–146, 175, 224, 251, 263, 268

complaint, 31, 39, 97, 130, 133, 148, 208, 211, 233, 234, 277

complexity, 28, 38, 65, 84, 85, 129, 130, 156, 159, 171, 198, 208–213, 217, 224,

233, 264, 275, 283, 287

complexity trap, 84

complex proposal, 113, 121, 140–142

compliance officer, 32

component manufacturer, 33

composed Project, 13

composite structures, 35

concurrent work, 24

configuration management, 156

confusion, 28, 90, 189, 225

congress, 12, 41, 70, 90

consortia, 3, 28–31, 78, 79, 126–129, 202, 203, 216, 223, 247–249, 282, 287

Consorzio Venezia Nuova (CVN), 78

constraint, 6, 103, 140, 156, 185, 225–228, 234, 254, 261, 262, 266, 269–271

construction, 7, 9, 17, 24, 28–31, 39, 53, 76–79, 82, 118, 120, 126–128, 194, 206, 229, 230, 236, 240

constructive change, 166, 181

Constructora Urbana, 31, 128

consultancy, 30

consumption materials, 33

continuous delivery, 15

contract, 1–4, 11, 13, 18–47, 50–58, 62, 65, 68–90, 93–121, 124–133, 138–140, 145–149, 153–211, 214–287

contract development, 90, 155, 159, 175, 176

contracting, 22, 25, 29, 38, 70, 71, 85, 98, 138, 140, 154, 155, 159, 162, 178, 183, 191–193, 197, 203, 206, 215–218, 238, 246, 247, 268, 275

contract-intensive project, 29

contract management, 38, 87

contract negotiation, 87

contractor, 1–4, 11, 13, 18, 19, 22–28, 31–39, 43–46, 50–58, 62, 65, 68–75, 78–88, 93, 94, 99–103, 108–112, 116, 119, 120, 124–129, 132, 145–148, 153–162, 165–177, 180–209, 214–256, 259–286

contract partner, 172, 175, 196, 217, 218, 222, 224, 247, 254

contract party, 18, 29, 40, 69, 79, 86, 140, 161–164, 172, 173, 183, 185, 232, 233

Contract SOW, 93
contract types, 38, 89, 159, 181–183, 191, 262
contractual conditions, 50
contractual disagreement, 165
contractual payment scheme, 58
contribution, 13, 39, 125–129, 207
control limit, 132, 133, 228
Conway, Melvin, 28
Conway's law, 28, 68, 165, 176, 249
Cook, Tim, 48, 49
cooperation, 29, 31, 57, 74–76, 104, 123, 139, 148, 149, 158, 178, 202, 206, 207, 212, 221, 224, 238, 246, 247, 251
copyright, 33
corporate income, 32
corporate restaurant, 161, 245
corporate strategy, 304
corporation, 3, 17, 26, 30–32, 49, 54, 55, 70, 80–86, 116, 126, 136, 145, 204, 218, 234, 240, 244, 247, 264, 266, 276
corruption, 77, 78, 82, 123, 175, 215, 216, 223, 234, 240, 241
Corruption Perception Index (CPI), 216, 223
cost, 1–3, 6, 11, 14, 18, 24, 32–35, 38–57, 68–71, 74, 78, 79, 83, 88, 89, 98, 99, 102, 103, 107–112, 118–121, 125–140, 145, 153, 157–162, 166–168, 171, 175, 176, 181–194, 197, 198, 201, 202, 205–213, 220, 223–228, 231, 235–240, 243, 246, 248, 251–256, 260–262, 266–272, 276, 277, 282–284
cost benefit, 54, 69, 70
cost center, 3, 11, 18, 24
cost engineering, 260–262, 269
costing, 2, 18, 43, 135–138
cost reimbursable contract, 160, 183, 186–188
counsel, 215, 224, 285
court, 18, 19, 26, 29, 83, 88, 111, 118, 119, 145, 147, 154, 155, 162, 164, 171, 172, 176, 181–185, 189, 194, 215, 222, 230, 240, 252, 254, 285
Covey, Stephen R., 69
CPI, See Corruption Perception Index

credit line, 3, 262, 277
creditor protection, 20
credulity, 167, 189
crisis management, 207, 251, 253, 278–280
critical misunderstandings, 90
culture, 10, 28, 29, 53, 65, 70, 72, 76, 77, 81, 104, 110, 120, 123, 155, 159, 163, 165, 168, 173, 177, 178, 181, 192, 196, 206, 215, 216, 221–223, 227, 240, 241, 244, 246, 249, 259, 260, 266–269, 276, 279, 287
customer, 1–4, 11–15, 18–61, 64–75, 80–87, 90, 93–149, 153, 156–162, 165–178, 181–211, 214–256, 259–287
customer interface, 32
customer-led consortium (CLC), 282
customer project, 3, 4, 11, 18–36, 39–44, 57, 58, 61, 65, 74, 90, 94, 95, 100, 102, 109, 111, 116, 121, 131, 153, 158, 170, 171, 181, 194, 195, 199, 203, 208, 209, 214, 224, 229, 253, 261–264, 267–276, 285–287
customer project management, 57, 58, 61
customers' mission, 38
CVN, See Consorzio Venezia Nuova

D
data leak, 83
deadline, 2, 14, 16, 34, 39, 42, 74, 82, 84, 96–99, 103, 104, 112, 116, 119–122, 126, 129, 138, 155, 156, 162, 169, 170, 175, 181, 184, 195, 211, 212, 226–228, 236, 242, 243, 260–271, 283
decomposed project, 13
defaulting contractor, 216, 223
Defoe, Danial, 13, 22
delegation, 48
deliverable, 1, 10–15, 18, 24, 30, 39, 42, 53, 71, 74, 84, 125, 126, 134–136, 158, 161, 175, 176, 179, 189, 191, 208, 212, 219, 220, 229, 235, 244–246, 261, 262, 266, 270, 273, 275
delivery date, 185, 189, 198, 248, 283
demolition, 54
department, 3, 17, 32, 42, 55, 71, 77, 87, 93, 104, 112, 116, 121, 239, 240, 244, 273, 277

dependencies, 2, 19, 24, 33, 61, 84, 138, 153, 157, 207, 214, 215, 222, 223, 233, 237, 259, 286
design work, 30
diligence, 90, 174, 279
direct contractor, 198, 201, 217, 270
disaster, 73–76, 99, 135, 220, 257
discretionary project, 14, 47
discrimination, 89, 118–120, 277
disposal, 53, 54
disruption, 24, 38, 47, 56, 68, 74, 87, 107, 150, 152, 158, 161, 180, 183, 184, 215, 224, 225, 228, 239, 258, 259, 267, 276, 281, 282
distributed team, 283
distribution, 21, 29, 33, 57, 61–64, 75, 83, 163, 204, 210, 213, 234, 258
distrust, 31, 56, 69, 90, 131, 158, 165–167, 175, 190, 206, 241, 257, 271
diversity, 4, 51, 177, 214, 215, 221, 259
diversity of skills, 51
documentary, 82, 89
Drucker, Peter F., 18
Dvir, Dov, 9
dynamics, 4, 7, 12, 21, 22, 28, 38, 43, 54, 58, 65, 90, 122, 132, 171, 179, 203, 219, 221, 232, 248, 249, 254, 256, 269, 271, 280, 281, 286, 287
dysfunctional question, 8

E

early adopter, 66, 205, 283
economic benefit, 109
economics, 57, 215
education, 29, 30, 33, 57, 65, 75, 275
educators, 65
effectiveness, 32, 43, 136, 171, 175, 264, 272, 275
efficiency, 32, 42, 55, 171, 205, 264
effort-reward imbalance (ERI), 110, 166
Eigen, Peter, 77
e-invoicing, 87, 276, 277
empathy, 15, 22, 28, 29, 69, 120, 140, 215, 221, 272, 285
empirical research, 57

employee, 3, 20, 24, 30, 38, 44, 45, 48, 52, 55, 56, 67, 80, 83–85, 94, 98, 110, 111, 116, 126, 136, 141, 146–148, 152, 166, 167, 175, 200, 204, 205, 214–216, 219–223, 230, 235, 241–246, 256, 263, 277, 278
enabling services, 24, 36, 87, 153, 156–162, 172, 200, 225, 228–230, 235, 245
end date, 10
engineering, 2, 8, 10, 13, 30, 39, 52, 94, 107–111, 130, 135, 236, 245, 253, 254, 260–273, 287
engineers' project, 13
enquiry, 97, 104, 241
enterprise project management, 32, 42
enterprise resource planning (ERP), 67, 86, 87
environmental conditions, 6, 119, 179
environmental standards, 54
equipment, 6, 11, 17, 31, 35, 38, 47, 54–57, 70, 99, 149, 151, 183, 194, 197–200, 229, 230, 245, 280
ERI, *See* effort-reward imbalance
ERP, *See* enterprise resource planning
error, 4, 17, 28, 45, 71, 72, 84, 125, 151, 162, 166, 167, 173, 176, 201, 204–207, 212, 213, 222, 243, 246, 266, 277
error fixing, 45, 71, 72, 207
error tolerance, 166, 173
escalation, 177, 286
ESPN, 82
Europe, 45, 52, 64, 85, 127, 163, 164, 177, 186, 246
Europrop International, 127
evolutionary deliveries, 15
exhaustion, 110, 166
exploratory project, 12
external assets, 56, 57, 70, 214
external resources, 43, 56, 287

F

failure, 4–9, 12, 19, 21, 24, 25, 28, 29, 38, 43, 44, 50, 54, 68, 69, 75, 79, 84, 90, 104, 122, 128, 132, 158, 162, 165–167, 179, 180, 201, 207, 216, 223, 226, 232, 238, 242, 254–257, 267, 271, 277, 280,
continued on next page

Failure, continued
    281, 284, 287
Fédération Internationale de Football
    Association, 82
fees, 20, 24, 26, 33, 47, 50, 52, 55, 72, 73,
    87, 88, 110, 124–128, 136–139, 143,
    145, 151, 160, 161, 164, 166, 173–175,
    182–194, 198, 204, 207, 216, 220–223,
    228, 234, 239, 263, 270, 282–287
field change, 36
fields of business, 32
FIFA, *See* Fédération Internationale de
    Football Association
final payment, 1
financial value, 41
fire protection system, 76
fitness tracker, 51
Five whys, 8
fixed price contract, 160, 182, 183, 186, 189,
    195, 284
flexibility, 33, 38, 89, 126, 183, 197, 198,
    227, 260
focused project, 10
focus group, 233
force-field analysis, 100–102
Ford, Henry, 52
forecasting, 32
forming, 2, 6, 11–14, 18, 21, 22, 25, 26,
    29–31, 47, 54, 57, 58, 64, 75, 108, 110,
    123, 133, 139, 159, 162, 202, 203, 237,
    243, 252–254, 260, 264, 273, 278, 285
Forschungs- und Innovationszentrum, 52
fragmentation, 26–29, 65, 68–70, 90
fragmenting forces, 29, 30
framework agreement, 109–111
freebie project, 33–35
Freedomhouse, 223
freelancer, 38, 39, 65, 87, 88, 251, 258, 286
friendly dog effect, 143
functional organization, 14, 31, 32, 38
functional question, 8
funding, 2, 11, 17, 67, 127, 226, 228, 271

G

gardeners' project, 13, 14
gentlemen's agreement, 154, 155

globalizing world economy, 65
gobbledygook, 141
golden handshake, 71, 83
golden rule, 146, 147
goLive, 15
good faith, 161–165, 168, 172–178, 182, 185,
    189, 192, 256, 261, 272, 276, 285
governance, 6, 8, 32, 42, 56, 120, 260
government agencies, 39, 54, 286
Green, Andy, 73
greenfield project, 9, 10, 229, 236
greengrocer's apostrophe, 151
growth potential, 52, 71
Grupo Unidos por el Canal (GUPC), 31, 79,
    128
GS Yuasa, 28
gullibility, 167
GUPC, *See* Grupo Unidos por el Canal

H

Hart, Oliver, 153, 178, 181
helpfulness, 144, 169, 172
high-impact project, 11
high-level management, 19
hit rate, 42, 98, 102–109, 170
horse title, 141
human resources, 99, 183, 214, 219, 273

I

IfB, *See* invitation to bid
IKEA, 81
impact, 4, 9, 11, 26, 55, 71, 76, 80, 94,
    95, 99, 136, 152, 163, 171, 172, 182,
    194–196, 205, 212, 221–225, 232–235,
    240, 252, 256, 264, 266, 271, 272,
    277–279, 285
Impregilo, 31, 78, 128
incentives, 97, 136, 182–186, 190, 266
incumbent parties, 36
India, 60, 82
individual trust, 168
Industria de Turbo Propulsores, 127
Industry 4.0, 80
informal relationships, 124
infrastructure, 7–10, 20, 30, 39, 50, 51, 55,

72, 76, 79, 82, 88, 89, 96, 115, 118–121,
    126, 127, 198, 202, 222, 226–229, 242,
    244, 267, 276, 283
innovator, 205, 283
insolvency, 20, 21, 79, 84, 109, 125, 216,
    223, 226, 251
insurance, 3, 53, 118, 161, 170, 214, 223,
    228, 244, 245, 276
integrating forces, 276
integration management, 22
intellectual property, 19, 33, 49, 84, 205
intention, 3, 7, 10, 15, 29, 35, 46, 73, 83, 89,
    93, 122, 127, 128, 134, 138, 142, 162,
    163, 167, 170, 181, 192, 196, 216, 224,
    225, 246, 252, 261, 266
interdependency, 207
interface, 11, 32, 48, 65, 83, 84, 123, 161,
    165, 169, 172, 198, 204, 208, 209, 216,
    257, 258, 269
internal customers, 11, 18, 19
internal logistics systems, 33
internal projects, 3, 4, 11, 14, 18, 21–26, 42,
    56–58, 93, 158, 238, 261, 276, 287
internal rate of return (IRR), 14, 53
internal requestor, 11
internal resources, 237
internal SOW, 93
internal structures, 48, 52, 68
internal vendor, 18, 19, 259
international contract, 85, 160, 176, 178, 215
International Organization for
    Standardization (ISO), 6
International Project Management
    Association (IPMA), 8, 22
internet booking, 55
Internet of Things (IoT), 71, 80
interview, 8, 56, 147, 149
intranet, 161
investment, 20, 22, 25, 33–35, 48, 49, 53,
    67, 89, 104, 105, 109, 118, 123, 126,
    127, 136, 137, 145, 156, 180, 207, 247,
    255, 264, 270, 283–285
invitation to bid (IfB), 44, 87, 91–94,
    114–116
invitation to pitch (ItP), 4, 6, 11, 17, 93, 114

invitation to tender, 92, 114
invitation to treat (ItT), 2, 13, 18, 44, 48,
    56, 68, 70, 81–96, 101, 108–122, 131,
    139, 140, 143–147, 150, 154–160, 164,
    168, 169, 173, 177–179, 185, 195, 198,
    207, 214, 221, 229, 239, 244, 247, 253,
    257, 285
invoice, 36, 38, 120, 183, 186, 188, 194,
    200, 240, 244, 246, 252, 269, 276–278
iOS, 43, 49, 83, 87, 116, 141, 175, 186–189, 213
IoT, *See* Internet of Things
iPad, 81
IP address, 64
iPhone, 49, 50, 81
IPMA, *See* International Project
    Management Association
iPod, 81
IRR, *See* internal rate of return
ISO, *See* International Organization for
    Standardization
iterative incremental, 12
ItP, *See* invitation to pitch
IT services, 32
ItT, *See* invitation to treat

J
JAM, *See* just about managing
joint venture, 3, 30, 31, 126, 127, 202, 203, 247
just about managing (JAM), 20, 98, 108,
    134, 278

K
Kafala, 82
key performance indicator (KPI), 266
kick-off meeting, 158
KPI, *See* key performance indicator

L
laggard, 284, 285
late majority, 284
lawsuit, 18, 26, 49, 85, 88, 118, 166, 176,
    194, 196, 225, 277
lawyer, 10, 26, 29, 77, 115, 138, 145, 156,
    176, 177, 241, 277
layperson, 164

LDs, *See* liquidated damages
leading venturer, 31, 128
learning process, 6, 54, 70, 99, 232, 269
legacy, 9, 35, 68, 135, 221, 229
legacy systems, 35, 68, 135, 229
legal action, 19, 26, 85, 88, 118, 154, 177, 185, 200, 215
legal advice, 115, 153, 162
legal entities, 18
legal perspective, 28, 157, 176, 180, 204, 214, 215
legal remedy of conflicts, 26
legal system, 10, 29, 65, 70, 84, 85, 115, 138, 145, 146, 155, 159–164, 168, 176, 178, 184, 215, 221, 222, 259, 276, 279, 287
legislation, 20, 99, 118, 125, 162, 264
Lesotho, 77, 215
Lesseps, Ferdinand de, 79
lessons-learned database, 2
letter contract, 155
letter of intent (LOI), 100, 154, 155
liability, 20, 42, 74, 111, 128, 156, 165, 177, 192, 214, 276
license, 33, 56, 57, 107, 124, 127, 129, 182, 200, 201, 219
lifecycle, 6, 14, 17, 53, 57, 58, 69, 103, 126, 127, 135–138, 156–159, 171, 226, 231, 240, 245, 246, 261, 262, 272
lifecycle cost, 103, 135–138, 226, 240
LinkedIn, 21, 60, 87, 152
Linnaean taxonomy, 4
liquidated damages (LDs), 97, 182–186
LOI, *See* letter of intent
long-term orientation, 168, 275
long-term planning, 17, 43
long-term strategies, 146
Lotus Elise, 66
low-balling, 107, 138, 260
low-impact projects, 11
low margins, 32
loyalty, 20, 24, 165, 216

M

Macintosh, 81
Mail Online, 81

maintenance, 13, 198, 273
make option, 58, 64, 66, 70, 130
make-or-buy decision, 34, 44, 46, 58–66, 84, 85, 91, 103, 175, 262, 275
makeshift, 70, 264
malware, 80, 217, 224
management, 1–28, 32–34, 38–50, 53–61, 65–71, 74, 75, 81–87, 90, 95–100, 103–116, 119–122, 128, 131, 132, 138, 147, 152–173, 178–181, 189, 192, 195–207, 211–240, 244–248, 251–256, 259–282, 285–287
management attention, 11, 26, 32, 47–50, 53, 56, 57, 70, 71, 152, 200–203, 212, 217, 220, 244, 264, 266, 285
mandatory project, 14, 22
Mark 1 project, 8, 9, 54
market conditions, 53, 65
market research, 29
market window theory, 283, 284
Mark *n* project, 8, 9, 54
Mars Climate Orbiter, 28
Mars Polar Lander, 28
matrix organization, 50
media, 2–4, 35, 39, 41, 50, 54, 59, 73, 78, 93, 109–111, 118, 135, 146, 147, 152, 173, 190, 191, 199, 205, 217, 223, 229, 233, 235, 239, 242, 243, 252, 264, 277–280
mediation, 199, 233
meetings, 1, 56, 89, 90, 158, 207, 235, 239, 252
melon project, 167
memorandum of understanding (MOU), 45, 47, 50, 52, 71, 79, 135, 141, 145, 151, 154, 157, 166, 168, 176, 180, 183–186, 194, 197, 206, 212, 217, 219, 223, 227, 230, 232, 239–244, 269, 270, 276, 278, 285
Mendeleev, Dmitri Ivanovich, 4
mental resources, 47, 167
methodology, 3, 32, 148, 252, 280, 281
MG, 52
micromanagement, 167
micro-survey, 57–61

Middle East, 64
miscommunications, 28
mishap, 29
"Mission Success First", 29, 72, 123, 140,
    165, 167, 172–178, 181, 192, 196, 215,
    220–226, 241, 248, 249, 257, 266, 269,
    272, 279, 286
misunderstandings, 25, 29, 74, 90, 123, 143,
    152, 153, 156, 167, 196, 212, 222, 227,
    239, 248, 249
MOdulo Sperimentale Elettromeccanico
    (MOSE), 78
modus vivendi, 31
money, 2, 11, 18–24, 29, 33, 37–39, 42, 47,
    50, 53, 54, 57, 58, 64–67, 71, 77, 83, 90,
    95, 104, 107, 111, 112, 115, 116, 121,
    124, 129, 132, 135, 138, 140, 145, 149,
    158, 162, 165–168, 188, 191, 200, 204,
    211, 214, 219, 227, 244, 246, 256, 264,
    266, 269–274, 278, 284, 285
Montesquieu, Baron de, 162
Morgan, Charles, 167
MOSE, See MOdulo Sperimentale
    Elettromeccanico
Motorola Mobility, 49
MOU, See memorandum of understanding
MTU Aero Engines, 127
multiple-deadline project, 16
multiple exclamation marks, 151
multiple-handover project, 15
multi-shot project, 16
multi-tier PSNs, 22, 35
Münchener Verkehrsverbund GmbH
    (MVV), 142
mutual distrust, 167, 175, 257
mutual obligations, 138, 146, 161
mutual success cultures, 29
MVV, See Münchener Verkehrsverbund
    GmbH

N

naming, 34, 125, 201, 203, 238, 239
NASA, 28, 153, 257
NDA, See non-disclosure agreement
negative press, 67

Nepal, 82
net present value (NVP), 14, 53
net total benefits of ownership, 136
new prospects, 29
Newsweek, 78
NGO, See non-governmental organization
no-deadline project, 16
nominating, 201, 203, 238, 239
non-binding offer, 115–118
non-compete clause, 145
non-disclosure agreement (NDA), 145–147,
    175
non-governmental organization (NGO), 81,
    225
norming, 123
North America, 64
novelty, 8, 9, 283
NPV, See net present value
Nul, Jan de, 31, 128

O

objective, 6, 22, 31, 35, 40, 44, 55, 59, 70,
    88, 129, 142, 148, 171, 185, 189, 192,
    193, 202, 224–228, 234, 251, 252, 260,
    270, 273
obsolescence management, 53
OFFA, See offer force-field analysis
offer, 13, 33, 37, 39, 42, 57, 70, 86–122,
    128–153, 156, 169–174, 194, 202, 206,
    223, 235, 238–241, 254, 260, 262,
    268–271, 276–278
offer development process, 96
offer force-field analysis (OFFA), 101
Ogilvy, David, 70
on-boarding meeting, 158
one-shot project, 16
online collaboration, 32
open typologies, 17
operational business, 33, 34, 86, 262
operational disruption, 38, 47, 56, 74, 107,
    152, 183, 184, 224, 225, 228, 267
operationalization, 32, 42
operations, 14, 17, 18, 28, 31–35, 53–56, 67,
    68, 109, 111, 124, 134, 152, 185, 198,
    204, 225, 234, 238, 240, 243, 244, 267

organizational process assets, 3, 32
outdating, 13
outlays, 53, 95, 161
outsourced project, 24
outsourced work package, 24, 25
outsourcing, 19, 50, 53, 54, 60–65, 70, 71,
    84, 211, 221, 285

P

panacea, 4
Panama Canal expansion, 30, 79
paradigm change, 51
paradise syndrome, 216, 223
partners, 2, 10, 26, 37, 38, 52, 88, 90, 120,
    123, 126, 140, 146, 151, 154, 155, 164,
    165, 171–178, 192, 196, 204–207, 214,
    217, 224, 239, 244, 253–256, 269, 275,
    276
party, 1, 2, 13, 18, 19, 26, 29, 33–40, 54,
    65, 68–74, 78, 79, 83–88, 94, 95, 111,
    115–118, 121, 127–132, 140, 146–149,
    153–167, 171–186, 191–196, 200,
    206–208, 211, 218–221, 224, 228–235,
    238–241, 244–246, 253–258, 261, 263,
    268–272, 275, 276, 279, 280
patent, 49, 56, 83, 216
payment, 1, 20, 56–58, 71, 72, 81, 87, 88,
    95, 111, 115–118, 130–134, 148, 156,
    158, 161, 162, 182, 184, 188, 191, 192,
    212, 229, 235, 244–247, 252–256, 261,
    276–278, 285, 286
PBM, *See* Project Business Management
penalty, 14, 97, 146, 177, 182–186
Penzhorn, Guido, 77, 215
perfect price, 130, 133, 134
performance, 1, 18, 19, 24, 48, 56, 67, 70,
    73, 90, 110, 115–119, 133, 161, 162, 171,
    176, 182–186, 189, 191, 200, 201, 207,
    223, 238, 243, 254, 257, 262, 266, 272,
    281, 282, 285
performance bonds, 118, 119, 161
performing, 2, 11–14, 18, 21, 22, 25, 26,
    29–31, 47, 54, 57, 58, 64, 75, 108, 110,
    123, 133, 139, 159, 162, 202, 203, 237,
    252–254, 260, 264, 273, 278, 285

performing organization, 2, 11–14, 18, 21,
    22, 54, 58, 162, 278
periodic table, 4
peripheral costs, 135
Philippines, 82
philosopher's stone, 4
Piquette Avenue Plant, 52
pitch, 93, 114, 152, 268
plan, 2, 7, 12–17, 22, 29, 34, 43, 47, 52–54,
    67, 68, 73–76, 79, 84, 111, 112, 119–
    121, 124–128, 135, 136, 146, 151, 156,
    158, 170, 173, 179–183, 186, 195–198,
    212–218, 228, 236–240, 243, 248,
    252–257, 268, 272, 283
PMBOK® Guide, 164
PMI, *See* Project Management Institute
PMO, *See* project management office
PMP®, *See* Project Management Professional®
point of total assumption (PTA), 1, 7, 13, 22,
    29, 43, 74, 77, 98, 115–117, 123, 125,
    131–134, 142, 143, 152, 176, 179, 181,
    186–189, 192–194, 199–203, 211–223,
    233, 234, 244–248, 260, 279
political estimates, 167
portfolio, 13, 21, 22, 29, 32, 41, 50, 111,
    212, 244, 253, 254, 269
portfolio management, 21, 111, 212, 269
post-contract services, 29
post-project operations, 31
PQQ, *See* pre-qualification questionnaire
practitioner, 4, 5, 8, 17, 97, 106, 157, 192
predictability, 16, 48, 139, 164, 165, 280
predictable project, 12
predictive approach, 12, 179, 181
premature, 283
pre-qualification questionnaire (PQQ),
    91–93, 113, 122
presentation, 12, 73, 114, 115, 122, 144–152,
    223, 268, 281, 282
Press Freedom Index, 223
prettification, 165
price, 2, 26, 33, 52, 66–68, 71, 74–77, 81,
    87–94, 103, 106–108, 111, 114, 118, 119,
    125, 126, 129–142, 148, 160, 170, 175,
    182–195, 198, 202–206, 219, 220, 236,

239, 254, 256, 263, 268, 270, 273, 278, 282–284

price ceiling, 160, 183, 188, 189, 192, 193

pricing, 130–138, 147, 268

prime contractor, 1, 37, 38, 79, 124–129, 175, 183, 186, 197–204, 208, 217–219, 229, 235–238, 242–244, 247, 252, 254, 260, 270, 276–280

prioritization, 15, 257, 286

privacy, 84, 191, 234

process description, 161

procrastination, 286

procurement, 11, 22, 25, 37, 42, 44, 57, 66, 77, 85–99, 103, 106, 111–115, 118, 119, 129–132, 138, 156, 159, 164, 174, 175, 201, 202, 214, 216, 224, 228, 239, 245–247, 272, 275, 277

procurement SOW, 93

product business, 35, 37

product development, 20, 52, 74

production lines, 34, 243

profile page, 86, 87

profit, 1, 3, 11, 18–22, 29–35, 38, 41–44, 48, 53, 58, 65, 69, 71, 85, 95, 98, 99, 104–112, 119, 125–128, 132–136, 149, 166, 168, 183, 186, 194, 197, 202, 219, 251–255, 262, 268, 273–279, 283, 285

profit center, 3, 11, 18, 29, 31, 35, 65, 273, 274

program, 21, 22, 25, 50, 52, 68, 80, 83, 136, 161, 202, 210–214, 242, 281, 282

program management, 21

progressive elaboration, 12, 194, 195

project, 1–90, 93–185, 189–287

Project Business Management (PBM), 1–4, 11, 18, 40, 44, 57, 65, 85, 97, 108, 109, 112, 119, 154–158, 161–163, 169, 171, 178, 215, 218, 221, 225, 232, 246, 247, 251, 254, 256, 269, 278–282, 285, 287

project environment, 50, 52, 61, 62, 172

project goal, 224, 260

project location, 63

project management, 1–17, 20–22, 32, 42, 43, 46, 54, 57–61, 65, 68, 69, 81, 87, 90, 96, 99, 106, 115, 120, 122, 131, 132,

138, 152, 159, 161, 169, 178–180, 189, 192, 196, 197, 205, 207, 211–213, 224, 225, 230–237, 240, 246, 248, 251, 252, 256, 260, 267–275, 279–282, 285, 287

project management business office (PMBO), 90, 164

project management experience, 8

project management expert, 7

Project Management Handbook, 32

Project Management Institute (PMI), 6, 8, 22, 78, 90, 126, 164

project management methodology, 32, 252

Project Management Office (PMO), 3, 32, 42, 260

project management team, 248

project manager, 1–17, 21, 22, 25, 26, 29, 32, 38, 42–47, 56, 59, 63, 65, 69, 70, 74–77, 80, 87, 96, 97, 104, 107, 119, 123, 145, 148, 154–158, 162, 165, 167, 175–181, 191–197, 200–203, 208, 211–217, 221–225, 229, 232, 237–243, 252–254, 258–274, 279–286

project portfolio management, 21

project procurement management, 90, 164

project provider, 31

project sponsor, 4, 47, 195

project success, 29, 57, 104, 158, 159, 172, 192, 202, 203, 225, 235, 251, 272

project supply network (PSN), 10, 11, 21, 22, 28–30, 35, 38, 65, 69, 72, 74, 81, 87, 129, 162, 169–173, 181, 189–192, 196–199, 202–209, 212–219, 222–258, 270, 271, 274–281, 285–287

project team, 4, 10, 15, 35, 39, 56, 65, 70, 74–76, 79, 83–86, 93, 174, 206, 222, 226, 239, 253, 263, 280

property, 19, 33, 49, 84, 125, 161, 168, 205

proposal, 29, 32, 42, 44, 90, 92, 95–98, 104–109, 112–114, 117, 120–122, 130, 139–143, 147, 148, 152, 220, 240, 252, 254, 264–269

proposal management, 42, 90, 95, 98, 108, 120, 252

proposal manager, 113, 220

prospective contractor, 46

prospective customer, 46, 66, 93, 94, 103, 107, 141, 144, 175
protective behavior, 165, 171
protest-proof process, 89
prototype, 66, 74, 75
provisions, 24, 36, 87, 153, 156–162, 172, 200, 225, 228, 230, 235, 245
proximity, 34, 107, 126, 135, 218, 268
PSN, *See* project supply network
psychopath, 75, 215, 222
PTA, *See* point of total assumption
publication, 4, 12
public procurement, 44, 92, 94, 118
public tendering, 88
purchasing department, 239

Q

Q&A session, 89
Qatar, 82
qualification program, 161
quotation, 91, 109–113, 130, 268

R

rapport, 56, 69, 148, 169, 170, 238, 269, 280
rating, 13, 22, 28–30, 49, 50, 53–55, 68, 73, 87, 98–101, 107, 112, 123, 127, 135, 171, 174, 190, 191, 205, 223, 226, 253, 256, 270, 276, 281
rationale, 57
razor-and-blade project, 33
realism, 139
recruiter, 38
reference customer, 69, 106, 107, 128, 203, 206
reference list, 86, 146
relation, 4, 20, 26–29, 35–37, 43–46, 49, 55, 56, 69–72, 75, 82–86, 100, 104, 105, 108–112, 123–125, 129, 131, 138–140, 146–149, 153, 156–162, 165–172, 175, 176, 181, 189–195, 198–206, 214, 217–220, 229, 233, 235, 238, 239, 244, 245, 253, 254, 257, 258, 262, 269, 271, 275, 276, 286, 287
relational perspective, 28
remoteness, 71

repetition, 97, 107, 179, 261, 279
Reporters sans Frontieres, 223
reputation, 25, 32, 39, 72, 80–82, 99, 107, 136, 142, 216, 224
request, 11, 15, 17, 25, 34, 38, 44, 59, 74, 76, 85–97, 100, 105–114, 118–124, 133, 134, 155, 178, 179, 182, 184, 193–198, 208, 211, 213, 219, 220, 229, 244, 271
request for information (RfI), 87, 91–93, 113
request for proposal (RFP), 44, 87, 92–94, 114, 122
request for quotation (RfQ), 87, 91, 94, 113
requirement, 1, 6, 12, 17, 24, 25, 43, 44, 47, 48, 65–68, 86, 93, 94, 98–101, 104, 105, 111, 121, 126–129, 138, 145, 152, 156, 162–165, 171, 175, 179, 180, 186, 194, 195, 203, 210, 219, 222, 225–228, 234, 244, 260–265, 268, 270, 273, 274, 282
requirements, 1, 6, 12, 17, 24, 25, 43, 44, 48, 65–68, 86, 93, 94, 98–101, 104, 105, 111, 121, 126–129, 138, 152, 156, 165, 171, 175, 179, 180, 186, 194, 195, 203, 210, 219, 222, 225–228, 234, 244, 260–265, 270, 273, 274, 282
requirements management, 68
research, 5–9, 13, 29, 31, 49, 57–65, 71, 77, 78, 82, 104, 119, 142, 150, 155, 216, 223, 233, 240, 257, 287
research project, 7, 13, 142
resistance, 7, 55, 56, 211, 232
resource, 2, 6, 11, 14, 17–21, 25, 26, 30–32, 38, 42–50, 53, 56, 57, 67, 70–73, 80, 82, 86, 88, 96–106, 112, 116–122, 125, 128, 129, 137, 138, 147, 152, 155, 158, 167, 171, 175, 180–183, 191, 194, 200, 203–208, 213, 214, 219, 220, 223, 224, 227–230, 235–240, 247–249, 252, 254, 264, 267–275, 278, 282, 285, 287
response, 5, 12–16, 21, 55, 59–66, 81, 85, 87, 90–95, 100, 103–107, 113–118, 122, 133, 137, 142, 148, 167, 179, 192, 205, 213, 231, 243, 248, 258
responsibility, 4, 10, 24, 25, 39, 48, 54, 65, 72, 80, 81, 94, 96, 106, 136, 164,

200–203, 211, 215, 225, 229, 238, 260, 271–275
responsiveness, 20, 43, 103, 110, 150, 193, 223
return on investment, 53
review board, 32, 233
RfI, *See* request for information
RFP, *See* request for proposal
RfQ, *See* request for quotation
risk, 2, 9, 18, 26, 29, 33–35, 38, 39, 44, 48, 50, 53–57, 66, 69, 71, 75, 76, 80–85, 94–96, 100, 103, 104, 108, 111, 112, 116–129, 133, 138, 139, 145–149, 157, 159, 166, 171–176, 182–188, 193–195, 201–203, 206, 208, 211–218, 221–224, 228, 232–237, 241, 244, 247, 248, 255, 256, 259–266, 269, 270, 273, 275, 278–281, 286, 287
ROI, *See* return on investment
role model, 43, 82, 86, 87, 174, 207, 256, 283
rolling award fee, 160, 189–192, 207, 221, 222, 270, 282, 284, 287
rolling wave, 13, 179, 180, 194, 213
Rolls-Royce, 127

S
Sabre, 55
Sacyr Vallehermoso, 31, 128
safety manager, 32
SAIC, *See* stakeholder attitudes influence chart
Salini Impregilo, 31, 128
Samsung, 49
Sarbanes-Oxley Act (SOX), 264
satellite project, 12
scalability, 33, 55
science communities, 55
scope, 37, 43, 61, 67, 68, 126, 129, 138, 158, 160, 171, 183, 195, 224, 228, 239, 254, 256, 259, 268, 272
search engine optimization (SEO), 41, 262, 275
seller, 11, 19, 24, 36–38, 44–46, 56–58, 66, 67, 81, 85–100, 103–107, 112–122,

130–153, 156–159, 172–175, 178, 182, 197, 214–218, 221, 223, 229, 231, 235, 238, 244, 247, 249, 262, 270–275, 280, 285
seller response, 87, 90, 93–95, 105, 113
SEO, *See* search engine optimization
service department, 273
settlement, 26, 83, 162, 194, 230, 234, 245
Shakespeare, William, 10
shareholder, 1, 25, 48, 264
Shenhar, Aaron, 9
Shinkansen, 28
Shipley Proposal Guide, 95
shortlist, 91, 113, 122
Siemens–Alsthom, 28
siloed project, 10
siloing, 10, 28, 29, 65
single-deadline project, 16
single-handover project, 15
single point of contact, 25
SitPM, *See* Situational Project Management
situational awareness, 221, 282
Situational Project Management (SitPM), 4, 6, 11, 17, 21, 43, 90, 122, 132, 232, 256, 271, 280
skill, 13, 28, 43, 48, 51, 54–57, 72, 76, 95, 99, 104, 119, 129, 156, 170, 193, 202, 203, 206, 215, 221, 253, 266–268, 274, 280 282, 286
slavery, 19
smart watch, 51
Snecma, 127
social network, 21, 50, 60, 87, 147, 161
sociopath, 75, 215, 222
software development, 1, 15, 49, 83, 282
solid project, 10, 28
SoP, *See* standard operating procedure
sophistry, 167
source code, 83
SOW, 92–94, 99, 104, 122
SOX, *See* Sarbanes-Oxley Act
special magazine, 88
specification, 11, 34, 68, 116, 132, 175, 194, 196, 226–228, 244
specification limit, 132, 228

speed blindness, 72–75
sphere, 24, 28, 141, 257, 271, 286
sponsoring organization, 36, 53, 54
staged deliveries, 15, 17, 244, 261
stakeholder, 2, 9, 10, 17, 25, 39, 40, 47, 48,
    74, 94, 101, 104, 119, 162, 178–180,
    192, 195, 211, 215, 226, 231–234, 239,
    240, 261–264, 267, 274
stakeholder attitudes influence chart (SAIC),
    40, 105
stakeholder register, 40
standalone project, 12
standardized cost reporting, 32
standard operating procedure (SoP), 4, 15,
    53, 162, 167, 181, 227, 228, 242, 243,
    246, 278
stare decisis, 162
start date, 10
start of production, 15, 34, 227, 228, 242,
    243
statement of work, 92, 93, 104, 119, 122, 141
step-by-step approach, 17, 186
storming, 31, 123
strategic interest, 30
strategic option, 70
stress, 41, 66, 71, 102, 111, 123, 140, 167,
    222, 277, 282
strong matrix, 14, 31
Stuttgart, 21, 7, 10
subcontractor, 2, 22, 26, 37–39, 43, 75, 79,
    80, 84, 125–129, 175, 183, 186, 188,
    191, 194, 197–204, 208, 217–220, 224,
    229, 235–238, 242–244, 247, 251–254,
    259, 260, 263, 267, 270, 275–278, 286
submission deadline, 98, 120, 121
subsidiary company, 19, 20
sub-team, 10, 28, 68
success, 1–12, 16–21, 24, 28–31, 34, 35,
    38–43, 50, 54–57, 65, 69, 72–76, 81,
    86, 90, 94, 96, 101–112, 118, 122, 123,
    128–132, 137, 140, 143, 145, 149, 152,
    158, 159, 164–169, 172–181, 191, 192,
    195–199, 202–208, 215, 216, 220–226,
    232–235, 241–243, 248–260, 263,
    266–273, 276, 279–282, 285–287

supportive project, 33, 35
survey weariness, 59
survival, 6, 65, 73, 165
Sword of Damocles, 88

T
Taiwan High Speed Railway Consortium, 28
tangible benefit, 14, 253
target cost contract, 160, 183, 186–189, 192
TBO, *See* total benefit of ownership
TCO, *See* total cost of ownership
team, 3, 4, 8–17, 25–40, 52, 56, 65–71,
    74–76, 79–90, 93, 96–99, 104, 108,
    109, 116, 121–126, 129, 130, 140, 147,
    149, 152–158, 161, 165–180, 198–207,
    213, 219, 222, 223, 226, 228, 232, 234,
    237–241, 248, 249, 253, 256, 257, 263,
    267, 272, 280–287
team communications, 149, 161
teaming, 31, 33, 37, 38, 122–124, 129, 140,
    165, 171–174, 177, 178, 200–203, 206,
    207
teaming agreement, 31, 38, 122–124, 129,
    200
teaming networks, 33
teaming partner, 37, 123, 129, 165, 171–174,
    177, 178, 206, 207
team spirit, 29, 173, 205, 257, 286
teamware, 32
technical requirements, 1, 261
technology, 17, 45, 52, 70, 74, 87, 109, 115,
    142, 143, 175, 205, 216
Technology Acceptance Model, 142
telefax, 55
template, 3, 19, 32, 42, 90, 97, 98, 206
temporary endeavor, 6
tender, 88, 92, 114, 121, 139, 140
terms of service (TOS), 73, 81, 93, 151
Tesla, 66, 67
Tesla Roadster, 66
TfNSW, *See* transport for NSW
Thales Group, 28
The Guardian, 82
third-party logistics provider, 33
third-party vendor, 19, 68

threshold, 48, 94, 123
time zones, 65, 70, 159, 287
Titanic, 73
T&M contract, 160, 181, 186, 225
tools and techniques, 55
TOS, *See* terms of service
total benefit of ownership (TBO), 226
total cost of ownership (TCO), 136, 226
total net benefit of ownership, 103
TRAC, 1–4, 7, 11–14, 18–47, 50–58, 62, 65, 68–121, 124–133, 138–140, 143–149, 153–287
trainer, 33, 65, 88, 111, 120, 145, 146, 156, 238, 240, 251, 252, 281
translation software, 152
Transparency International, 77, 216, 223, 241
transport for NSW (TfNSW), 67
travel agencies, 55
trend, 58–64, 86, 171
trust, 15, 25, 31, 56, 69, 75–78, 83, 84, 90, 124, 128, 131, 132, 139, 146, 154, 157, 158, 165–170, 174–176, 190–192, 198, 203, 206, 216, 223, 226, 241, 244, 254–257, 263, 267–271, 277–280, 286
trustworthiness, 69, 75, 131, 154, 168, 169, 206, 223, 279, 286
Tuckman, Bruce, 123
Tuckman Team Development Phases, 123
turnover, 41
typological dimension, 9, 11, 17
typology, 4–9, 17, 57
typology of projects, 4, 8, 57

U
unambiguity of language, 90
uniqueness, 5, 6, 25
urgency, 47, 100, 286

V
value perception, 39, 130, 131

value stream, 13
venturer, 31, 126–129, 202, 203, 216, 223, 247
verbal contract, 115, 157
Viadeo, 87, 152
VIP position, 31
Volkswagen, 28

W
walk-away limit, 132–134
wall of silence, 77
warning limit, 132
warranty, 31, 115, 125, 148, 181, 182, 244, 245, 262, 273
Waste Of Money, Brain, And Time (WOMBAT), 264
waterfall, 12, 180
WBS, *See* work breakdown structure
weak matrix, 14
wear and tear, 13
Wills, Harold C., 52
winning bids and proposals, 95
winning the contract, 100–102, 116
winning the project, 100, 101
WOMBAT, *See* Waste Of Money, Brain, And Time
word cloud, 45, 46
workaround, 264
work breakdown structure (WBS), 13, 22, 180
workload, 39, 42, 48, 60, 62, 71, 84, 103, 120, 158, 167, 208, 247
workload Assignment, 62
work package, 13, 22–25, 129, 203, 228, 237
World Bank, 77
World Championship, 82

X
Xing, 28, 45, 60, 71, 72, 84, 87, 152, 199, 207, 208